the GOLDEN RELATIONSHIP

ART MATH NATURE

Book 1:

UNIVERSAL PATTERNS

by Martha Boles
and Rochelle Newman

Design & Illustrations by Sylvia T. Burnside

pythagorean press

Second Edition

All graphic work in this book was done by human hands without the aid of sophisticated computer equipment.

Library of Congress Catalog Card Number: 87-60685

ISBN 0-9614504-1-X

Published by:

pythagorean press
P.O. Box 162 Bradford, MA 01830

Manufactured by ZBR Publications, Inc.
3 Industrial Way, Wilmington, MA 01887

Cover photos by Richard Newman

Morris Kline once said to a group of mathematics teachers: "96.6% of the people are different from us. They don't like mathematics." It is to this large segment of the population that we dedicate this book.

Acknowledgements

We wish to acknowledge that cooperation is better than competition and that love is more powerful than money. All through this project we have received much support and encouragement, but without the commitment of the following we would only have had a good idea.

Bradford College for giving us the financial aid that allowed us to begin,

Northern Essex Community College and Bradford College for building bridges between public and private education,

Judy Moran who provided much of the initial stimulus for this book,

Reuben Hersh for his helpful criticism,

Marcia McCullough and Jody Zafris for giving us many hours' worth of skillful typing,

Dover Publications who published little known works relating to this project and who provided copyright free visuals which we have used,

the students of Bradford College and Northern Essex Community College who caught our enthusiasm, reinforced our ideas, and showed us that those ideas could be translated into beautiful works which they were willing to share so that they could be incorporated into this book,

Richard Broadhurst, Karen Pope, John Roberts and Mary Saunders who were generous with either their time or their ideas or both,

Richard Newman for his photographic work.

Contents

Constructions

Chapter I: Basic Constructions

Chapter II: Unique Relationships

Chapter III: Special Triangles

Chapter IV: Dynamic Rectangles

Chapter V: Fibonacci Numbers

Chapter VI: Spirals

PREFACE

"To search for relationships between things apparently disconnected" is the Aristotelian Quest. Our desire is a variation on this quest: To show relationships between things which appear disconnected. The quest is a way to identify wholeness.

This world is of a single piece; yet, we invent nets to trap it for our inspection. Then, we mistake our nets for the reality of the piece. In these nets we catch the fishes of the intellect but the sea of wholeness forever eludes our grasp. So we forget our original intent and then mistake the nets for the sea.

Three of these nets we have named Nature, Mathematics and Art. We then conclude since they are different because we call them by different names, they are forever separate with nothing bonding them together. It is not the nets that are at fault but rather our misunderstanding of their function as nets. They do catch the fishes but never the sea, and it is the sea that we ultimately desire.

Our aim in this book is to explore some of the connections--to look at Nature and then to see how the disciplines of Art and Mathematics catch the same fishes of this natural world. It is both the holes in the net and the net itself that we would like to examine. Our concern is in relationships, connections, interfaces, overlaps.

One of us comes from the discipline of Mathematics with an interest in Art; the other comes from the discipline of Art with an interest in Mathematics. Our skills show that both Art and Mathematics are necessary for a fuller understanding and enjoyment of the world which we call Nature.

Though Nature is both complex and simple simultaneously, she is comprehensible to the human mind and consistent with her laws. She suggests ultimate order, which is perfection, on one hand; then offers us deviation and disorder on the other. She is ever involved in ceaseless activity. She never gives us the mathematician's precision but rather offers us the artist's abundance and variety. She seldom works with a simple one-force system but rather delights in the interplay of several forces. She must be understood by viewing the interaction of systems synergetically; the whole always being greater than the sum of the parts.

> It is this silent swerving from accuracy by an inch, that is the uncanny element in everything. It seems a sort of secret treason in the universe...Everywhere

> in things there is this element of the quiet and incalculable.
>
> The real trouble with this world of ours is not that it is an unreasonable world, nor even that it is a reasonable one. The commonest kind of trouble is that it is nearly reasonable, but not quite. It looks just a little more mathematical and regular than it is; its exactitude is obvious, but its inexactitude is hidden; its wildness lies in wait.
>
> Gilbert Chesterton

Nature is all inclusive; however, for purposes of discussion let us call Nature that which is given, not the man-made nor the synthetic. Nature is a verb--partaking of, creating, re-creating--the universal interconnectedness and interdependence of all phenomena in an ever dynamic cosmic dance. Each individual dancer is part of an environment which, in turn, is part of a larger environment which is part of a planetary system which, in turn, is part of a solar system which itself is part of a galaxy which is part of the Cosmos. Thus, the dancer and the dance are one. We cannot decompose this Nature into discreet, isolated, independently existing bits, but must look at these bits in process and in relation to one another. This planet, this "spaceship earth," this biosphere can sustain, maintain, generate life and adapt to changing conditions like the individual organisms that are a part of this system. Thus, the planet is a living organism and must be considered such. It is our responsibility to understand this dynamism to work in consonance with it. Then, arrogance shall give way to the wisdom that we must be responsible participants who are part of the larger processes.

Despite all our attempts at understanding, the essential enigma will remain. For whenever we use one of our nets to examine a part of Nature we exclude other parts; therefore, none of our explanations can be total Truth but rather only partial truth limited to a particular time and place. As Jacob Bronowski said, "No formal system embraces all the questions that can be asked."

Our senses give us only a limited range of Nature's reality. Beyond and below the boundaries of our senses are other aspects of the same reality. The world of the fly is not the same as that of the elephant since size is affected by gravity which in turn affects both function and structure.

But at the minutest level there is similarity. We are just now beginning to understand that ours is a modular world built out of multiple replication of the simplest structures which we call atoms. These atoms, which are mostly space, are composed of electrons, protons and neutrons ever moving as individual members, but stable in their patterns. All atoms have the same basic internal structure. What makes them different from each other is the number of electrons and protons contained within them. There are only 92 different elements, each with their own atomic structure, which become building blocks of our natural world. Atoms join together to form molecules which in turn cluster to form larger units of matter. The variety that Nature provides is gotten by permutations of these building blocks.

There is individuality; there is community
There is microcosm; there is macrocosm
There is visible; there is invisible
There is randomness; there is organization
There is reason; there is intuition
There is mind; there is body
There is inside; there is outside
There is motion; there is rest
There is equilibrium; there is fluctuation
There is competition; there is cooperation
There is chance; there is choice
There is uncertainty; there is predictability
There is perfection; there is flaw
There is process; there is product
There is symmetry; there is asymmetry
There is theme; there is variation
There is diversity; there is unity
There is the infinite filled with the finite

Despite the essential difficulties in understanding the whole of Nature, both Mathematics and Art reflect a human desire to comprehend aspects of the world. Both come after the fact of Nature and each never fully explains her, but in their complementarity they work to give wholeness.

Both Art and Mathematics are "games," constructs of the mind with recognizable limits. Depending upon the cultural climate and the "problem" in need of solution, there is an initial premise as to how the game must be played; what the object of the game is; the rules necessary to play by; the techniques needed to play with; the board one must play on. And both disciplines look for relationships among the game pieces.

To understand relationships is to search for patterns that are identifiable and repeatable. In Mathematics, however, the interest is the relationship without the reference to things. There is no concern for the affect that an object might have upon its perceiver, while in Art the interest in the relationship is <u>because</u> of the things. The "thingness" of the object is as strong a concern as its connections. And it is in part the emotional qualities that engage the artist.

If it is the aim of Mathematics to seek the simplest explanations of the most complex of facts, is it then the aim of Art to express the joy in experiencing the multitude of relationships?

Art looks to the specific to speak about the general; Mathematics uses the general to explain the specific. One is idiosyncratic, the other, impersonal. As D'Arcy Thompson has pointed out, mathematically we can define the form of a wave or a heap of sand, but never ask a mathematician to define the form of any particular wave of the sea, nor the actual form of any mountain peak or hill. That is the task of the artist and not only will he give you the particular wave or mountain peak, but it will be his particular perceptions of that wave or peak unlike any other.

In this world of overspecialization, much of education deals with discreet bits of information rather than large systems. People, therefore are not trained to find connections. Without connections, value systems are difficult to develop. In the evolution of civilization, Art and Mathematics are disciplines that have been seen as polarities without connection. Yet, in fact, they are the left and right hand of cultural advance: One is the realm of metaphor, the other, the realm of logic. Our humanness depends on a place for the fusion of fact and fancy, emotion and logic. Their union allows the human spirit freedom.

Student Work
Chapter 1, project 1

Carol Gagnon

Introduction

THIS BOOK IS ABOUT SPACE.

SPACE IS THE ARENA OF *ALL* LIFE'S INTERACTIONS.

SPACE IS THE COMMON DENOMINATOR OF MATHEMATICS, ART AND NATURE.

NATURE *IS* RELATIONSHIPS IN SPACE.

GEOMETRY *DEFINES* RELATIONSHIPS IN SPACE.

ART *CREATES* RELATIONSHIPS IN SPACE.

Space, however, cannot be perceived without our initially forming a space conception--an idea that first defines what space is. The experience of physical space exists for everyone but the *idea* of space is an abstraction - a way to step outside the experience and then attempt to define that experience through words and symbols.

To the mathematician the term space is chosen to be undefined, as are point, line, and plane. The reason these terms remain undefined is that mathematicians are concerned with the development of logical symbolic structures, one of which is geometry.

In playing the game of geometry, the undefined terms are the first game pieces and one of the rules of the game dictates that new pieces (terms) may be added only if they can be defined by terms which are either undefined or have been previously defined. This rule prevents the game from becoming circular in its logical structure.

Another rule in this game allows the player to formulate statements which will be accepted as true without proof. These statements are called postulates and axioms. Yet another rule requires that these be kept to a minimum and never be used when a theorem, which can be proved, may be used instead.

This game was originally played informally by the Classical Greek philosophers. However, it was Euclid, circa the third century B.C., who wrote *Elements*, the rule book for plane geometry that deals with only two dimensions. New strategies were developed by later players who altered one of the postulates giving rise to new geometries that were no longer played on a flat board.

Mathematics is the tool by which we attempt to understand how the natural world operates. As our understanding of this world increases, the mathematical models need to change in order to adapt to that new understanding. Ancient geometry originated from the practical need to survey and measure the land which was perceived as flat. Current models have expanded to fit the concept of a universe composed of curved surfaces ever changing in relationship to each other.

Thinking about space requires a high degree of abstraction. This is difficult to do since this "nothingness" of space is everywhere and yet nowhere is it visible. It is intangible and yet we must give it tangibility in order to examine it, but not all qualities of space are accessible to sense perception. We cannot comprehend large-scale or small-scale space without the aid of the telescope or microscope.

Most of our experience of the external world comes through to us via our senses with vision and touch playing the most significant roles. When we speak of space, we must also talk of its corollary form--the objects that are placed within space.

How humans place objects in relation to one another is instinctive and operates at a subconscious level. Art objects are symbolic representations of spatial experiences. The various periods throughout the history of art show us changing concepts. Man's views toward space go beyond the physical into the realms of the emotional and the metaphysical through art. How we humans envision our position in the cosmos can be comprehended through the graphic and sculptural projections of a particular cultural period.

Out there, in the cosmos, space is continuous and there is multidirectionality; all directions are equally valid; all reference points equally important. There is no reference to a "me."

In here, we live in a three-dimensional world. It has length and width and depth. We continually move through our space and our movement plays a decisive role in the perception of space.

By touching and seeing, form becomes accessible. With our ever roving eyes, our ever moving bodies, our ever moving limbs, we experience a dynamic, not static, environment which is space.

As physical creatures in this three-dimensional space, our understanding is tempered by both our physical and perceptual selves. We no longer walk on all fours. Because of this upright posture, verticality has become the dominant direction along with its complement, the horizontal. A person standing on the ground plane with an outstretched arm could allow for the development of the abstract idea of the "right angle" just as a person pivoting on one leg could lead the way for the development of the concept of "circle." Notions of measure and distance in space grow from our human scale and the understanding of our physical place in this space. Each person is a point of orientation.

In Art, the vertical became the universal organizing principle that has reigned supreme since the neolithic period up to the present day. But we must remember, it is a relative position. It is not the only way to portray order. This vertical has come to stand for the line of movement and of striving upward while, in contrast, the horizontal suggests repose and balance.

Our two eyes, located in the front of our head, apprehend our position in space. We move our heads right and left giving us 180 degrees of vision before we must turn around to deal with "back" and "front." Our limited directions of movement forces us to define space in a particular fashion.

What if we were birds in the air or fishes in the sea? Gravity pulls upon our bodies so that the ideas of "up" and "down" are determined by that force. "Here" and "there" are notions of location. "Near" and "far" involve the concept of time, movement through space or duration.

Thus, our brains and perceptual processes evolved to suit our own particular brand of space. This medium became our message. We did not learn to really "see" our space until the non-Euclidean mathematicians and Albert Einstein pointed out that there were other spaces whose forms and patterns would differ from ours in very significant ways. Mathematically, we have spaces with more than three dimensions that allow for examination of certain problems that could not be solved within the structure of Euclidean geometry. The artist-designer, however, works within a relatively limited spatial range. Two and three dimensional space, at this time, is the most useable for generating designs.

All of Nature's interactions occur in a spatial field and there are structural constraints that allow for only a limited number of basic patterns. All of these occur as a result of Nature's practicality since everything she does has a purpose. Spirals, meanders, explosions, branchings, and crackings solve different spatial problems with the greatest efficiency. Using these, she works within the interaction of many complex forces, and conditions are never simple. We find these recurring patterns, with infinite variation, at the microscopic and macroscopic levels.

Nature seems to abhor an undisturbed space, a flat plane, a vacuum. These she tries to fill with whatever is most convenient and economical. The chief angles used in her geometry are 90°, 45°, 60°, and 120° for three way angles. Certain proportional relationships appear again and again. There is a concern for minimum energy expenditure, minimum material use and minimum area.

Nature's aim is to design functionally, which mandates the economic use of materials, energy, work, weight and space, yet to produce maximum strength and enclosure. These patterns swerve from the absolutely perfect geometric model which is only an abstraction created by the human mind to parallel experience. Polar force and gravity interact within space to provide deviation from perfection.

In creating forms in space, Nature doesn't indulge in sloppy craftsmanship. All details of design are fully worked out with a fusion of function, technique and material. She gives the same amount of attention to small and large, organic and inorganic forms alike, and no aspect of her design is unimportant. Her packing and partitioning of space are ultimately practical and efficient, yet beauty is the by-product of her design because the form follows function.

Whatever we may say is obviously inadequate to the reality of Nature, for, as Jacob Bronowski has said, "The world is totally connected. Whatever explanation we invent at any moment is a partial connection, and its richness derives from the richness of such connections as we are able to make."

But, through this book, we hope to encourage a reverence for all aspects of her being.

> Everything made by human hands looks terrible under magnification--crude, rough, and unsymmetrical. But in nature every bit of life is lovely. And the more magnification we use, the more details are brought out, perfectly formed, like endless sets of boxes within boxes.
>
> Dr. Roman Vishniac

Photo by Richard Newman

Francisco Colom

Student Work
Chapter 1, project 3

Further Reading

Bronowski, Jacob. The Origins of Knowledge and Imagination. New Haven: Yale University Press, 1978.

Bronowski, Jacob. Science and Human Values. New York: Harper and Row, 1965.

Capra, Fritjof. The Tao of Physics. New York: Bantam Books, 1980.

Capra, Fritjof. The Turning Point. New York: Simon and Schuster, 1982.

Feininger, Andreas. The Anatomy of Nature. New York: Dover Publications, Inc., 1979.

Feininger, Andreas. Nature Close-Up: A Fantastic Journey Into Reality. New York: Dover Publications, Inc., 1981.

Jammer, Max. The History of Theories of Space in Physics. 2n ed. Cambridge: Harvard University Press, 1969.

Judson, Horace Freeland. The Search for Solutions. New York: Holt, Rinehart and Winston, 1980.

Kline, Morris. Mathematics in Western Culture. New York: Oxford University Press, 1953.

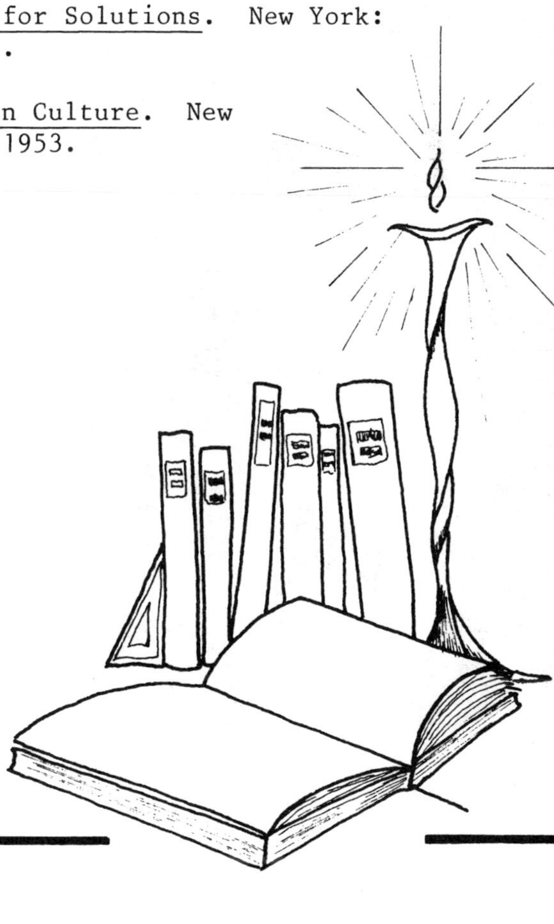

Basic Constructions

1

Mathematics flourishes in an environment where there is unlimited freedom of inquiry regardless of whether or not the results have immediate practical application. Such was the intellectual climate of classical Greece. The culture was that of a fabric tightly interwoven with the strands of philosophy, mathematics and art. Each enhanced and clarified the other as well as defining the position of man in his relationship to Nature. Unlike all preceding civilizations, the Greeks were the first to claim that Nature was comprehensible to man through the use of reason and through the application of mathematics.

Though the elements of geometry were known by the Egyptians and Babylonians, and transmitted to the Greeks, it was the latter that created a mathematics that far surpassed that of their predecessors and which has lasted to this day. For beings endowed with a reasoning faculty, there was a method of gaining knowledge that was different from induction based on experience. Deduction allowed one to draw conclusions based on few premises.

In spite of limits to this method of reasoning, the Greeks insisted that all mathematical conclusions be obtained by the deductive process alone. They were naturally drawn to this approach by their involvement with philosophy's mental activities dealing with abstract questions and broad generalizations. The plane geometry of Euclid with its ten axioms is the supreme example of this method. The hundreds of theorems grew from these few original statements.

This love of rationality led to their concept of beauty in the visual arts. In sculpture and architecture they sought the ideal, the eternal, the perfect, the complete, the consistent, the pure. What they sacrificed was the ephemeral, the fleeting, the changing, the imperfect. Theirs was a static world of closed form. It was man's reason, and not his unreliable senses, that was to govern their way of thinking. With the use of reason mathematics, too, became involved with the abstract whereby mental concepts rather than real, physical things were essential. A theorem could be used in a great many cases, thus was removed from examining a specific concrete thing and need only apply to an ideal form. These were permanent and perfect and incorruptible unlike the experiences of the senses. Today we understand that our senses give us access to a limited range of experience only. It is through reason and deduction that scientists have been able to discover the electrical and atomic aspects of our universe which reside in realms outside our senses.

Plato, in philosophy, spoke of eternal ideas that existed in a sphere outside that of human experience which was a mere approximation of these ideal truths. Greek sculpture, too, with its concern for the ideal, eternal and harmonious type was interested in the proportions that were inherent within each form and which expressed the harmony of all things.

When the Greeks realized that the harmonies of the Universe could not always be explained in terms of whole number ratios, as they originally had thought, plane and solid geometry became the subjects of mathematical emphasis. They were investigated because figures and shapes had constant properties. Instead of dealing with numbers, everything was conceived of in terms of geometry whereby a length was designated as number 1 and then other numbers were represented in terms of this length. Thus, "answers" could be expressed in terms of figures rather than numbers.

Need arose then for a method of accurately constructing the figures being examined. Liking a challenge, the Greeks set very tight limits on which tools were permissible for construction. Most used were the compass and straightedge, and, with a few notable exceptions, almost all of the figures they dealt with could be constructed with these tools.

Interestingly enough, there is speculation that limitations on construction tools came about because the Greeks, preferring mental activities, shunned working with their hands, an occupation for slaves. Occasionally they extended the challenge, setting further restrictions on their tools, such as using a compass that only had one setting - the so called "rusty compass". Without going to this extreme, we invite the reader to experience the challenge of the Greeks while examining the harmonious relationship that exists between the worlds of Art, Mathematics and Nature.

Linda Lazzaro

Student Work
Chapter 1, project 3

In order to better understand the concepts explored in this book, hands-on experience is essential. At the end of each chapter, we have provided problems and projects that the reader can do. Each enables the reader to physically engage the ideas presented. The problems are simpler and require less time to complete. The projects are designed to give aesthetic satisfaction as well.

Because of its precision, we have chosen to use mathematical language. Realizing that the layman frequently places a different meaning from the mathematician upon a term, we will point out the discrepancies in interpretation whenever possible.

In the back of the book we have provided a glossary which not only defines terms mathematically but also gives diagrams and symbols for these terms. Acquaint yourself with the glossary because terms and symbols will frequently be used without definition in the body of the text. Appendix A explains the use of symbols that you will come across throughout this book. Some useful formulas are given in Appendix C. Appendix B provides further explanation of some of the mathematical concepts presented, while Appendix D gives information on art materials and techniques.

Since we are investigating the properties of physical space and the ways in which we may apply Nature's laws to gain aesthetic divisions of that space, a knowledge of certain geometric construction techniques is necessary. Using a T-square and a draftsman's triangle frequently simplifies construction problems since with these tools perpendicular and parallel lines are quite easily drawn; however, many constructions cannot be handled with a T-square and triangle. These constructions frequently require knowledge of basic compass and straightedge techniques for their execution. We have, therefore, presented all constructions as compass and straightedge problems.

Quality tools directly influence the quality of the finished product. We suggest that you use a good compass; a metal straightedge that is elevated to prevent smudging; an HB pencil with a good sharp point or a mechanical pencil with HB lead; and a Pink Pearl eraser. Heavy weight bond paper is an appropriate type to use for the problems, but projects may require heavier paper. Other tools and supplies will be suggested as needed.

Before you start, make sure the tips of your compass are aligned and that you have a smooth surface to work on. Keep in mind as you work that larger constructions are usually easier to do and are generally more accurate.

In the section that follows, eight basic constructions are presented with points labeled as an aid to understanding the steps. Labels are not an essential part of constructions but you may want to use them at first to gain better understanding of the techniques involved. Later on, when using constructions as a basis for a project design, you will find labelling unnecessary.

> Creation directly from principles, and not through the imitation of appearances, is the real way to freedom for an artist. Originality is the product of knowledge, not guesswork.
>
> Joseph Schillinger

The only permissible **TOOLS** for the following constructions are:

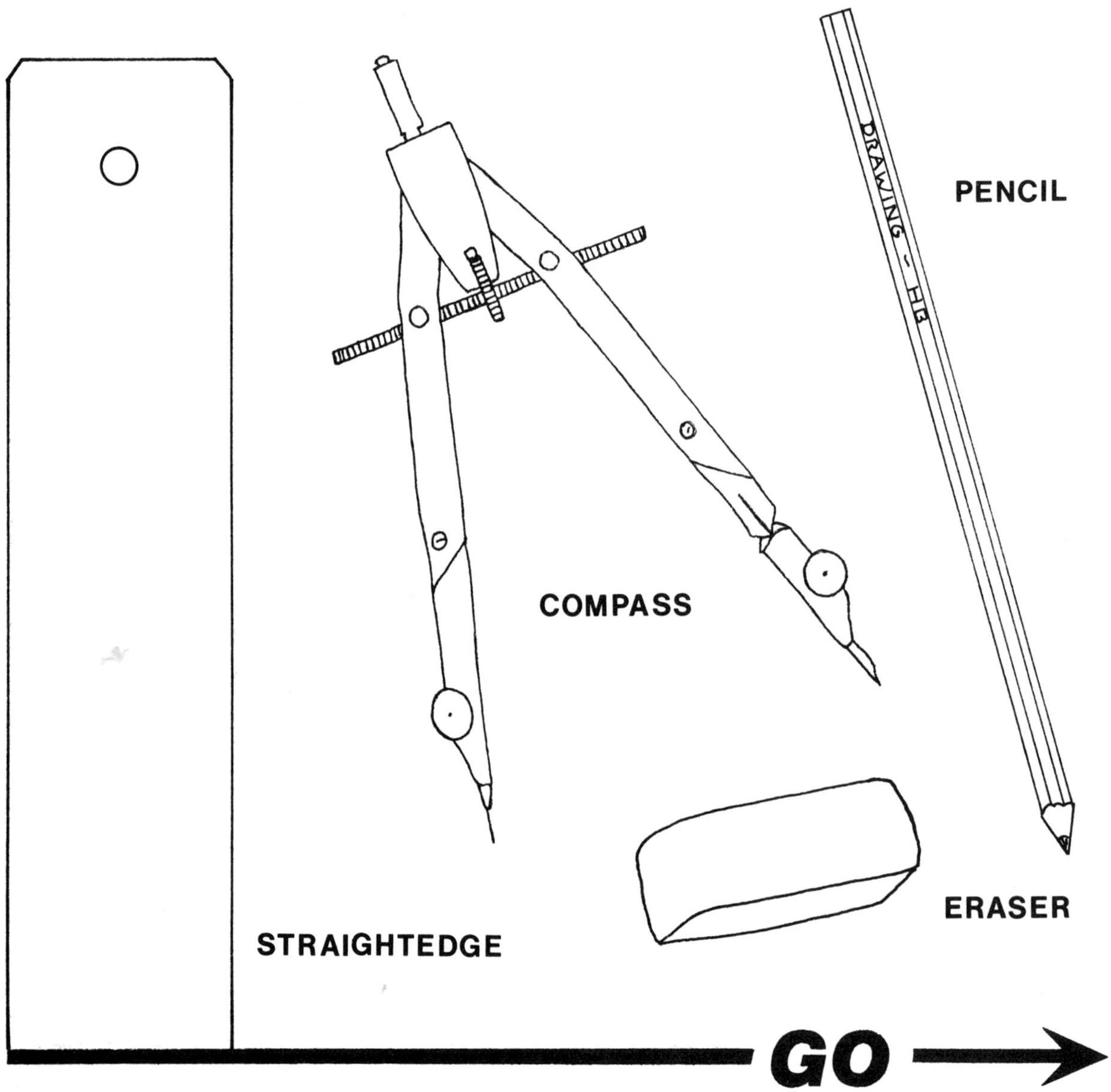

CONSTRUCT A LINE SEGMENT CONGRUENT TO A GIVEN LINE SEGMENT

Given $\overline{AB}$ construct $\overline{CD}$ such that $\overline{AB} \cong \overline{CD}$.

1. Draw a line segment* longer than $\overline{AB}$.
2. Choose a point C on the longer segment to be one endpoint of the constructed segment.
3. Place the tips of the compass on A and B respectively.
4. Without changing the compass setting, place the metal tip on C and cut an arc that intersects the longer segment.
5. Label the point of intersection D.

Now $\overline{AB} \cong \overline{CD}$. Note that the compass is used for measuring in Euclidean constructions. Equal lengths are obtained when the setting is left unchanged.

* All underlined words can be found in the glossary and all mathematical symbols are explained in Appendix A.

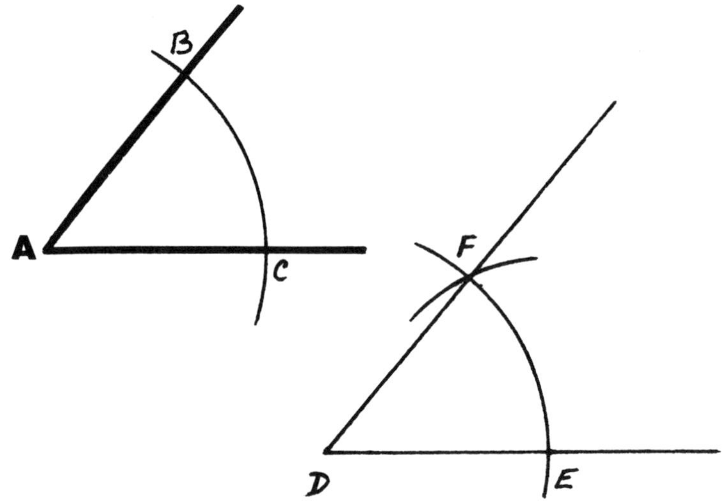

CONSTRUCT AN ANGLE CONGRUENT TO A GIVEN ANGLE

Given $\angle A$, construct $\angle D$ such that $\angle A \cong \angle D$.

1. Draw a line segment and label one end-point D.
2. Put the metal tip of the compass on A and cut an <u>arc</u> that intersects both sides of $\angle A$.
3. Label these points of intersection B and C.
4. Without changing the compass setting, place the metal tip on D and cut an arc longer than $\overset{\frown}{BC}$ so that it intersects the line segment.
5. Label this point of intersection E.
6. Put the metal tip of the compass on C and adjust the setting so that the pencil tip rests on B.
7. Without changing the setting, put the metal tip on E, and cut an arc that intersects the arc through E.
8. Label this point of intersection F.
9. Draw $\overrightarrow{DF}$.

Now $\angle A \cong \angle D$.

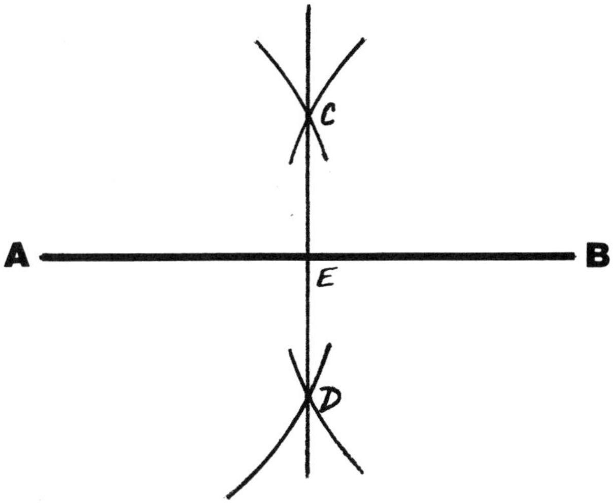

BISECT A GIVEN LINE SEGMENT

Given $\overline{AB}$, divide the segment into two congruent segments.

1. Open the compass to more than half the distance from A to B.
2. Place the metal tip on A and cut arcs above and below $\overline{AB}$.
3. Without changing the compass setting, place the metal tip on B and cut arcs above and below $\overline{AB}$ so that they intersect the other two arcs.
4. Label these points of intersection C and D.
5. Draw $\overleftrightarrow{CD}$.
6. Label E the point of intersection of $\overline{AB}$ and $\overleftrightarrow{CD}$.

Now $\overline{AE} \cong \overline{EB}$.

$\overleftrightarrow{CD}$ bisects $\overline{AB}$ and is called the bisector of $\overline{AB}$.

E is the midpoint of $\overline{AB}$.

Also, $\overleftrightarrow{CD} \perp \overline{AB}$ and therefore $\overleftrightarrow{CD}$ is the perpendicular bisector of $\overline{AB}$.

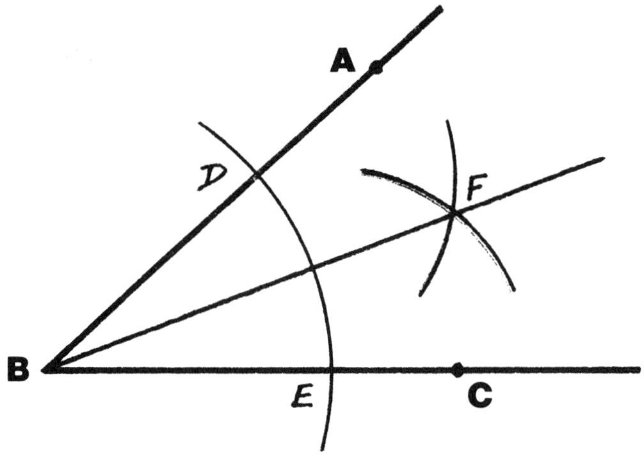

BISECT A GIVEN ANGLE

Given ∠ABC, Divide the angle into two congruent angles.

1. Place the metal tip of the compass on B and cut an arc that intersects both sides of ∠ABC.
2. Label these points D and E.
3. Open the compass to more than half the distance from D to E.
4. Place the metal point on D and cut an arc in the interior of ∠ABC.
5. Without changing the compass setting, place the metal tip on E and cut an arc that intersects the one drawn in Step 4.
6. Label the point of intersection of the two arcs F.
7. Draw $\overrightarrow{BF}$.

Now ∠ABF ≅ ∠FBC.
$\overrightarrow{BF}$ bisects ∠ABC and is called the bisector of ∠ABC.

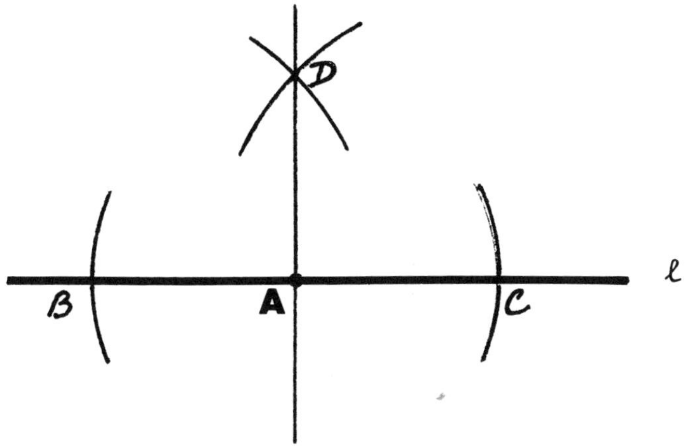

CONSTRUCT A PERPENDICULAR TO A LINE THROUGH A POINT ON THE LINE

Given point A on line ℓ, construct a line through A, perpendicular to line ℓ.

1. Place the metal tip of the compass on A and cut an arc that intersects ℓ on either side of A.
2. Label the points of intersection B and C.
3. Open the compass to more than half the distance from B to C.
4. Place the metal tip on B and cut an arc on one side of the line.
5. Without changing the compass setting, place the metal tip on C and cut an arc that intersects the one drawn in Step 4.
6. Label the point of intersection D.
7. Draw $\overleftrightarrow{AD}$.

Now $\overleftrightarrow{AD}$ is perpendicular to line ℓ.

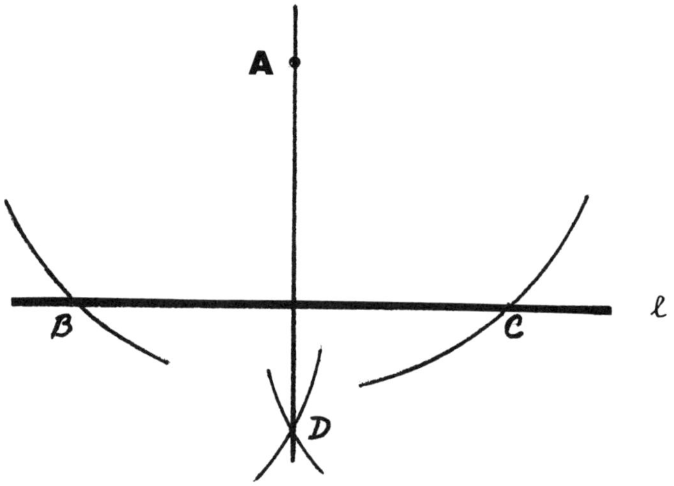

CONSTRUCT A PERPENDICULAR TO A LINE THROUGH A POINT NOT ON THE LINE

6

Given line ℓ and point A not on ℓ, construct a line through A, perpendicular to line ℓ.

1. Open the compass to more than the distance from A to ℓ.
2. With the metal tip on A cut an arc that intersects ℓ in two places.
3. Label the points of intersection B and C.
4. Place the metal tip on B and cut an arc on the side of ℓ opposite A.
5. Without changing the setting, place the metal tip on C and cut an arc that intersects the one drawn in Step 4.
6. Label the point of intersection D.
7. Draw $\overleftrightarrow{AD}$.

Now $\overleftrightarrow{AD}$ is perpendicular to line ℓ.

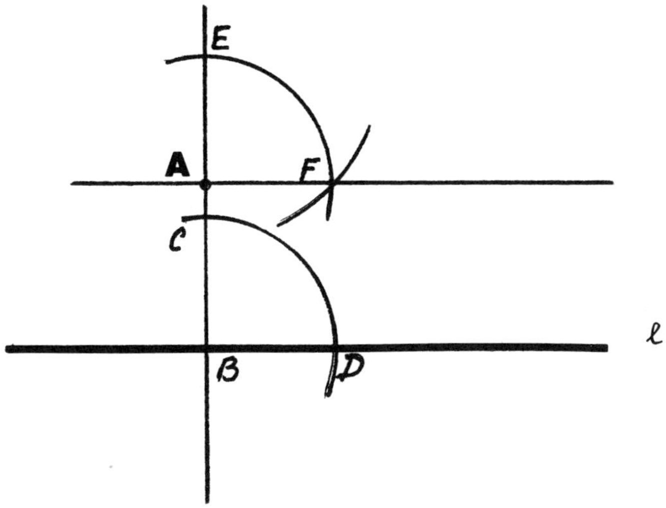

CONSTRUCT A LINE PARALLEL TO A GIVEN LINE THROUGH A GIVEN POINT

Given line ℓ and point A, construct a line through A <u>parallel</u> to line ℓ.

1. Draw a line through A that intersects ℓ.
2. Label the point of intersection B.
3. Place the metal tip of the compass on B and cut an arc that intersects both $\overleftrightarrow{AB}$ and ℓ.
4. Label the point of intersection C and D.
5. Without changing the compass setting, place the metal tip on A and cut an arc on the same side of $\overleftrightarrow{AB}$ as the arc drawn in Step 3.
6. Label E the point of intersection of $\overleftrightarrow{AB}$ and the arc drawn in Step 5.
7. Place one of the compass tips on D and the other on C.
8. Without changing the setting, place the metal tip on E and cut an arc that intersects the arc drawn on Step 5.
9. Label the point of intersection F.
10. Draw $\overleftrightarrow{AF}$.

Now $\overleftrightarrow{AF}$ is parallel to line ℓ.

Frequently it is useful to be able to divide a line segment into any number of congruent segments. If the number of segments desired is a power of two, i.e., 2, 4, 8, 16, 32, etc., it is probably easiest to bisect the segments and continue bisecting the newly formed segments.

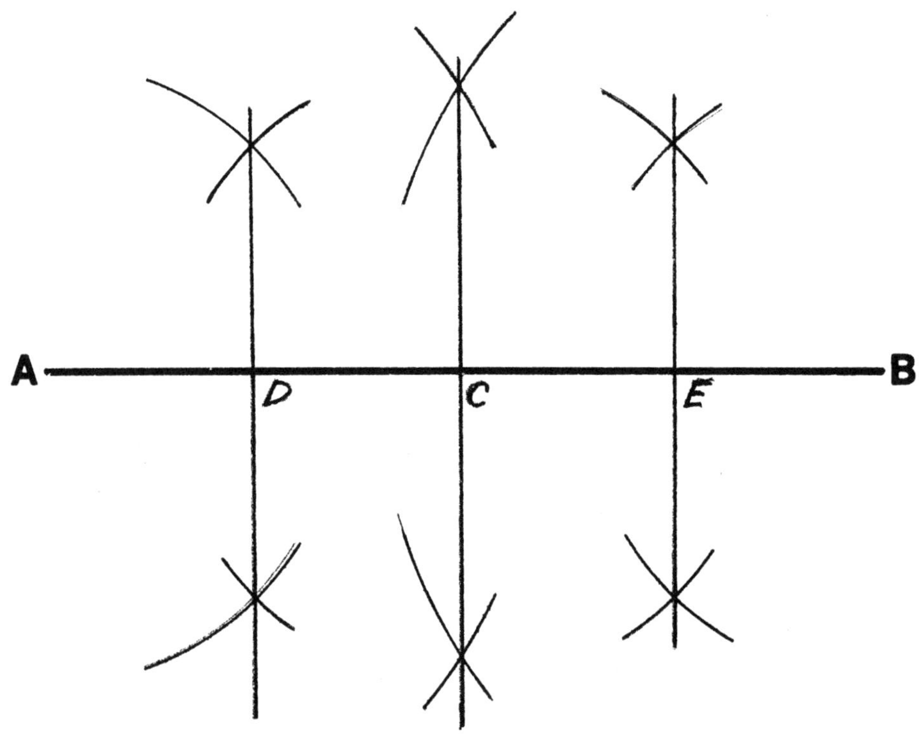

In the diagram, $\overline{AD} \cong \overline{DC} \cong \overline{CE} \cong \overline{EB}$

If the number of segments is not a power of two, the following construction must be used. We have chosen to divide the line segment into three congruent segments for purposes of demonstration. The same construction, however, may be used for any number of divisions. But, Step 2 must be adjusted to allow for the desired number of segments, that is, if the segment were to be divided into six congruent segments, six congruent segments must be marked off on the ray.

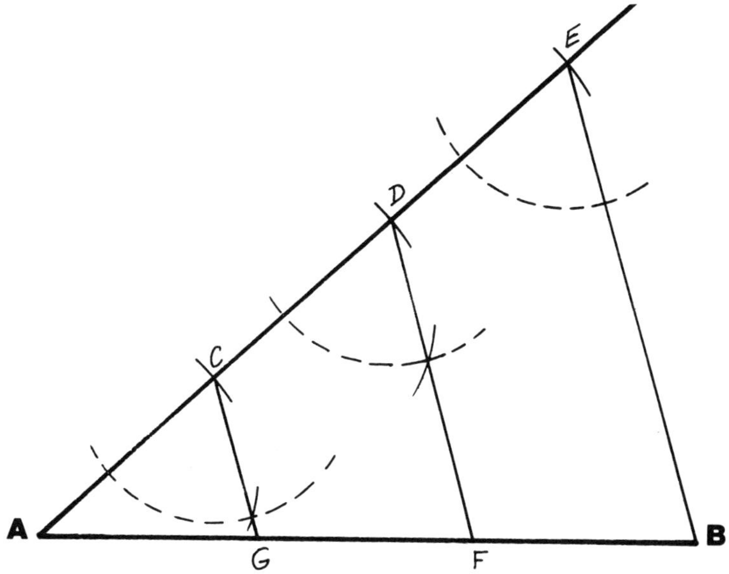

DIVIDE A LINE SEGMENT INTO A GIVEN NUMBER OF CONGRUENT SEGMENTS

8

Given $\overline{AB}$, divide the segment into three congruent segments.

1. Draw a <u>ray</u> from A to form an <u>acute</u> angle with $\overline{AB}$.
2. Mark off three congruent segments on the ray by first placing the metal tip on A and cutting an arc on the ray. Repeat without changing the compass setting, each time placing the metal tip on the most recently cut arc.
3. Label the points of intersection C, D and E, respectively.
4. Draw $\overline{EB}$.
5. Construct an angle at D (with side $\overrightarrow{DC}$) congruent to $\angle DEB$ and extend the side so that it intersects $\overline{AB}$. Refer to construction 2.
6. Label the point of intersection F.
7. Repeat Step 5 at C.
8. Label the point of intersection G.

<u>Note</u>: Once $\overline{FB}$ is determined, this measurement could be used to mark off the remaining congruent segments on $\overline{AB}$. This makes it necessary to copy only one angle.

Now $\overline{AG} \cong \overline{GF} \cong \overline{FB}$

Keep the lead point of your compass finely chiseled, to a depth of about 3/8"; a dull point produces inaccuracies!

PROBLEM **1** Construct a triangle with sides congruent to segments a, b, and c respectively.

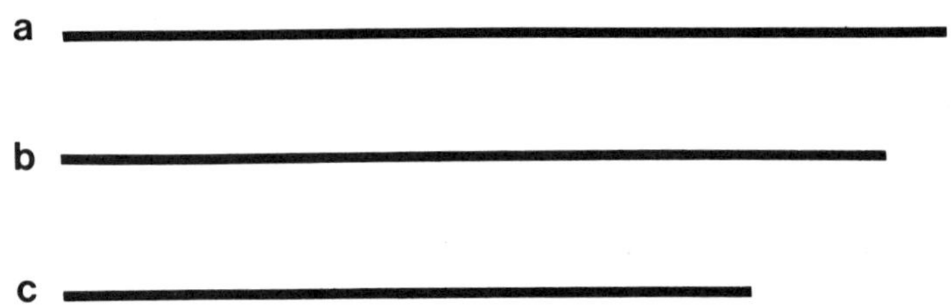

Hint: Refer to construction 1.

PROBLEM **2**

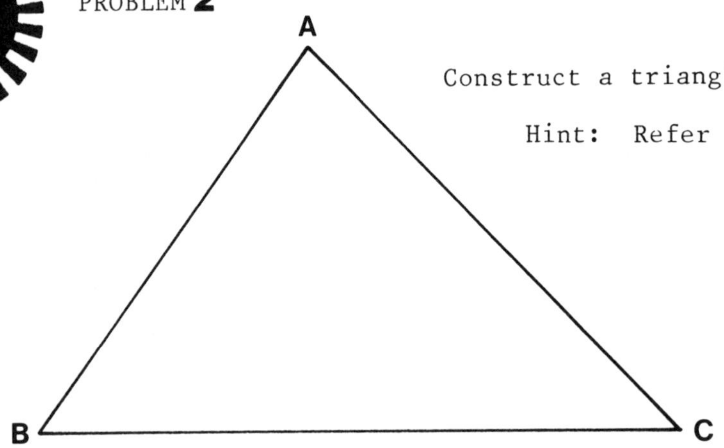

Construct a triangle congruent to $\triangle ABC$.

Hint: Refer to previous problem.

PROBLEM **3**

a. Divide $\overline{RS}$ into eight congruent segments.

Hint: Refer to Construction 3.

b. Construct a segment congruent to $\overline{RS}$ and divide it into seven congruent segments.

Hint: Refer to construction 8.

PROBLEM
4

Construct all three altitudes of $\triangle$ ABC.

Hint: Refer to Construction 6.

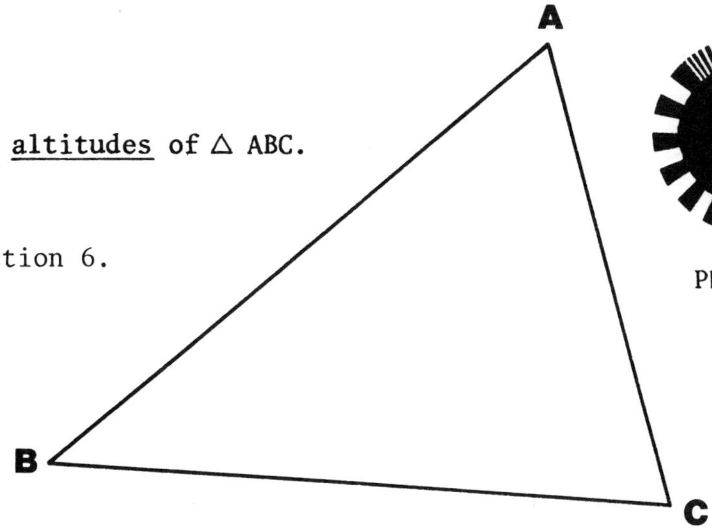

PROBLEM
5

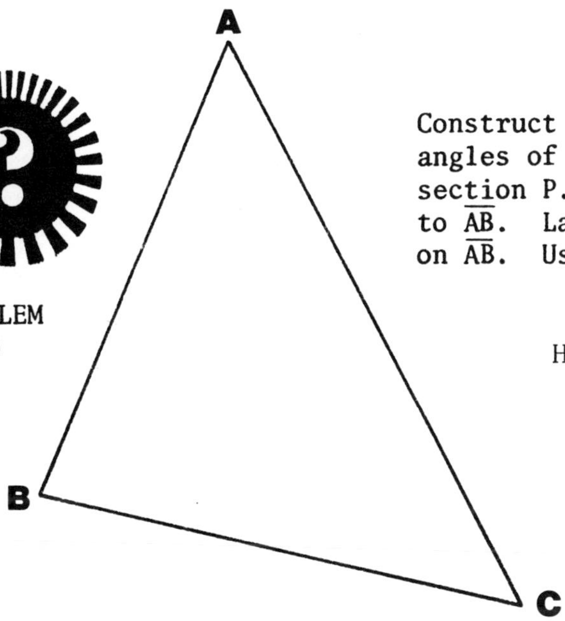

Construct the angle bisector of each of two angles of $\triangle$ ABC. Label the point of intersection P. Construct a perpendicular from P to $\overline{AB}$. Label Q, the point of interesection on $\overline{AB}$. Using $\overline{PQ}$ as a radius, draw $\odot$P.

Hint: Refer to Constructions 4 and 6.

PROBLEM
6

Construct the perpendicular bisector of each of two sides of $\triangle$ ABC. Label the point of intersection O. Using $\overline{OA}$ as a radius, draw $\odot$O.

Hint: Refer to Construction 3.

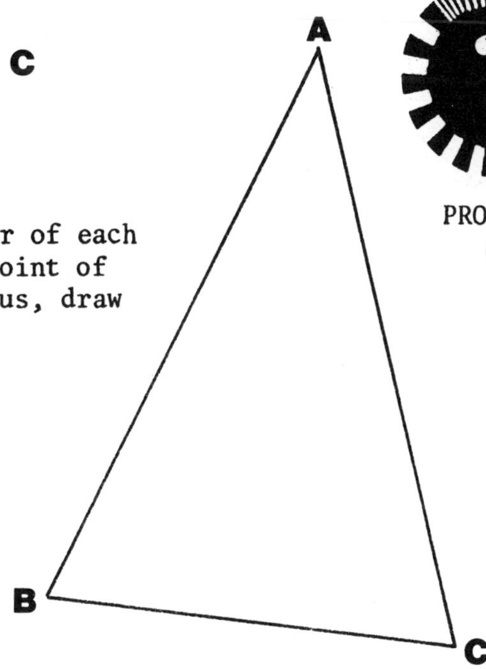

PROBLEM 7

Construct a <u>rectangle</u> with width equal to MN and length equal to OP.

Hint: Refer to constructions 5 and 1.

PROBLEM 8

Construct two different sized triangles each of which has two angles congruent to the two given angles.

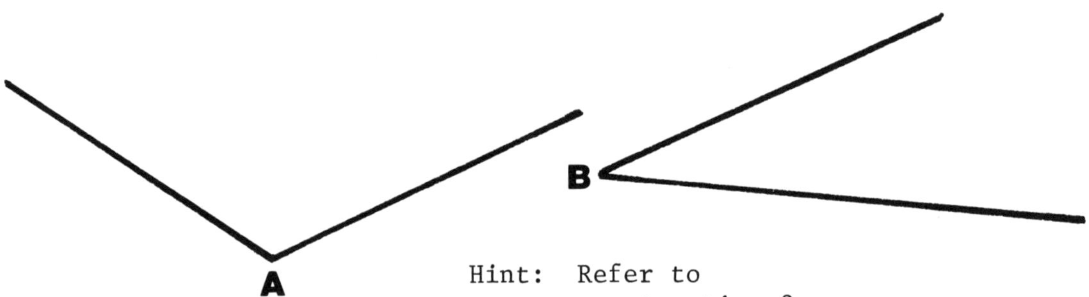

Hint: Refer to construction 2

PROBLEM 9

On $\overleftrightarrow{AB}$ construct lines parallel to $\overleftrightarrow{CD}$ through points A and B respectively.

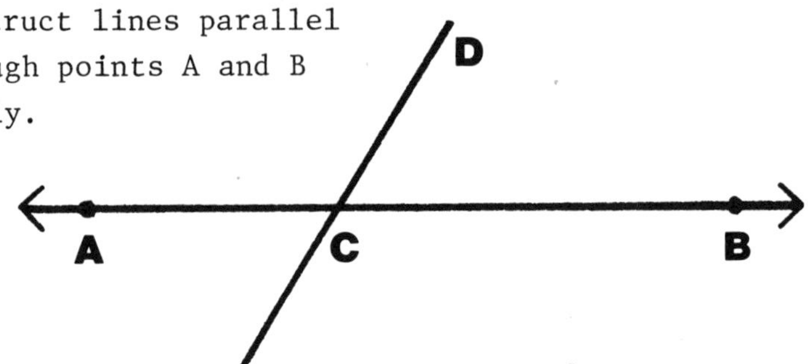

Hint: Refer to construction 7.

PROJECT **1**

Choose one of the following geometric designs and duplicate it in a square measuring four inches on a side. Repeat the construction a minimum of three times making variations on the original each time. Put the four (or more) blocks in finished form by doing them in black and white or color and by being concerned with the arrangement.

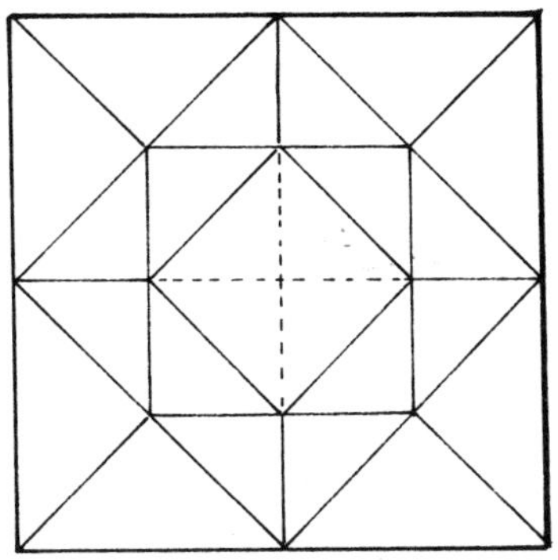

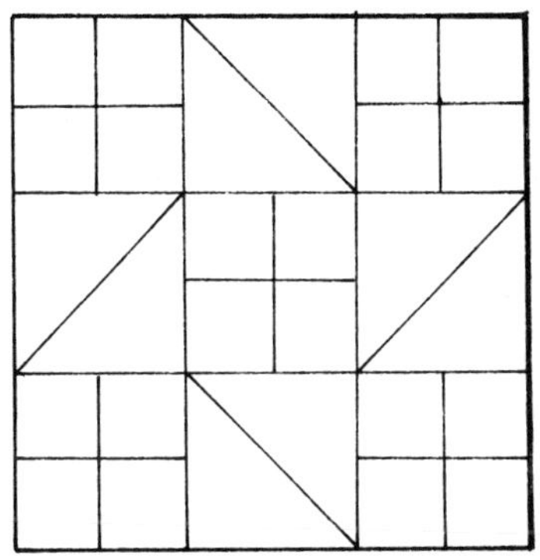

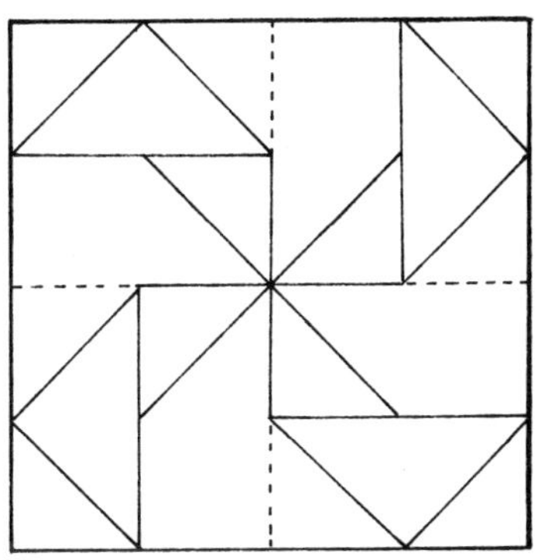

PROJECT **2**

On heavy paper or illustration board, construct a square that measures 12" x 12". Then, with compass and straightedge, construct a figure similar to one of the ones drawn below. Give the figure color by using materials of your choice (fabric, colored pencils, paints, inks, construction paper, etc.).

Check Appendix B for the properties of a square.

PROJECT 3

On heavy paper or illustration board, construct a rectangle that measures 8" x 13". (Check Appendix B for the properties of a rectangle.) Create a design with the following restrictions:

1. The design must be done within the limits of this constructed rectangle.
2. It must use at least <u>three</u> of the basic constructions you have just mastered.
3. It must be done in black and white only.

PROJECT 4

Using the above instructions for project 3 as a basis, create a solution using color.

PROJECT 5

Using multiples of one of the squares given in PROJECT 2 (a photocopier would be useful here), create a paper "quilt" of three blocks by four blocks. It may be in black and white only or use a color scheme of your choice.

Sharon Maniates

Student Work
Chapter 1, project 1

Further Reading

Bruni, James V. Experiencing Geometry. Belmont, California: Wadsworth Publishing Company, Inc., 1977.

Gardner, Martin. "Geometric Constructions with a Compass and Straightedge, and Also with a Compass Alone." Scientific American. Vol. 221, no. 3, September, 1969.

Seymour, Dale and Reuben Schadler. Creative Constructions. Palo Alto, California: Creative Publications, 1974.

Unique Relationships

2

Proportion is a principle in Nature which is a purely mathematical one and to be rightly interpreted by man through the means of geometry; therefore geometry is not only the gateway to science but it is also a noble portal opening wide into the realms of art.

"Nature's Harmonic Unity"
Samuel Colman

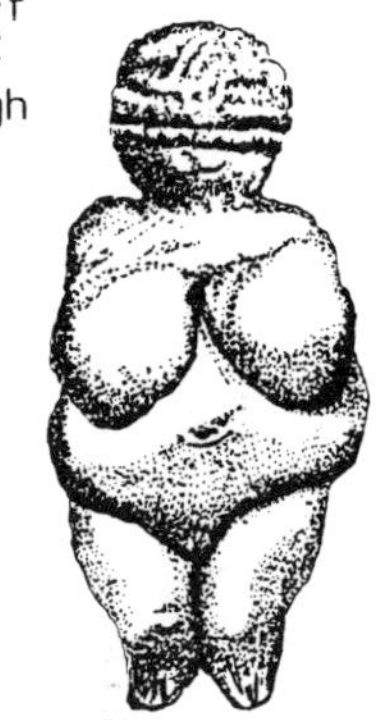

Venus of Willendorf
15,000 - 10,000 BC
Stone. 4 3/8" High

Division of space is a problem that the artist meets continually. The question is how does one go about dividing the space so that there are harmonious relationships between the smaller elements of area and the whole? There must be some way to determine the corect relationships beyond intuition. The understanding of the concept of proportion leads to more rational solutions to the problem. Reason and intuition must complement each other.

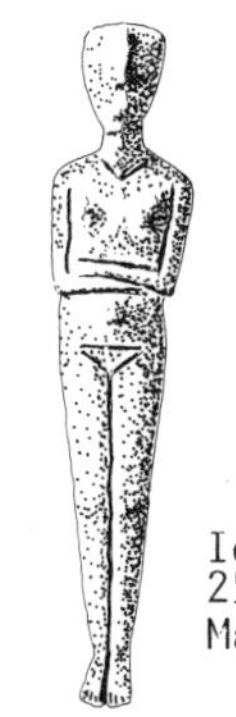

Idol from Amorgos
2500 -1100 BC
Marble. 30" High

One of the most obvious areas of concern is man's concept of human body proportions. Each age redefines the "Body beautiful"; however, many of our ideas about correct proportions come down to us from Classical Greece. When working with the nude form the Greek artist was fundamentally interested in the ideal; articulating a concept of a perfect shape and adapting natural form to this ideal. The famous Greek sculptor Polycleitos was interested in the ideas of completeness, balance and clarity communicated through the nude form of the athlete. It is believed that he wrote about a system of proportions in a book presumed lost.

Mantiklos Apollo
7th C. Bronze
8" High

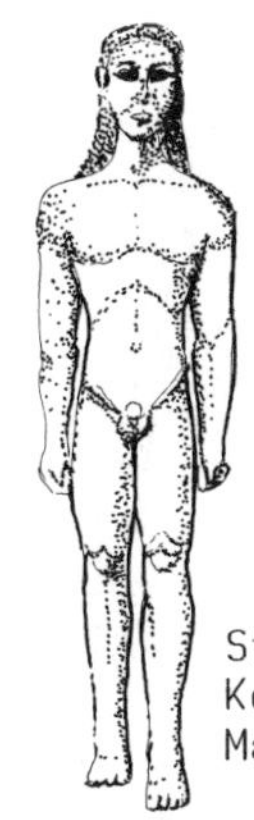

Standing Youth
Kouros. c 600 BC
Marble. 6 ½" High

Standing Woman by
Gaston Lachaise
1912 - 1927
Bronze. 70" High

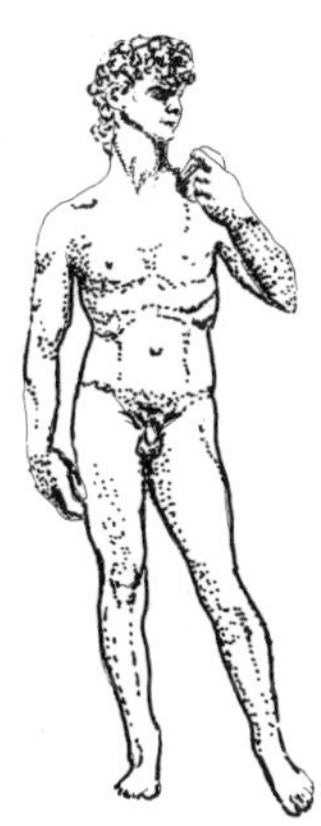

David by
Michelangelo
1501 - 1504
Marble. 13' 5"

Phidias, whom we shall refer to again, determined that the ideal was 7½ heads with the length of the person's head as the measuring unit. His concept is still used by artists today with:

a. the waist determined at 3 heads
b. the hips at the halfway point of 3 3/4 heads
c. the legs being considered one-half the length of the entire body
d. the knees at 5 heads
e. the ankle at 7 heads
f. ankle to foot at ½ heads

For Praxiteles, the Hellenistic Greek sculptor, ca. 340 B.C., and the Italian Michelangelo, 16th Century A.D., the ideal became 8 heads.

The difference between the two concepts is illustrated in the diagram.

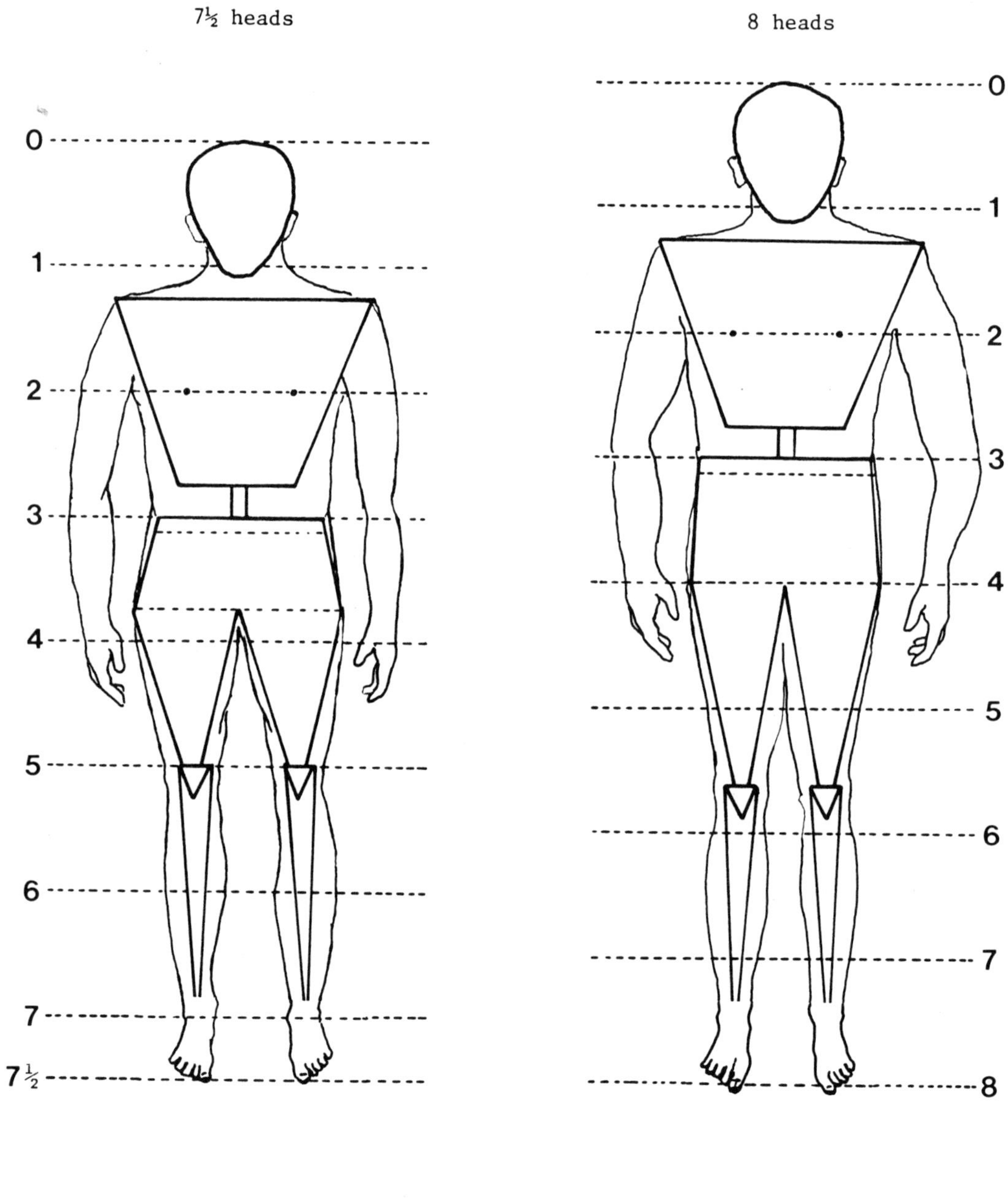

What does proportion mean?

This is a term we use frequently in our everyday lives; one that brings to mind the qualities of harmony, balance, and fitness. We sense that something is "out of proportion" when it feels "wrong" and we tend to describe that condition as ugly.

The artist talks about pleasing proportions suggesting the relationship of a part to the whole, but never articulates a precise definition except as expressed through a given artwork.

Language tends to break down in discussing this idea. It is no wonder that the expression "beauty is in the eye of the beholder" has become a catchphrase or a last defense.

Even the dictionary "defines" proportion to be:

1. the comparative relation between things or magnitudes;
2. PROPORTIONS, dimensions or size;
3. any portion or part in its relation to the whole;
4. symmetry or balance.

The definition is not only circular but moves away from specifics and becomes more and more general.

Through the course of our research we have discovered that Nature tends to rely repeatedly on the same relationships. Therefore, we are more strongly committed to the idea that Nature provides the model for beauty, and mathematics describes it. The dictionary gives us the feeling that proportion is a relationship between two things. Mathematics tells us precisely what that relationship is.

Before defining proportion, it is necessary to define the term ratio. A ratio is the comparison of numbers to each other. This comparison can be expressed in a number of ways. If, for example, we are comparing the number 2 to the number 3, we may write 2 to 3 or 2:3 or simply 2/3. Hence, every fraction is a ratio in which the numerator is compared to the denominator. There are times when a ratio compares three or more numbers such as is the case when discussing the relative lengths of the sides of a triangle. For such cases we use the notation a:b:c.

A proportion is simply an equation in which two ratios are set equal to each other, such as 2/3 = 4/6 or 2:3::4:6. This same equation may be expressed without the mathematical symbols by writing 2 is to 3 as 4 is to 6. When we write $\frac{a}{b} = \frac{c}{d}$, a and d are called the extremes of the proportion and b and c are called the means. A mean proportion is one in which the means are equal. The names of the terms become much more meaningful when we write a:b::c:d. The extremes, then, are the outside terms and the means are the middle terms.

One mean proportion which appears with amazing frequency in Nature and mathematics and has been used throughout the centuries by artists is called the Divine Proportion. This proportion is derived from the division of a line segment into two segments with the special property that the ratio of the whole segment to the longer part is the same as the ratio of the longer part to the shorter part.

Photo by Richard Newman

Michael Grutkowski

Student Work
Chapter 2, project 2

THE DIVINE PROPORTION

$$\frac{AB}{AC} = \frac{AC}{CB}$$

or

$$\frac{a+b}{a} = \frac{a}{b}$$

The ratio expressed by either side of this **equation** is called the Golden Ratio or the Golden Mean.

The point which divides a line segment into two segments in the Golden Ratio is unique with the exception of order; that is, the longer segment may be at either end of the original segment.

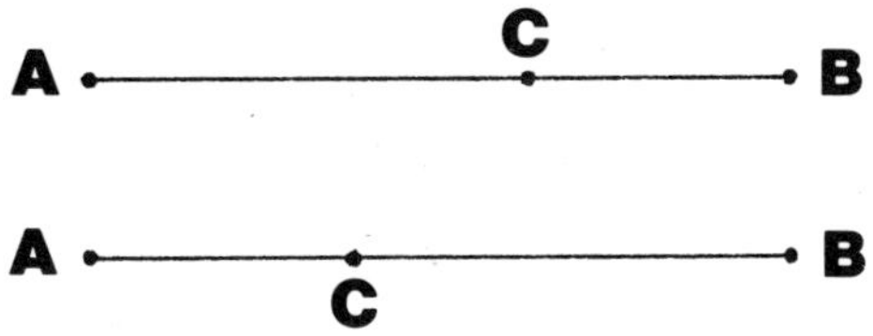

The point C in the above illustration is called the Golden Section or the Golden Cut and may be found geometrically quite easily by using the following technique with compass and straightedge.

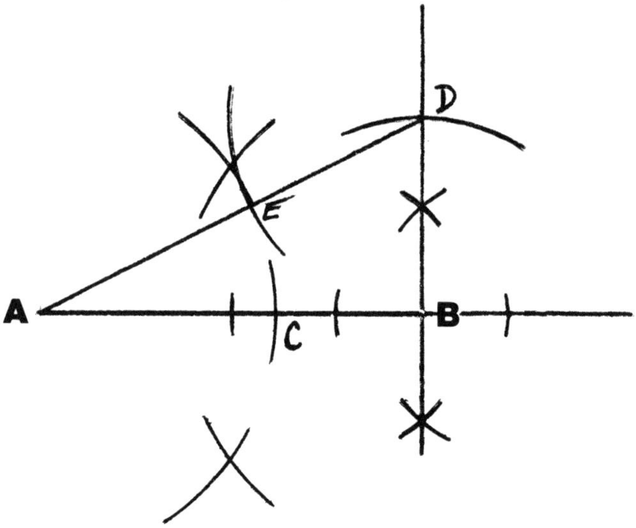

CONSTRUCT THE GOLDEN SECTION OF A LINE SEGMENT

Given $\overline{AB}$, find the point C such that

$\frac{AB}{AC} = \frac{AC}{BC}$

1. Extend $\overleftrightarrow{AB}$ through B.
2. Construct a perpendicular to $\overleftrightarrow{AB}$ through B.
3. Open your compass so that it measures $\frac{1}{2}$AB. (Use the segment bisector construction to divide $\overline{AB}$ in half.)
4. Without changing the compass setting, put the metal tip on B and cut an arc that intersects the line perpendicular to $\overleftrightarrow{AB}$.
5. Label the point of intersection D.
6. Draw $\overline{AD}$.
7. Put the metal tip on D and the pencil tip on B. Then cut an arc that intersects $\overline{AD}$.
8. Label the point of intersection E.
9. Put the metal tip on A and the pencil tip on E. Then cut an arc that intersects $\overline{AB}$.
10. Label the point of intersection C.

Now C is the Golden Section of $\overline{AB}$ and

$\frac{AB}{AC} = \frac{AC}{BC}$.

THE GOLDEN RATIO

Let C be the Golden Section of $\overline{AB}$. We know then that $\frac{AC}{CB}$ is the Golden Ratio. We can easily find a numerical value for this ratio if we let CB=1 and AC=x. Then $\frac{AC}{CB} = \frac{x}{1} = x$.

By using a little basic algebra it is a simple matter to find x. From the Divine Proportion we know that $\frac{x + 1}{x} = \frac{x}{1}$. Cross multiplying, we get the <u>quadratic equation</u>

$$x + 1 = x^2$$

or

$$x^2 - x - 1 = 0$$

We can solve for x using the quadratic formula. (Check the glossary if the quadratic formula may have slipped your mind!)

$$x = \frac{1 \pm \sqrt{(-1)^2 - 4(1)(-1)}}{2(1)} = \frac{1 \pm \sqrt{1 + 4}}{2} = \frac{1 \pm \sqrt{5}}{2}$$

There are two solutions here, $\frac{1 + \sqrt{5}}{2}$ and $\frac{1 - \sqrt{5}}{2}$.

However, $\frac{1 - \sqrt{5}}{2}$ leads to a negative number which does not have meaning when associated with the length of a line segment; therefore, we choose the positive solution.

$$\frac{1 + \sqrt{5}}{2} = 1.61803 \text{ correct to five decimal places.}$$

Because this number appears in the most surprising places and has such unique properties, mathematicians have given it a special name Ø. This symbol is the Greek letter Phi, the first letter in the name Phidias, the Classical Greek sculptor, who used the Golden Ratio so extensively in his work. It was chosen in his honor. Phidias was chief overseer of artistic enterprises under the Athenian leadership of Pericles in the Golden Age of Greece (5th century B.C.). He was considered responsible for the frieze panels, a continuous decorative band 525 feet in length, around the Parthenon, and was attributed with creating the huge ivory and gold statue of Athena, to whom the Parthenon was dedicated, which was to be located in the <u>cella</u> of that temple.

THE GOLDEN RATIO

$$\Phi = 1.61803$$

Since $\emptyset = \frac{1+\sqrt{5}}{2}$, $\frac{1}{\emptyset} = \frac{2}{1+\sqrt{5}}$ and $\frac{1}{\emptyset}$ is the reciprocal of $\emptyset$.

Correct to five decimal places $\frac{1}{\emptyset} = .61803$.

Note that $1.61803 - 1 = .61803$ or $\emptyset - 1 = \frac{1}{\emptyset}$. A proof of this property appears in Appendix B.

PHI IS THE ONLY NUMBER THAT BECOMES ITS OWN RECIPROCAL WHEN DECREASED BY ONE.

If you recall, when we solved the quadratic equation $x^2 - x - 1 = 0$, we got two solutions and chose the positive one to represent $\emptyset$. If we examine both roots we see that they bear an unusual relationship to each other. As we know, $\emptyset = \frac{1+\sqrt{5}}{2}$. Let us name the negative root $\emptyset'$ (Phi prime) so $\emptyset' = \frac{1-\sqrt{5}}{2}$. Multiplying we get:

$$\emptyset \cdot \emptyset' = \left(\frac{1+\sqrt{5}}{2}\right)\left(\frac{1-\sqrt{5}}{2}\right) = \frac{1-5}{4} = \frac{-4}{4} = -1$$

That is, $\emptyset$ and $\emptyset'$ are negative reciprocals since $\emptyset \cdot \emptyset' = -1$.

If we add we have:

$$\emptyset + \emptyset' = \frac{1+\sqrt{5}}{2} + \frac{1-\sqrt{5}}{2} = \frac{2}{2} = 1$$

and again the number 1 appears inextricably linked with $\emptyset$.

Let us consider the sequence whose terms are consecutive powers of ϕ.

$$1, \phi, \phi^2, \phi^3, \ldots, \phi^n, \ldots$$

This is a geometric sequence since each term is gotten by multiplying the preceding term by the same number, namely ϕ.

$$\begin{aligned} 1 \cdot \phi &= \phi \\ \phi \cdot \phi &= \phi^2 \\ \phi^2 \cdot \phi &= \phi^3 \\ \phi^n \cdot \phi &= \phi^{n+1} \end{aligned}$$

If we examine the terms more closely, we see that it is the case that each term is also the sum of the two preceding terms. Therefore, it is at the same time a summation sequence.

Since $\phi^2 - \phi - 1 = 0$, we have $\phi^2 = 1 + \phi$

$$\phi^2 = \phi^0 + \phi^1$$

then: $\phi^3 = \phi^2 . \phi = (1 + \phi)\phi = \phi + \phi^2 = \phi^1 + \phi^2$

and $\phi^4 = \phi^3 . \phi = (\phi + \phi^2)\phi = \phi^2 + \phi^3$

$$\phi^5 = \phi^4 . \phi = (\phi^2 + \phi^3)\phi = \phi^3 + \phi^4$$

This process could be carried on indefinitely (with no small amount of tedium involved!) and the result would continue to hold; that is,

$$\longrightarrow \quad \phi^n + \phi^{n+1} = \phi^{n+2} \quad \longleftarrow$$

To diminish masochistic impulses for the hard-to-convince, a proof is supplied in Appendix B, along with proofs of the other three properties.

In Chapter 5 we will be examining in detail the Fibonacci Sequence, the simplest of the summation sequences. We will look at the connection between ϕ and this very special sequence at that time.

PROPERTIES OF ϕ

1. $\phi - 1 = \frac{1}{\phi}$

2. $\phi \cdot \phi' = -1$

3. $\phi + \phi' = 1$

4. $\phi^n + \phi^{n+1} = \phi^{n+2}$

The construction on page 48 shows how to divide a line into ϕ proportions, a useful technique for the designer since each segment is related to the next harmoniously.

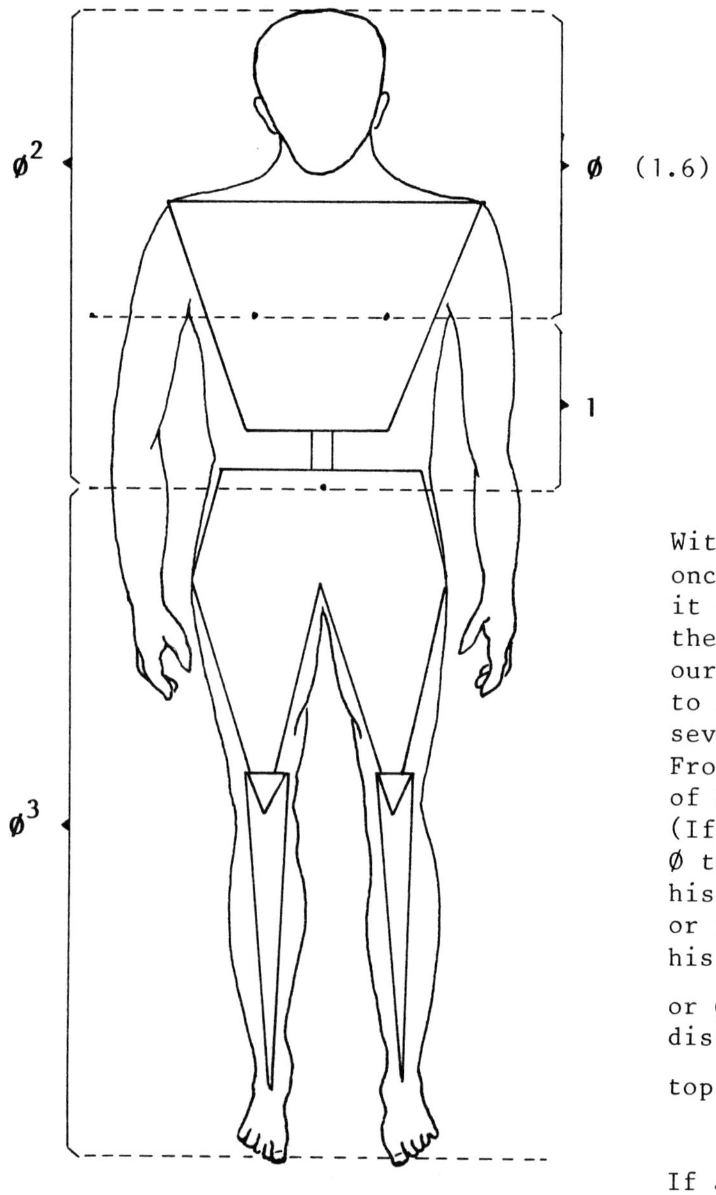

With this understanding of ϕ, let us once again look at the human body as it conforms to Phidias' concept of the ideal. If we take ten inches as our unit of measurement and apply it to a man 68 inches tall, we find several lengths in ϕ proportion. From the top of his head to the line of his breasts is 16 inches or ϕ. (If the unit is 10 then 16 becomes ϕ to one decimal place or 1.6). From his breasts to his navel is 10 inches or one unit. The measurement from his navel to the ground is 42 inches or ϕ^3. In the other direction, the distance between his navel and the top of his head is 26 inches or ϕ^2.

If a unit other than 10 inches is chosen, we can observe other occurrences of body measurements in the ratio of ϕ as illustrated in the following two examples of Classical Greek sculpture.

In the male figure, the unit is taken to be the distance between the top of the head and the line of the breasts. In the female figure, the unit is the distance from the shoulders to the navel.

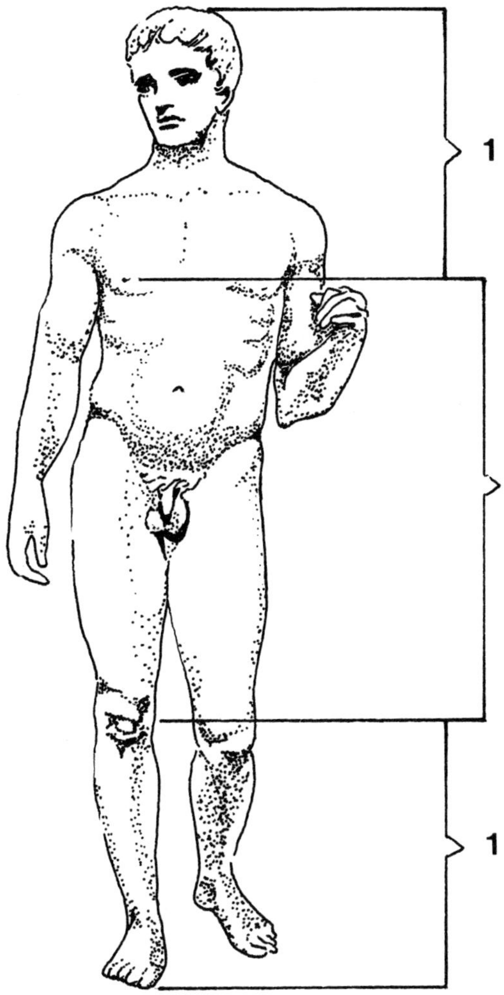

Doryphoros, the Spearbearer

Polycleitos
c. 450-440 B.C.

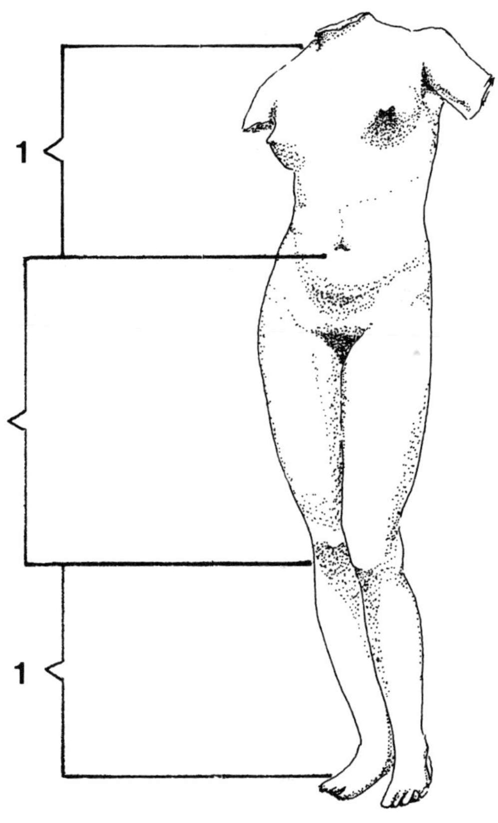

Aphrodite of Cyrene

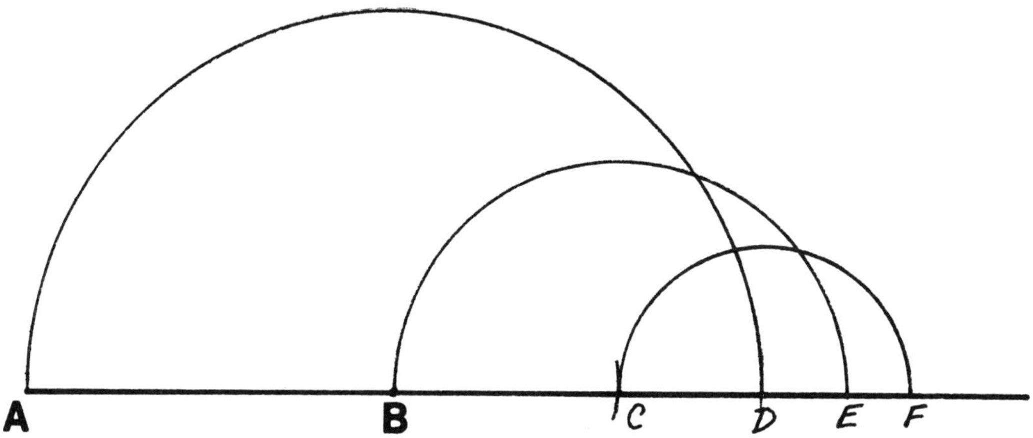

DIVIDE A LINE SEGMENT INTO ϕ PROPORTIONS

10

Given $\overleftrightarrow{AB}$, construct on $\overrightarrow{AB}$ successively smaller segments joined only by their endpoints such that the ratio of any consecutive pair is Ø.

1. Place the metal tip on B and the pencil tip on A and cut a semicircle.
2. Label the point of intersection D.
3. Use Construction 9 to find the Golden Section of $\overline{BD}$ and label it C.
4. Place the metal tip on C and the pencil tip on B and cut a semicircle.
5. Label the point of intersection E.
6. Place the metal tip on D and the pencil tip on C and cut a semicircle.
7. Label the point of intersection F.

Now $\frac{AB}{BC} = \frac{BC}{CD} = \frac{CD}{DE} = \frac{DE}{EF} = Ø$

Also,
$$AB = BC + CD$$
$$BC = CD + DE$$
$$CD = DE + EF$$

*NOTE: A close approximation of Ø-proportion subdivisions can be obtained using graph paper marked off in consecutive Fibonacci numbers. We refer the reader to Chapter 5 for further information.

The Pentagon

The Pentagon is introduced at this time rather than later on with all the other polygons because it is the structure par excellence found throughout the animate world. It provides the framework for the pentagram which clearly demonstrates the Golden ratio and is rich with visible harmonic properties.

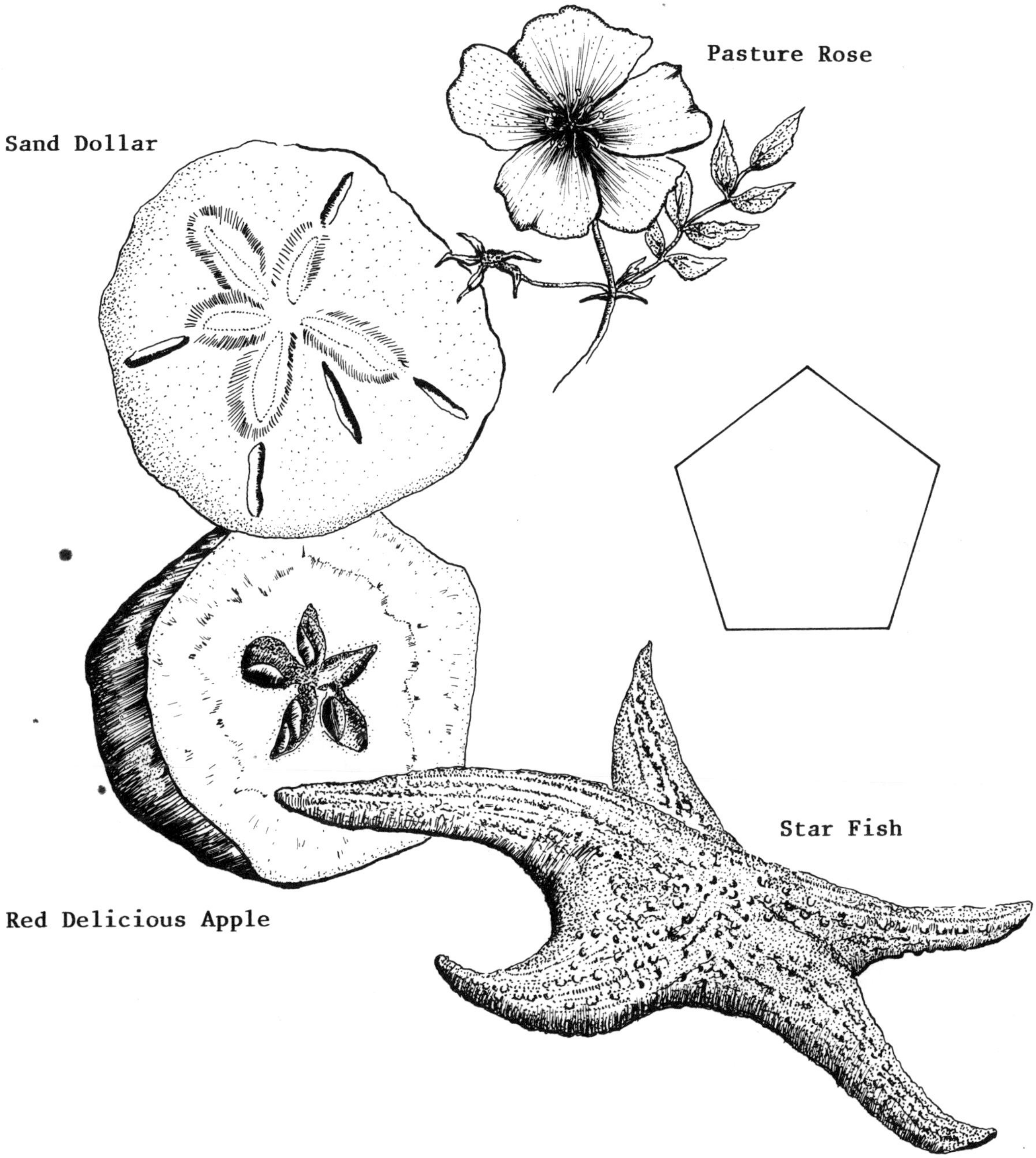

The pentagon is a five-sided simple closed curve composed of line segments that intersect only at their end points. It is called regular if all sides and angles are congruent.

The pentagram, also known as pentalpha, pentacle or pentangle, is the five-pointed star derived from connecting the non-consecutive vertices of a regular pentagon. The following diagram illustrates the appearance of the Divine Proportion in the different segments of the pentagram.

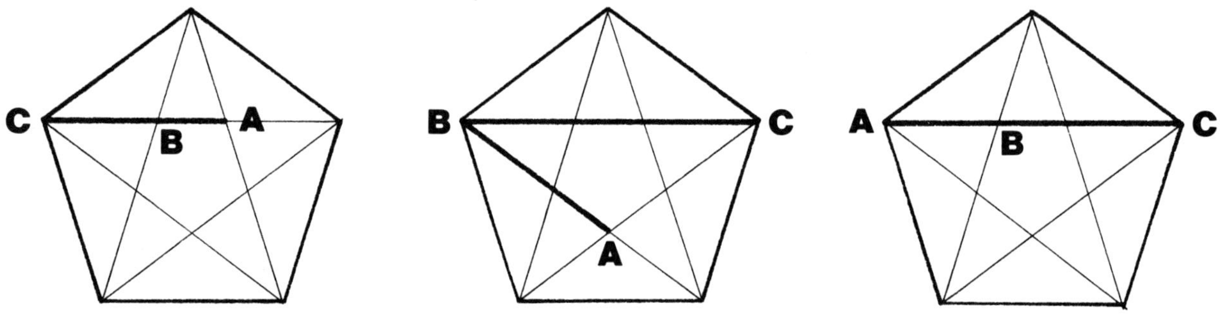

In each of the above: $\frac{AB + BC}{BC} = \frac{BC}{AB} = \phi = 1.61803$

The Divine Proportion also appears in the segments of the pentagram and its corresponding pentagon as shown below:

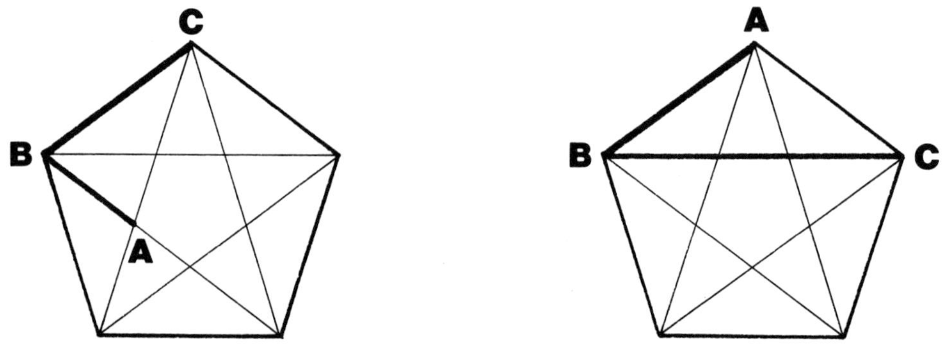

Again: $\frac{AB + BC}{BC} = \frac{BC}{AB} = \phi = 1.61803$

The pentagram is a fitting symbol of visually expressed harmony and order. It was the chosen symbol and seal of the Order of the Pythagoreans, a secret society of Ancient Greek mathematicians organized in the 6th century B.C. They were fascinated by natural harmony and order and believed it could be explained by numerical relationships.

> The power of the golden section to create harmony arises from its unique capacity to unite the different parts of a whole so that each preserves its own identity, and yet blends into the greater pattern of a single whole. The Golden Section's ratio is an irrational infinite number which can only be approximated. Yet such approximations are possible even within the limits of small whole numbers. This recognition filled the ancient Pythagoreans with awe: they sensed in it the secret power of a cosmic order. It gave rise to their belief in the mystical power of numbers. It also led to their endeavors to realize the harmonies of such proportions in the patterns of daily life, thereby elevating life to an art.
>
> The Power of Limits
> Gyorgy Doczi

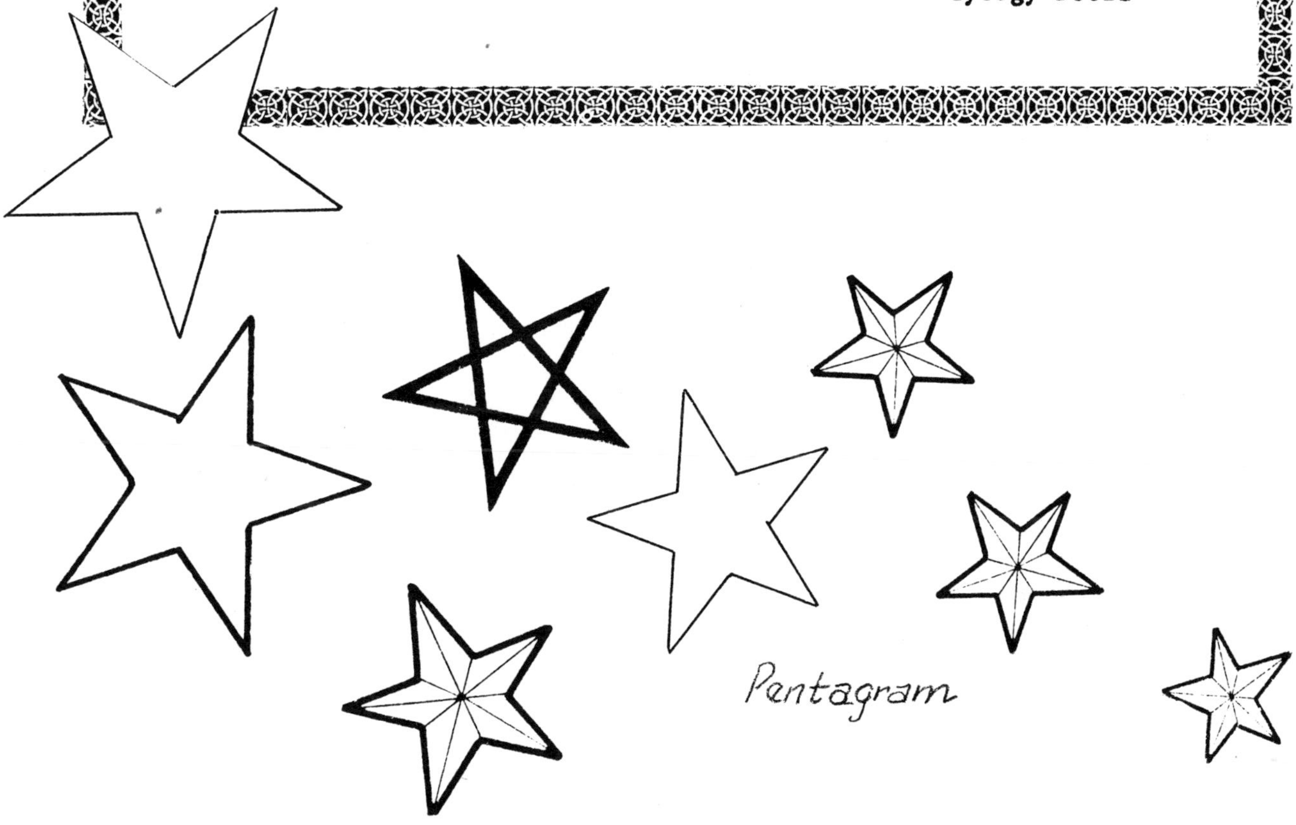

There are several ways to construct a regular pentagon. Two are illustrated here since the given conditions for the problem may vary. It is one thing to generate a pentagon if the length of a side is known and quite another to fit a pentagon into a given space.

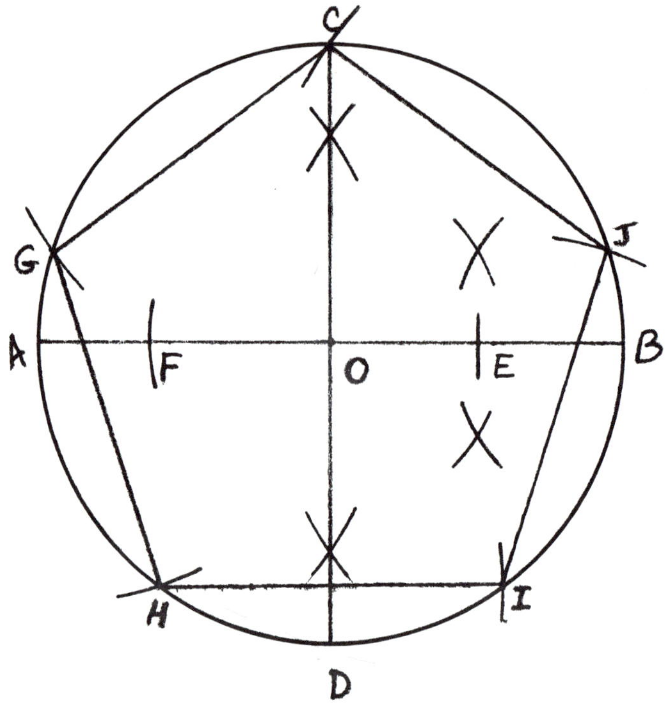

INSCRIBE A REGULAR PENTAGON IN A CIRCLE

Given $\odot$ O, inscribe within it a regular pentagon.

1. Draw a diameter and label its endpoints A and B, respectively.
2. Construct another diameter which is the perpendicular bisector of $\overline{AB}$.
3. Label the endpoints C and D, respectively.
4. Bisect $\overline{OB}$ and label its midpoint E.
5. Put the metal tip on E and the pencil tip on C and cut an arc that intersects $\overline{AB}$.
6. Label the point of intersection F.
7. Place the metal tip on C and the pencil tip on F and cut an arc that intersects the circle.
8. Label the point of intersection G.
9. Draw $\overline{CG}$.
10. Without changing the compass setting, put the metal tip on G and cut another arc that intersects the circle.
11. Label the point of intersection H.
12. Draw $\overline{GH}$.
13. Without changing the setting, continue around the circle placing the metal tip on H to cut intersection I, and on I to cut intersection J.
14. Draw $\overline{HI}$, $\overline{IJ}$ and $\overline{CJ}$.

*make sure your compass lead is sharp to insure accuracy

Now CGHIJ is a regular pentagon.

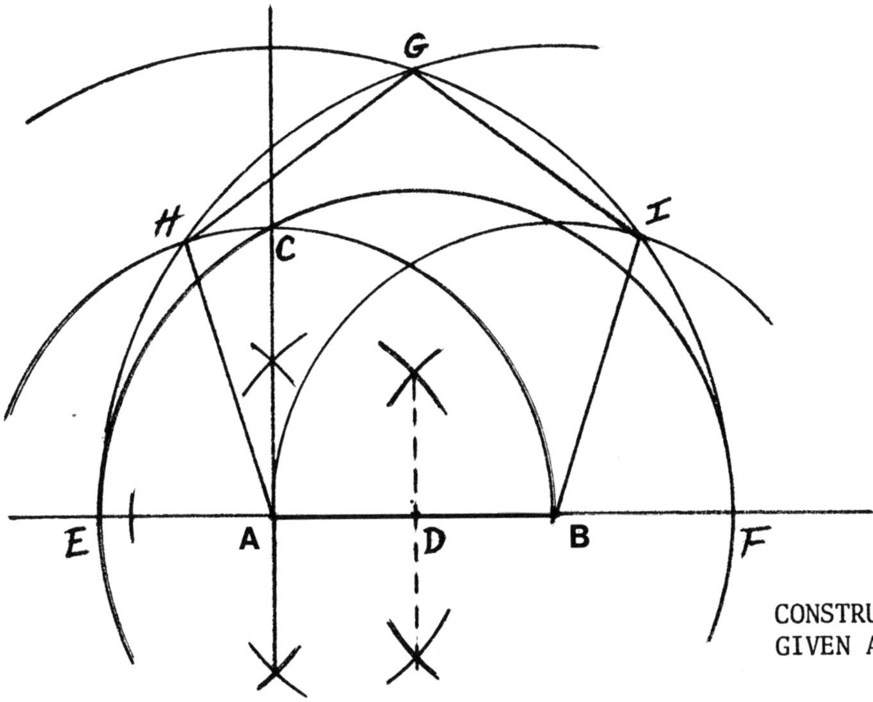

CONSTRUCT A REGULAR PENTAGON
GIVEN A SIDE

Given $\overleftrightarrow{AB}$ as a side, construct a regular pentagon.

1. Extend $\overleftrightarrow{AB}$ through A and B.
2. Construct a line perpendicular to $\overleftrightarrow{AB}$ at A.
3. Put the metal tip on A and the pencil tip on B and cut an arc from B well through the line perpendicular to $\overleftrightarrow{AB}$.
4. Label C the point of intersection of the arc and the line perpendicular to $\overleftrightarrow{AB}$.
5. Bisect $\overline{AB}$ and label the midpoint D.
6. Put the metal tip on D and the pencil tip on C and cut an arc that intersects $\overleftrightarrow{AB}$ on both sides of D.
7. Label the points of intersection E and F.
8. Put the metal tip on A and the pencil tip on F and cut an arc on the same side of $\overleftrightarrow{AB}$ as $\widehat{ECF}$.
9. Without changing the setting, place the metal tip on B and cut an arc that intersects the one drawn in Step 8.
10. Label G the point of intersection.
11. Label H the point of intersection of $\widehat{EG}$ and the arc drawn in Step 3.
12. Put the metal tip on B, the pencil tip on A and cut an arc that intersects $\widehat{GF}$.
13. Label I the point of intersection.
14. Draw $\overline{AH}$, $\overline{HG}$, $\overline{GI}$, and $\overline{IB}$.

Now AHGIB is a regular pentagon.

*Make sure your compass lead is sharp to insure accuracy.

The pentagon and the pentagram provide a framework for varied designs with harmonic proportions. Some examples follow:

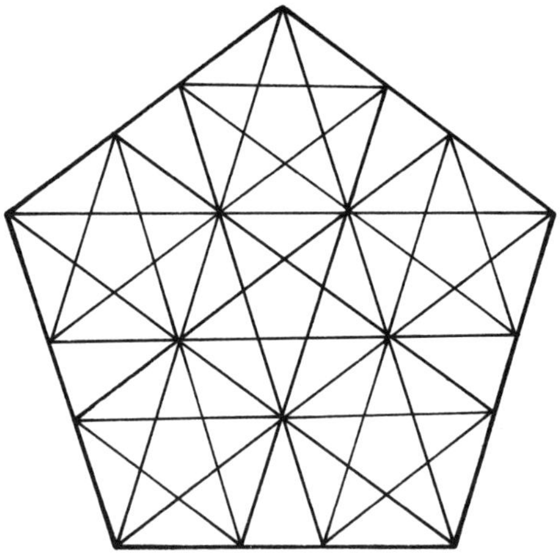

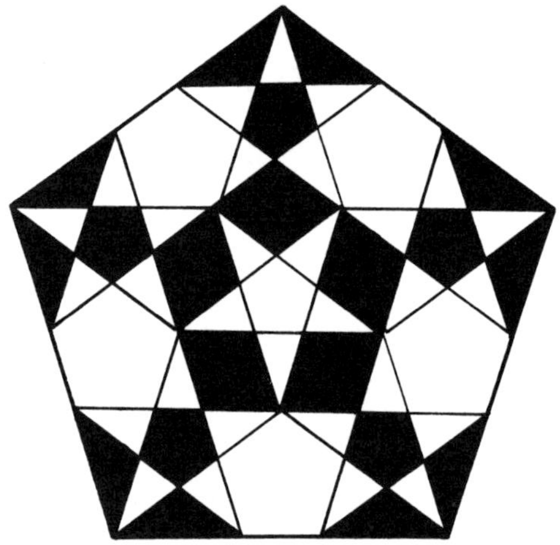

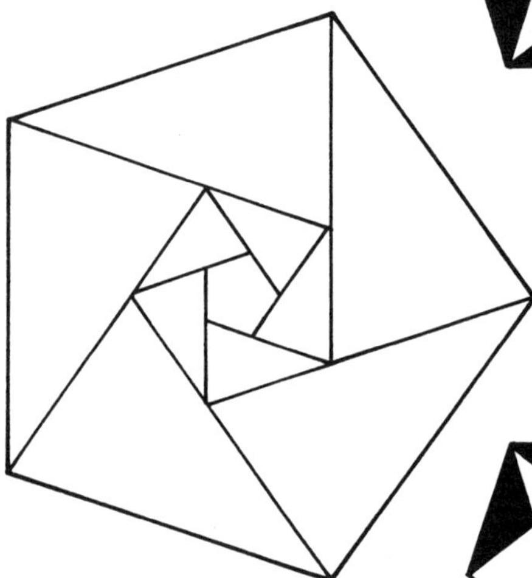

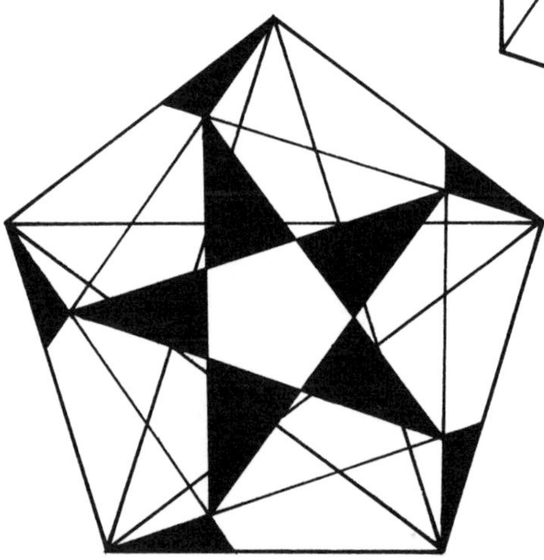

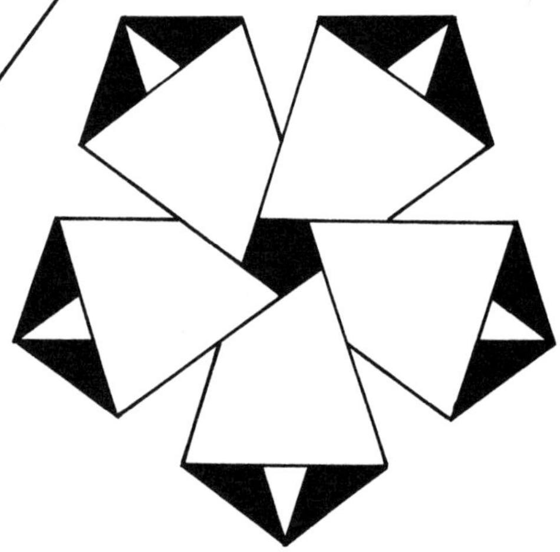

The decagon, or ten-sided polygon, can be easily constructed from a pentagon inscribed in a circle. If the perpendicular bisector of one side (QP) is drawn, the point at which it intersects the circle (R) bisects the arc cut off by the side of a pentagon. The remaining arcs may be bisected by measuring RP with the compass and using this measurement to mark off subsequent arcs around the circle.

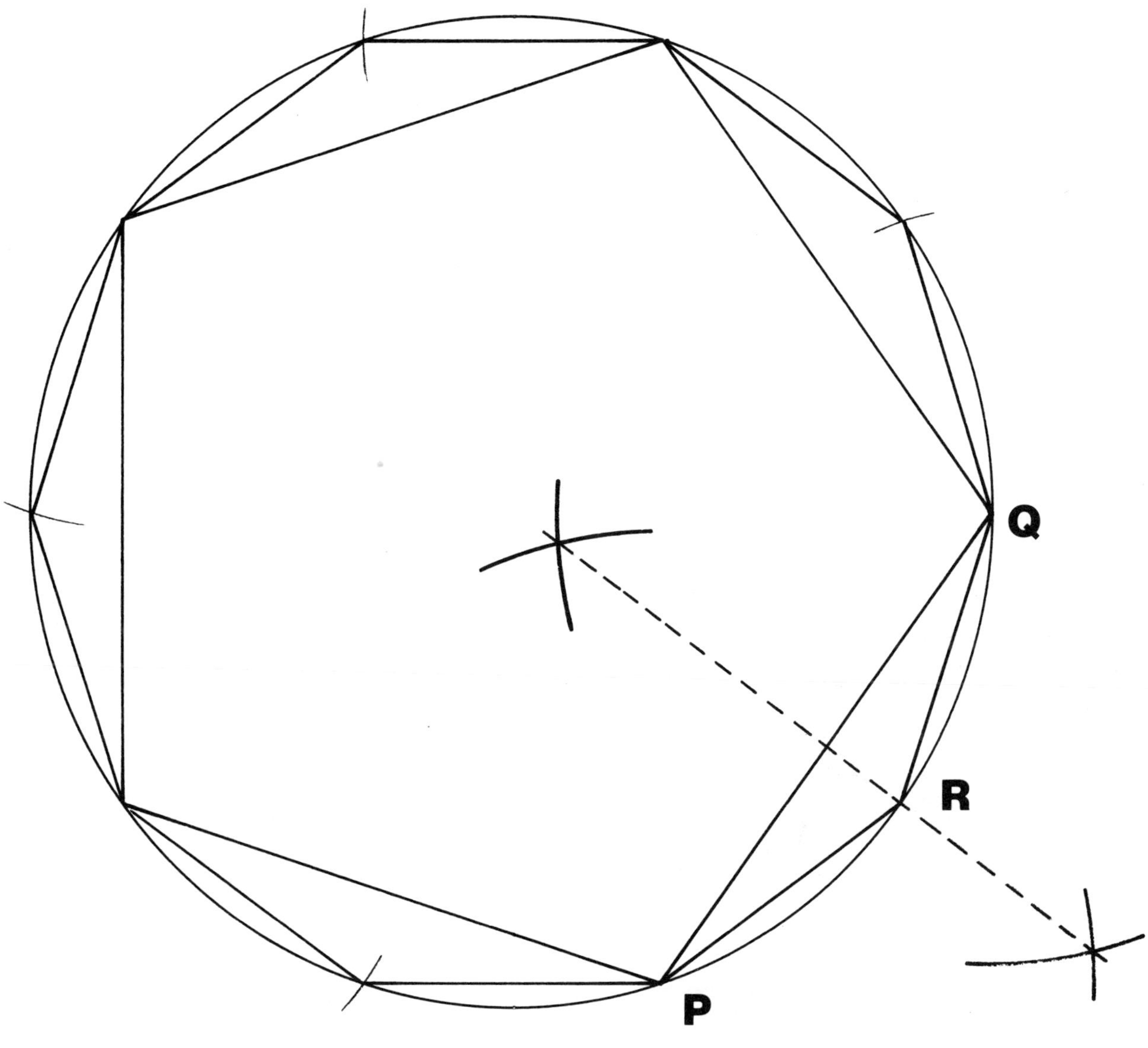

The decagon is useful since it inherits the harmonic properties of the pentagon, and it offers diverse and interesting design possibilities.

Each of the following designs may be derived from the intersecting inscribed pentagons shown in the first figure. Figures 3, 4, and 5 are variations of figure 2, gotten by eliminating line segments and shading appropriately.

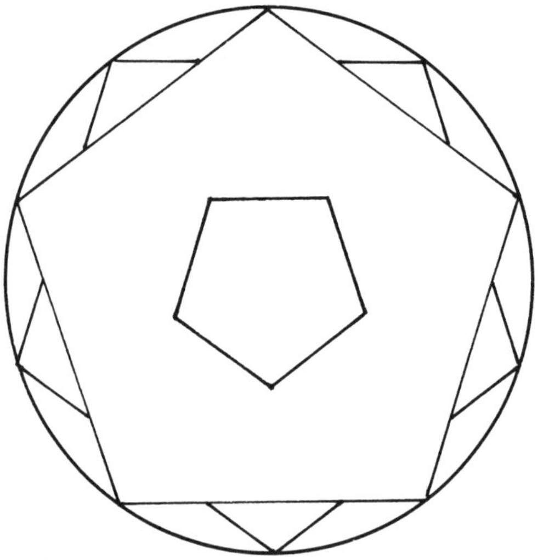

1

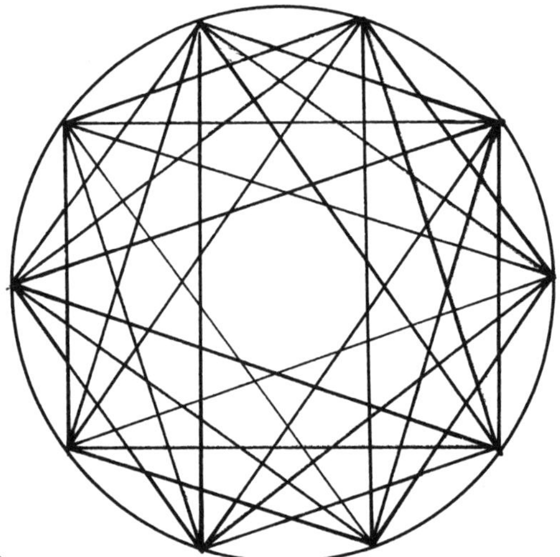

2

3

4

5

An effective variation on the decagon is shown in the following student work. The accompanying line diagram illustrates the principle that was used. The points are first marked on the circle. Then, line segments are drawn connecting every third point. Since ten divided by three yields three with a remainder of one, every three consecutive segments begin and end one unit apart. By continuing the process, each point on the circle is used up and eventually the path returns to the starting point.

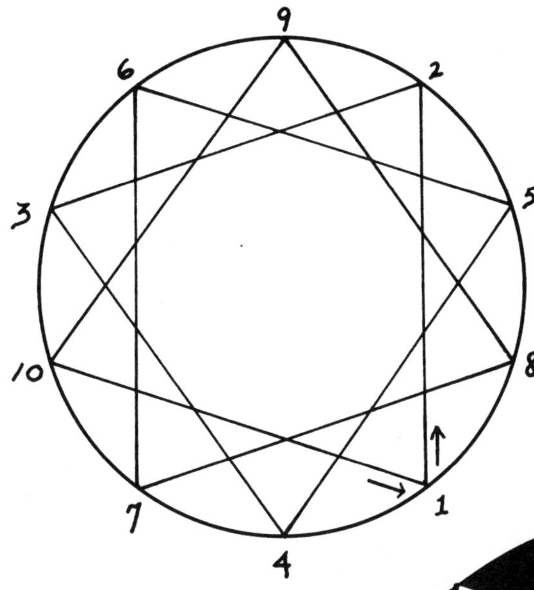

Photo by Richard Newman

Student Work   Lana Yonkers

Photo by Richard Newman

Julie Conway

Student Work
Chapter 1, project 3

THE GOLDEN RECTANGLE

We have examined the existence of the Golden Ratio $\emptyset$ in the pentagon and pentagram and now turn our attention to the rectangle that has become known as the Golden rectangle.

If the lengths of the sides of a rectangle are in the Golden Ratio, then the rectangle is a Golden Rectangle.

$$\frac{DC}{DA} = \frac{1.61803}{1} = \emptyset$$

The mathematician has explored this rectangle as an abstract intellectual exercise. The artist has used it as the structure for harmonious compositions. And the Golden Rectangle is an implied "frame" for many natural forms.

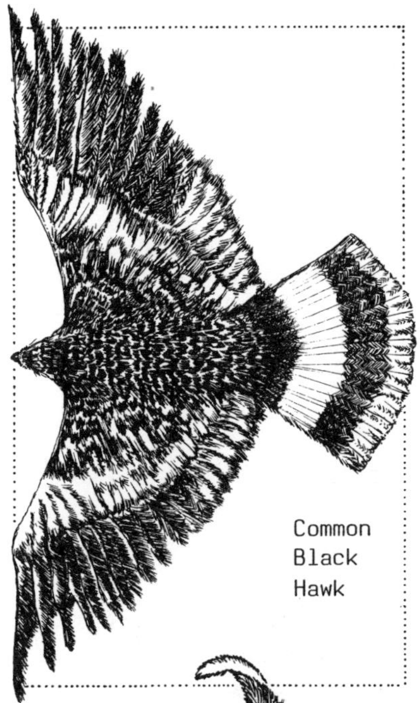

Common Black Hawk

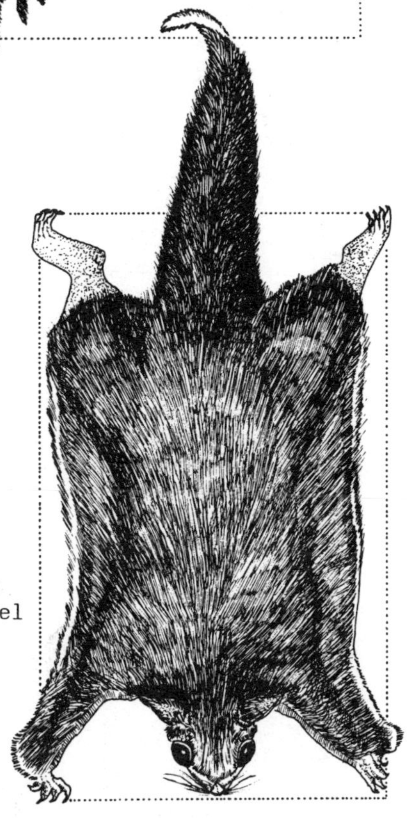

Flying Squirrel

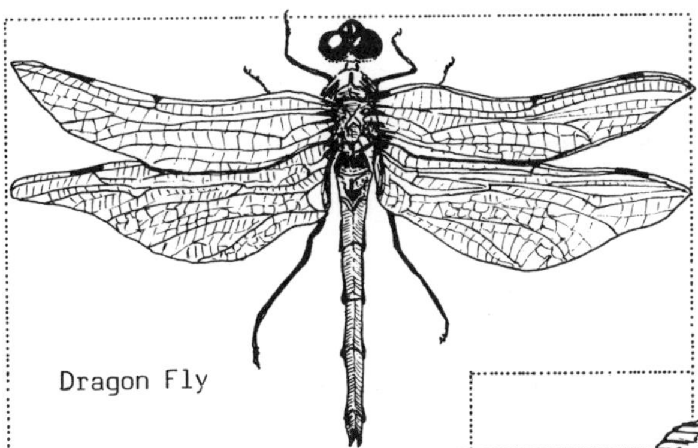

Dragon Fly

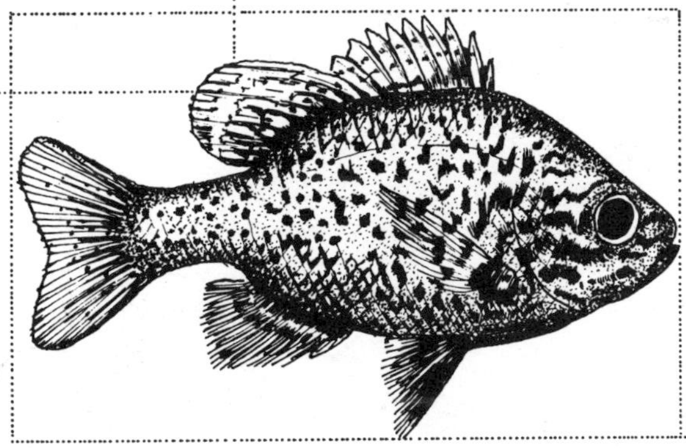

Sunfish

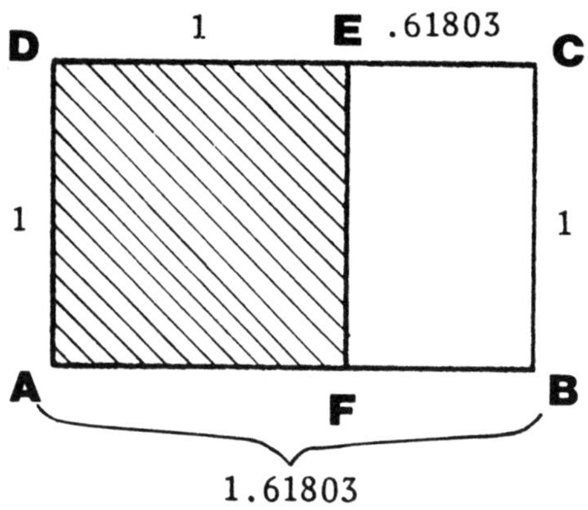

ABCD is a Golden Rectangle. We have assigned the numbers 1 and 1.61803 to the sides of this rectangle independent of real-world measurement standards such as inches, meters, etc. The width becomes the unit measure regardless of the size of the rectangle; therefore any Golden Rectangle may be said to contain a unit square.

If we cut off this square from the parent rectangle, what remains is rectangle BCEF with dimensions of 1 and .61803. Since $.61803 = \frac{1}{\phi}$; $\frac{1}{.61803} = \phi$ BCEF is again a Golden Rectangle and is similar to ABCD.

If we then cut a square from BCEF, we again are left with another smaller Golden Rectangle since $\frac{.61803}{.38197} = 1.61803 = \phi$.

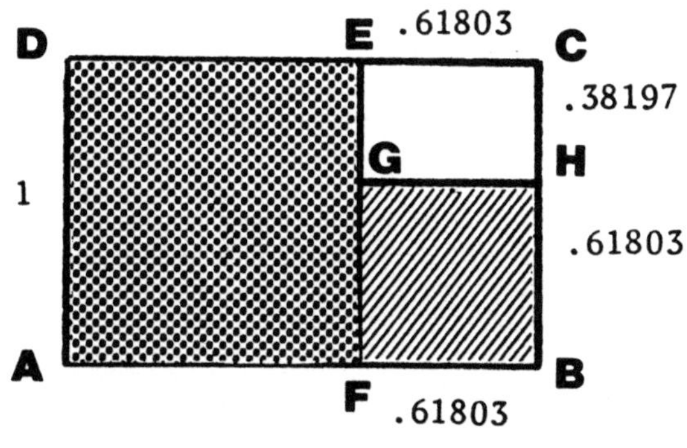

We could continue with this process indefinitely getting smaller and smaller Golden Rectangles.

To increase the size of our original rectangle, then, we need only construct a square on the side measuring 1.61803.

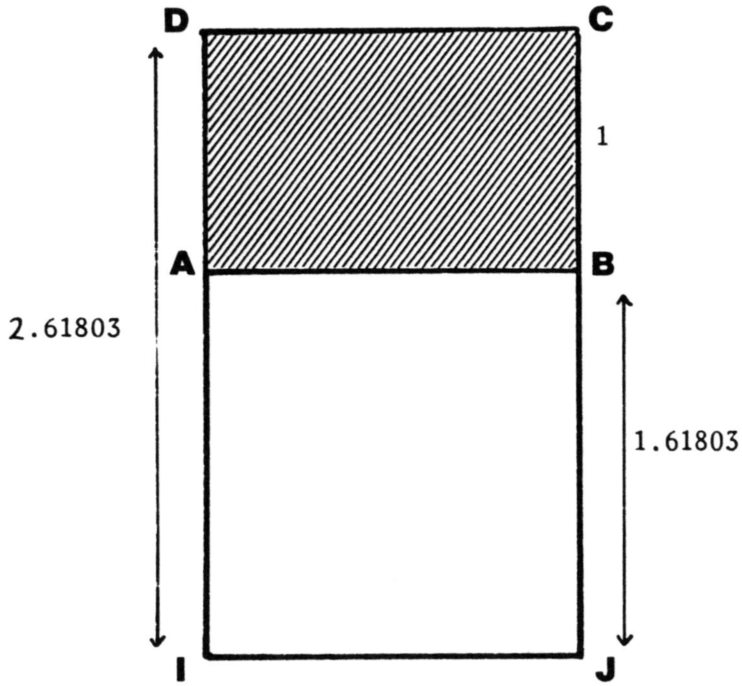

$$\frac{2.61803}{1.61803} = 1.61803 = \Phi$$

We could continue with this process indefinitely getting larger and larger Golden Rectangles.

The Golden Rectangle is the only rectangle from which a square can be cut and the remaining rectangle will always be similar to the original rectangle.

If you have run out of Golden Rectangle mixes, here is a recipe for whipping up a Golden Rectangle from scratch:

This construction is done in two parts. First, construct a square by the steps below. The side of this square will be the width of the Golden Rectangle.

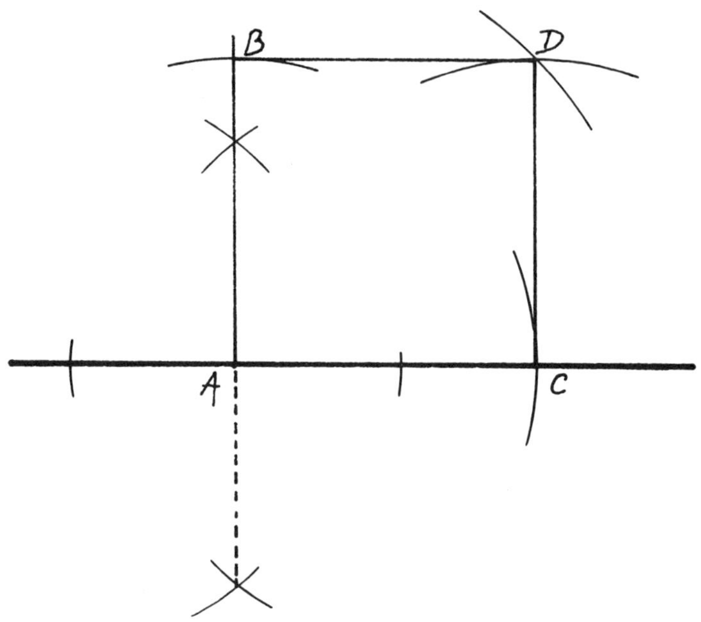

CONSTRUCT A GOLDEN RECTANGLE

1. Construct a pair of perpendicular lines.
2. Label the point of intersection A.
3. Open the compass so that it measures the desired length of the side of the square.
4. Without changing the compass setting, place the metal tip on A and cut one arc on each of the two perpendicular lines.
5. Label the points of intersection B and C.
6. Place the tips of the compass on B and C respectively.
7. Without changing the setting put the metal tip on A and cut an arc in the interior of $\angle BAC$.
8. Put the metal tip on C and the pencil tip on A.
9. At this setting and with the metal tip still on C, cut an arc that intersects the one drawn in Step 7.
10. Label the point of intersection D.
11. Draw $\overline{BD}$ and $\overline{DC}$.

Now ABDC is a square. This square will be used to construct the Golden Rectangle whose width is AB.

ABDC has been reproduced below without the construction marks in order to avoid confusion when undertaking the second part of the construction.

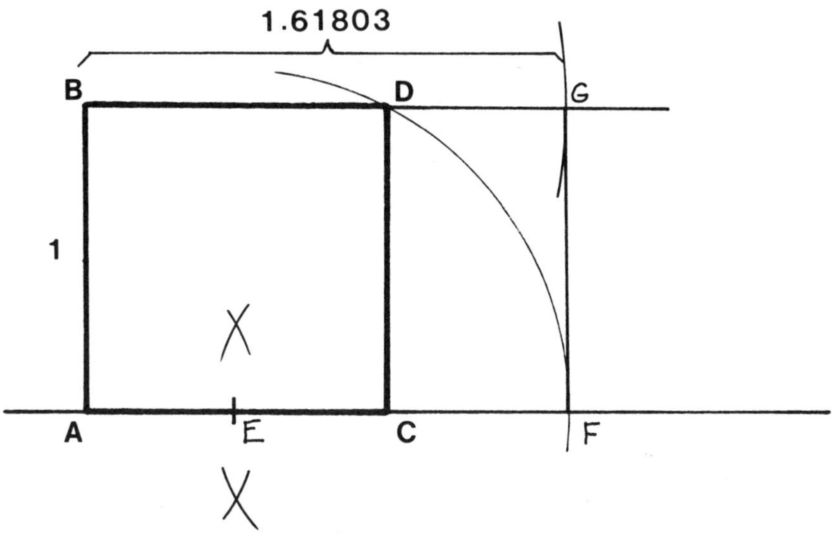

1. Extend $\overrightarrow{AC}$ through C.
2. Bisect $\overline{AC}$ to find its midpoint.
3. Label the midpoint E.
4. Put the metal tip on E and the pencil tip on D and cut an arc that intersects $\overleftrightarrow{AC}$.
5. Label the point of intersection F.
6. Extend $\overleftrightarrow{BD}$ through D.
7. Place the tips of the compass on A and F respectively.
8. At this setting put the metal tip on B and cut an arc on $\overleftrightarrow{BD}$.
9. Label the point of intersection G.
10. Draw $\overline{GF}$

Now ABGF is a Golden Rectangle and $\frac{AF}{AB}$ is the Golden Ratio.

Photo by Richard Newman

Julie Conway

Gift Box

Student Work
Chapter 2, project 2

Here is a useful construction for increasing or decreasing the sides of a rectangle without destroying its proportions. It works, not only for the Golden Rectangle, but also for those rectangles we will be working with later in chapter 4.

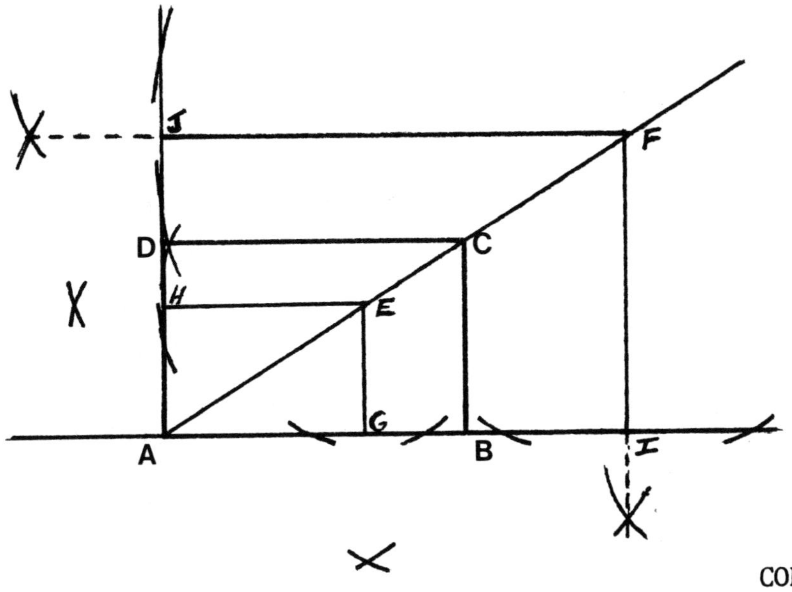

CONSTRUCT A RECTANGLE SIMILAR TO A GIVEN RECTANGLE

Given rectangle ABCD, construct a rectangle similar to ABCD.

A. To construct a smaller rectangle:

1. Draw $\overline{AC}$.
2. Choose any point on $\overline{AC}$ and label it E.
3. Through E, construct perpendiculars to $\overleftrightarrow{AB}$ and $\overleftrightarrow{AD}$.
4. Label points of intersection G and H.

B. To construct a larger rectangle:

1. Extend $\overrightarrow{AC}$ through C.
2. Extend the sides of the rectangle.
3. Choose a point on $\overrightarrow{AC}$ in the exterior of ABCD, say F, and construct perpendiculars to $\overleftrightarrow{AB}$ and $\overleftrightarrow{AD}$.
4. Label points of intersection I and J.

Now AGEH and AIFJ are similar to ABCD.

Probably you have realized that this construction would be simplified tremendously by using a T-square and triangle!

It is essential for students of the arts to understand and be able to use harmonious relationships in their compositions. Art is in part an intellectual discipline. The intellect determines the underlying structure for a work and the emotions provide the content.

Two of the great periods of art, Classical Greece of 5th century B.C. and Renaissance Italy of 15th and 16th century A.D., the latter being indebted to the former, sought an understanding of the world of mathematics so that their arts would have a sound structural base. Greek sculptors and architects and Renaissance artists, relied heavily on the Golden Rectangle for the armature of their works. Twentieth Century artists such as the architect Le Corbusier and the painter Mondrian have also been inspired by principles found in mathematics. In the book, The Painter's Secret Geometry, author Charles Bouleau examines a great many paintings in Western art in order to understand the mathematical structure beneath outward appearances. We refer you to this book for explanations of works of artists such as Duerer, daVinci, Vermeer, Seurat, and Mondrian.

Let us examine, now, one of the most famous pieces of architecture produced in the Western world. We shall see how the Golden Rectangle is an integral part of the form. But first, let us give you some background on that structure.

view of the acropolis with reconstruction drawing of the Parthenon.

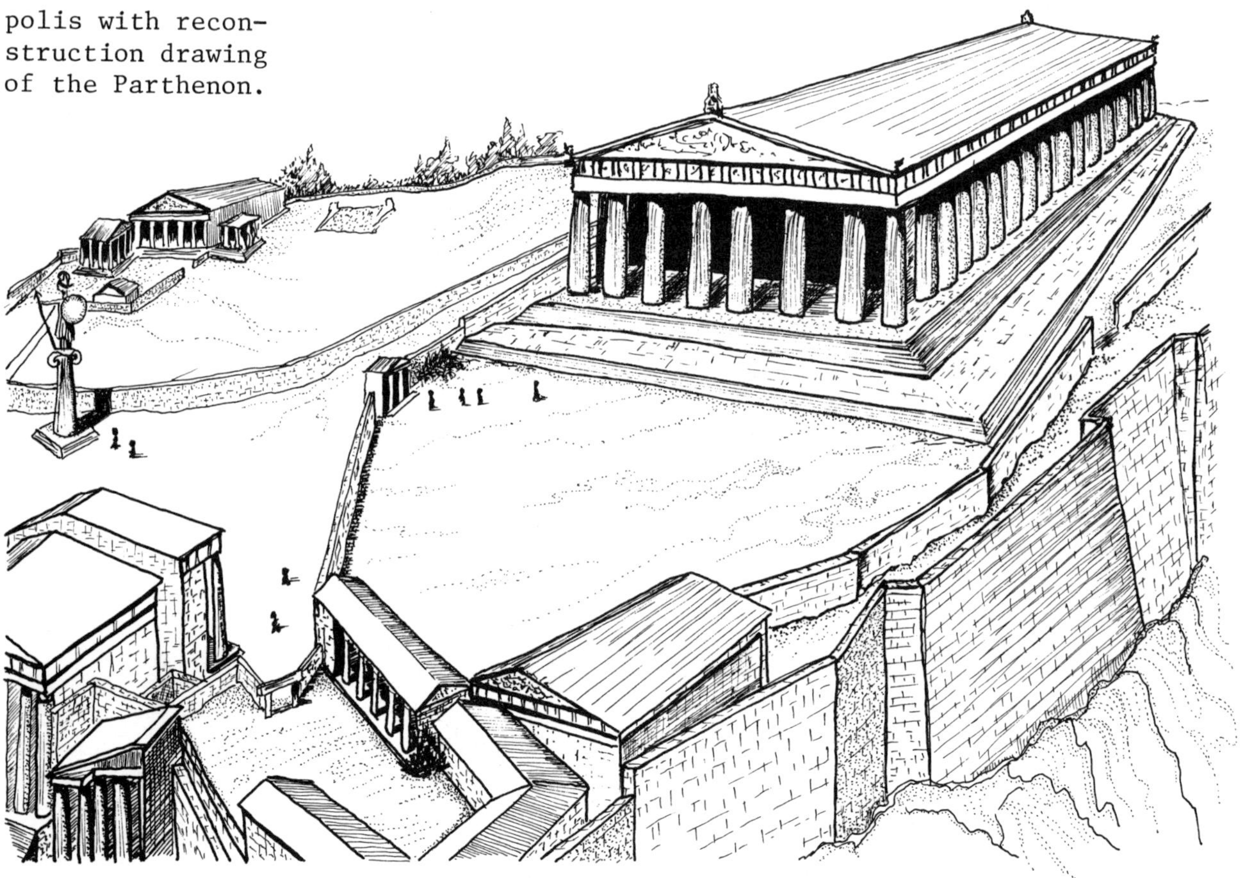

THE PARTHENON

On the highest part of the hilltop of the acropolis in Athens, Greece, standing among the other temples, is the Parthenon. It was built during the years 447-432 B.C. and it is still there today despite the ravages of human enemies and time. Its function was to serve both as a treasury and a dwelling house for the goddess Athena.

The Greeks are an outdoor people. They relate to sea and sky and mountains under a very brilliant light and the Parthenon was meant to be seen from many vantage points around the city. Every four years there was a festival in honor of Athena to whom the temple was dedicated. She became the protectress of the city when she won a contest against the god Poseidon. By prodding the ground with her lance she brought forth an olive tree as her gift to humanity. The sculptural elements on the building give the details of this contest as well as other aspects of her story. The temple was built under the aegis of Pericles during the Golden Age (450-400 B.C.). Ictinus and Callicrates, the architects, and Phidias, the sculptor, were the persons responsible for the aesthetics of the work.

In order to reach the Parthenon, the worshippers had to move along the Panathenaic way, the main avenue of the city, go up a single difficult steep climb and enter a gateway before coming upon the temple. Then, the celebration procession would begin on the west side of the building, run along the north and south sides in two separate groups, and converge at the centre of the east front to conclude the "ceremony of the robe" in which a new mantle of saffron and purple was placed on the statue of Athena.

According to Duane Preble in his book _Man Creates Man_ , the axis of the building was carefully calculated so that on the birthday of Athena the rising sun would come through the east doorway and fully illuminate the gold covered statue created by Phidias.

This building is the culmination of a long evolution of the particular Doric style of architecture, the simplest of the three classical modes. The other two are called the Ionic and Corinthian being named after the type of ornamentation at the top of the columns. The Parthenon is a special refinement on the essentially simple post-and-lintel method of construction. It is a rectangular marble structure that is 228' long by 104' wide with columns 34' in height. In the interior was another rectangular room which was windowless as opposed to the outer rectangle which was composed of a series of columns. This inner room housed both the cult statue and the treasury. The Parthenon was a totally marble creation except for the timbered roof under the marble tiles and the wooden doors and frames, these members being long gone.

At the short ends of the temple were eight columns and on the long sides there were 17 in number. There were porticos on both front and back and a space between the colonnade and the _cella_ in which to walk around, though essentially the building was meant to be used as only exterior space.

ABCD is a Golden Rectangle.

E an F are the Golden Sections of $\overline{AB}$ and $\overline{DC}$ respectively, and $\overline{EF}$ underscores the architrave.

The architrave and the entablature can be subdivided by Golden Sections G and H into two Golden Rectangles and a square.

By further dividing the vertical sides with Golden Sections I and J and drawing in the diagonals of the various rectangles formed (called a harmonic subdivision of the rectangle), we see that the columns are centered on points of intersection of the diagonals alone or the diagonals and sides of the smaller rectangles (see points K, L, M, and N).

The placement of the frieze and the pediment are also determined by such points of intersection (see points O and P).

Since there is symmetry about the vertical line through Q, corresponding points would be determined on the right hand side of the facade by drawing in the corresponding diagonals on that side.

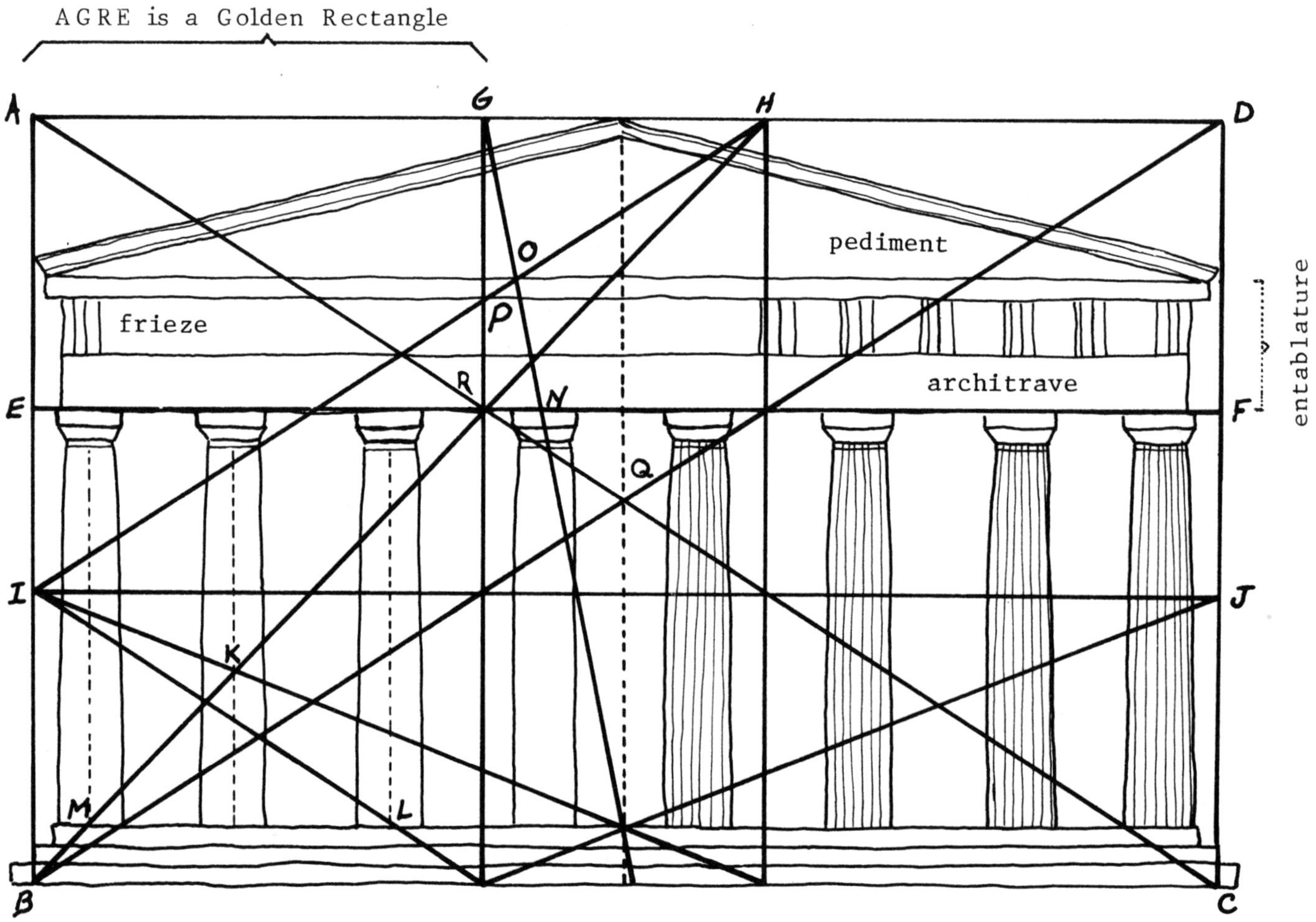

FACADE OF THE PARTHENON

Facade of the Parthenon

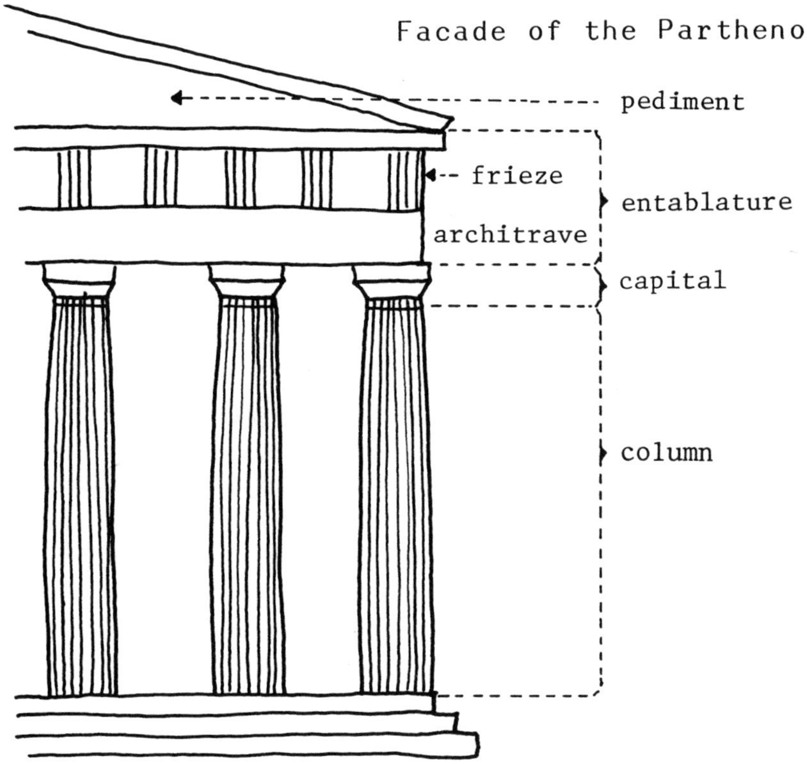

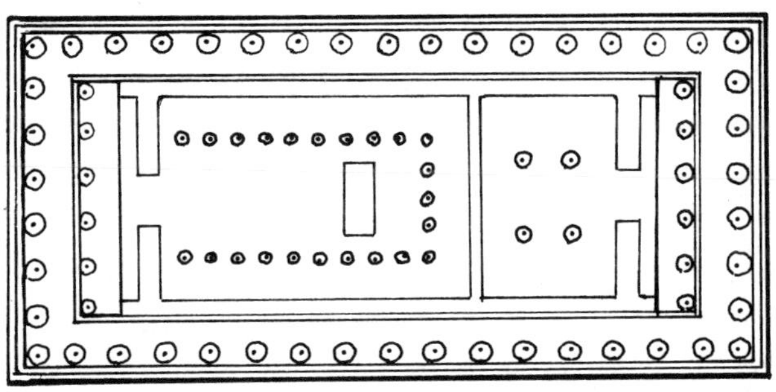

Floor plan of the Parthenon

In this piece of architecture, there is an orderly organization of the vertical members to the horizontal ones; the length and breadth to the height. The ratio of solid masses of columns to open voids was calculated by a unit of measure known as the module. This module was developed by an internal unit such as the radius or diameter of a column, not by an outside unit such as feet and inches.

We have already looked at the organizing principles of the design of this structure, so now let us take a look at a rather surprising element in this design. Despite the concern and interest in geometry, the structure hardly possessed a straight line. The desire seemed to be to create a sense of the organic filled with life rather than to produce a mechanical, only mathematically "correct" system.

Though the stepped courses at the base of the building appear to be equal in height, they are not. The lowest is the narrowest and the stylobate, the topmost, the widest.

The entire platform is not evenly horizontal but rises convexly in the center. The stylobate arches slightly from corner to corner having the center point on the short ends about 2 3/4 inches higher than the corners and the long sides rise about 4 inches in the center. According to some references the reason for this seems to be an optical one. If it were even, the platform would appear to be hollow when seen from the side. Above the column, the architrave echoes this curve. The entire base, therefore, optically resists the downward pull of the entire monument.

The columns have a slight swelling as they rise, 11/16 of an inch, to give the eye the sense of straining, as if these were human muscles meant to carry the heavy entablature. This swelling is called entasis.

While the columns themselves appear equally spaced and all perpendicular, they are not. The ones at the four corners are nearer to their neighbors than the others. This is because they are seen against open space rather than against the solid mass of the inner cella wall. They also serve as members of both the lateral and logitudinal colonnades. The columns are also 2" thicker here. They all incline to varying degrees toward the center, some off 3" plumb, so that the whole appears to converge at some point in the heavens. All of these suggest a conscious and considered reaching upward.

Greek architecture is an art of related lines and surfaces on the plane not of solids in space. Instead of true depth, what we have is successive and usually parallel planes in space, a presentation then of two-dimensions rather than felt volumes existing in space. As a result, only one angle, the right angle of 90 degrees, is permitted. Since this angle can only appear in parallel or perpendicular planes, we can only define or bound solid space but we cannot enclose it and have that enclosed space be perceived as a felt something. We needed to wait until the twentieth century before the architect could be concerned with "emptiness".

Unlike the civilizations that went before theirs, or even with those that were contemporaneous, the Greeks believed that Nature was orderly and that her laws could be comprehended rationally. With this came a belief that there is an ideal form for every class of objects and that such a true form is characterized by its geometric simplicity and by the commensurability of its component members. This idea could then be expressed through Art. Thus, the Parthenon is a rational structure and geometrically comprehensible. It is believed that it was designed "on paper" before it ever was built. Thus, we can study its proportions in terms of the original design and not be concerned with the departures resulting from special circumstances that might be due to time or geography or natural forces.

The man we are indebted to for the largest part of the following analysis is Jay Hambidge, author of The Parthenon and Other Greek Temples.

The advantage of designing "on paper" beforehand is that models of the whole building or any of its parts could be constructed at any scale without mathematical computations. They would be accurate as far as the proportions are concerned since in geometrical constructions you use only the compass and the straightedge. Then, the geometrical figure could be laid out full size on the ground and with the help of some sort of measuring tape, the actual dimensions of every member of the building could be accurately ascertained and recorded in Greek feet and their subdivisions. This would be an indispensable tool for writing the specifications for quarrying, finishing and setting in place of blocks of marble.

The Greeks, as we shall see in subsequent chapters, recognized a geometric commensurability. Two dimensions which would not be commensurable in linear units would be in the square. The theory of root rectangles which we shall examine in Chapter IV is the essence of the theory of proportion that Mr. Hambidge used in his analysis of the Parthenon; especially the principle of proportion inherent in the root-five rectangle. The proportions he found in looking at the temple could have been laid out with the help of pegs and cords as we shall see when we look at Egyptian architecture in the following chapter. There seems to be a constant recurrence in Greek works of art of a very limited number of proportions which are derived by simple constructions from the square. These are the root rectangles.

Here we shall only look at the facade of the temple and its relationship to the Golden Rectangle. In Chapter IV we shall examine it again in terms of the other Dynamic Rectangles.

On the preceding pages we have begun to examine the role of ϕ as it relates to the disciplines of mathematics and art through the Golden Rectangle. In his book, The Power of Limits, Gyorgy Doczi has examined in detail many natural forms and has found that certain proportions relating to ϕ appear again and again. He analyzes such diverse creatures as butterflies, fish, beetles, skates and humans. We refer the reader to this book for an in depth study.

In the following chapters we shall continue to examine ϕ as a common thread that weaves through the patterns of our existence.

Gayle Stinson

Student Work
Chapter 2, project 6

PROBLEM 1

How do you measure up to classical standards of beauty? Fill in the following chart with your measurements. You must be barefoot; the rest is optional.

Women

A. Floor to knee

B. Knee to navel

C. Navel to shoulder

B/A = _______

Men

A. Floor to knee

B. Knee to line of breasts

C. Line of breasts to top of head

B/C = ______

In the two classical scuptures pictured, these measurements appear in the Golden Ratio. The classical sculptors frequently used the navel as the Golden Section of the body.

Is your navel the Golden Section of your body?

A. Floor to navel = ________
B. Overall height = ________

B/A = ______

PROBLEM 2

Find the Golden Section of the following line segments.

PROBLEM **3**

Divide the following line segment into at least 4 segments in ϕ proportions if the first segment has measure PQ. Disregard any left over portion of the line segment.

PROBLEM **4**

A. Inscribe this figure in a circle whose radius is a minimum of two inches.

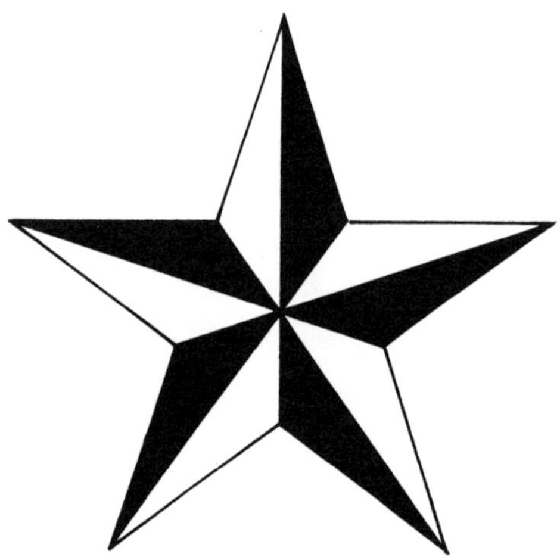

B. Construct the same figure without using the circle.

PROBLEM 5

Using your newly acquired information about Golden Rectangles, do an anlysis of common rectangular objects around the house. Examine such objects as playing cards, stationery, magazines, books, etc. Draw up a chart giving dimensions,type of article, rectangular analysis.

PROBLEM 6

Construct four different sized Golden Rectangles.

PROBLEM 7

Reconstruct one of the figures on page 54.

PROJECT **1**

Use an illustration board or heavy paper. Divide a line segment of any length into Ø proportions. (See Construction #10) Use this line segment as a "ruler" to first construct a rectangle and secondly to sub-divide the rectangle into Ø proportional areas.

Color in a manner of your choosing.

PROJECT **2**

Using the pentagon and pentagram and variations thereof, (refer to constructions 11 and 12), create a pattern in black and white that would be suitable for fabric or wrapping paper design. Repetitions of your pattern may be achieved by: stencil, silk screen, linoleum cut, wood block, rubber stamp, potato print, photocopy, etc. Your finished work should measure at least 8½" by 11" and may be done with the material of your choice.

PROJECT **3**

Make a marionette using the classic concept of body proportions involving a total length of 7½ heads as described at the beginning of the chapter. It may be as simple as a jointed paper figure (such as a shadow puppet) or as complex as a stringed wooden figure.

PROJECT **4** Sculptural Form

12 pentagons join together to form a three dimensional figure called the dodecahedron. The pattern below shows how to lay out the twelve pentagons in the plane. By pulling the pentagons together so that edge meets edge, the dodecahedron is formed. Create your own variation on this "Platonic Solid" by:

a. Making a template out of heavy paper following the directions in construction #11 or #12. Use this template to make all 12 units.

b. Divide each pentagonal face into interesting subdivisions and fill in with black and white or color.

c. Join all 12 units, or multiples of 12, to create a sculpture.

d. Use materials of your choice to make the finished form. Consider the possibilities of using:

1. felt material
2. patchwork units — these would need to be stuffed to
3. knitted units — fill out the forms
4. crocheted units
5. foamcore board
6. plexiglas
7. sheet copper, brass, or alumninum
8. colored acetates
9. stained glass pieces
10. clay units
11. any other materials that you might fancy

Cut on dotted lines.

Fold on solid lines.

You may not need all the flaps. See Appendix D for helpful hints in working with polyhedra.

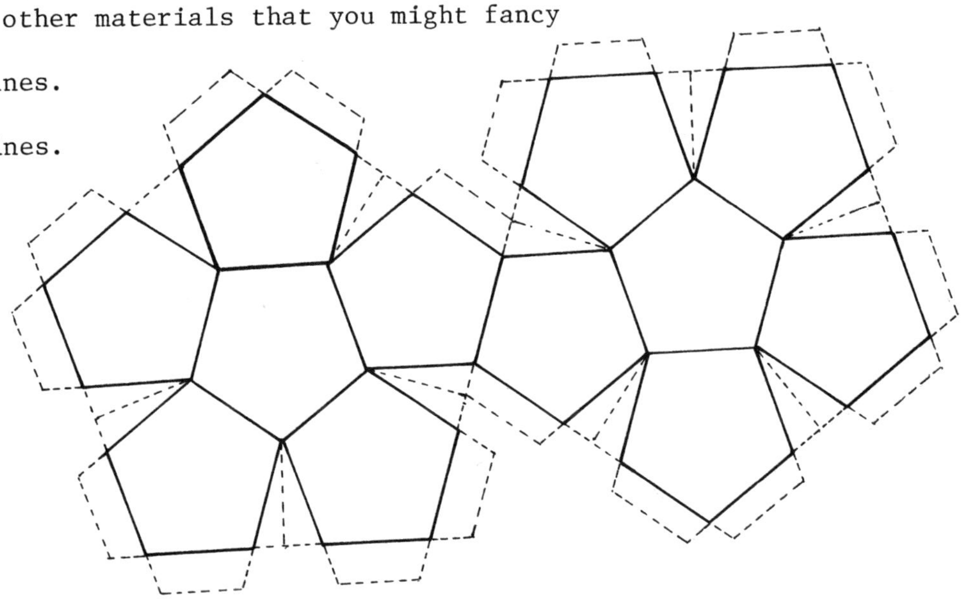

PROJECT 5

Using the following drawings of life forms for inspiration, create your own work in two or three dimensions.

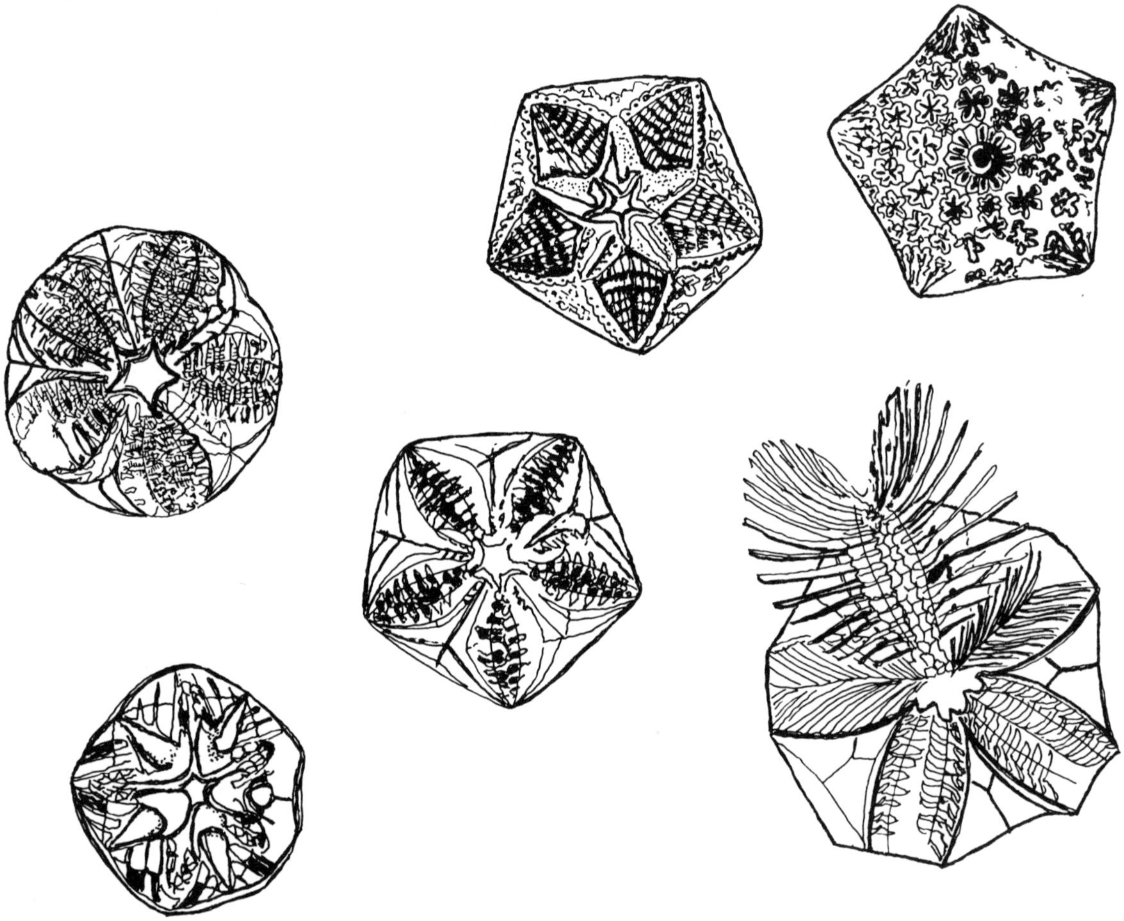

PROJECT 6

Using a maximum of four colors, Golden Rectangles and squares, create a masterpiece (!) in either two or three dimensions.

Further Reading

Bouleau, Charles. The Painter's Secret Geometry. New York: Harcourt Brace and World, 1963

Cook, theodore Andrea. The curves of life. New York: Dover Publications, Inc., 1979.

Doczi, Gyorgy. The Power of Limits. Boulder, Colarado: Shambala Publications, Inc., 1977

Ghyka, Matila. The Geometry of Art and Life. New York: Dover Publications, Inc., 1977.

Grillo, Paul Jacques. Form, Function, and Design. New York: Dover Publications, Inc., 1960

Hambridge, Jay. The Parthenon and Other Greek Temples. New Haven: Yale University Press, 1924.

Huntley, H.E. The Divine Proportion. New York: Dover Publications, Inc., 1970.

Seuphor, Michel. Piet Mondrian, Life and Work. London: Thames and Hudson, 1957.

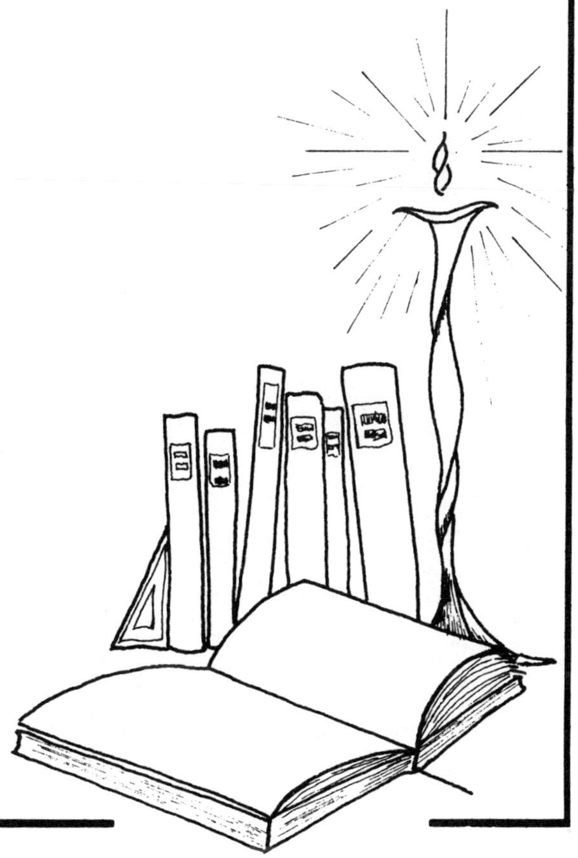

Special Triangles

3

One aspect of the study of space is the concept of the plane surface or what we can call a "slice out of space." Mathematically both the terms "plane" and "space" are undefined; however, for all practical purposes, humans experience and understand spatial relationships intuitively. But, in order to manipulate space in some fashion, one needs to give it boundaries.

The triangle is the simplest figure made up of straight line segments that encloses a portion of the plane. Neither one nor two line segments is able to do this. Structurally, it is not only the simplest but also the most stable as it is unable to be collapsed.

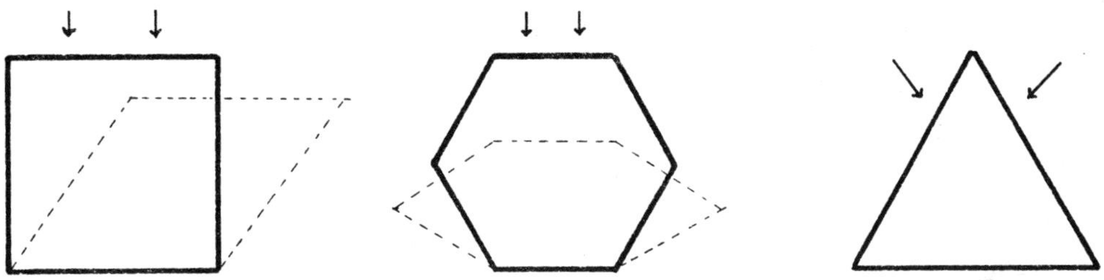

Thus, it is a wonderfully economical figure and certainly one with a long history of use.

In the Ancient World, especially for the Egyptian civilization, the need for bounding a portion of the plane was a problem of extreme importance. Each year, the River Nile overflowed its banks erasing the boundaries which marked the rich farm lands adjacent to the water's edge. Some method was needed to re-establish those boundaries. Thus, for the Egyptians, it appears that the use of geometry existed as a purely practical discipline long before the Greeks gave it the name meaning "earth measure." The Egyptians needed to have a working knowledge of points (boundary markers), lines (distances between markers) and areas (the field enclosed by the boundaries). They had discovered how to find the area of square, rectangular and triangular shapes. Not all fields, however, lent themselves to these simple regular shapes, but if the sides of the fields were straight, any field could be reduced to a series of triangular shapes whose areas could be found and then added together to give the total area. We call this process triangulation, a technique still used by surveyors today.

Through trial and error and the use of inductive reasoning, the Ancients acquired knowledge of some geometrical principles and began to develop simple working tools.

This knowledge and these tools were also needed for the construction of those architectural marvels, the pyramids, especially the three at Giza known as the "Great Pyramids" of Menkaure, Khafre, and Khufu during the Fourth Dynasty of Egypt. The pyramid Khufu, called Cheops by the Greeks, has come to be the symbol of excellence on many levels. It is exactly square, each side is 760'; height 482'; slope 51° 52'. It orients to the four cardinal points of the compass (N, S, E, W), differing no more than 3'6" from the exact north-south orientation.

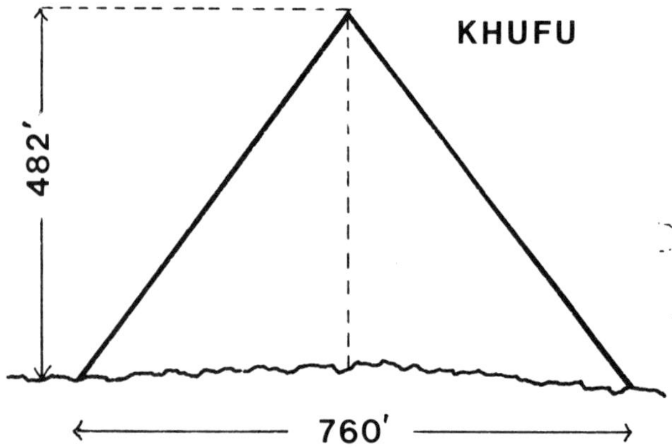

The builders had to be able to measure in the air as well as on the ground. Once the square base was set (no small task for a building so large!), the architects needed to be certain that the building blocks were at the proper angles to the ground since each pyramid was composed of three thicknesses of blocks set in successively smaller square layers, each course being centered on the previous one. A single pyramid could have as many as 123 courses.

In laying out the base of the pyramid, the architects needed to be certain that the corners of each block were at right angles since an inaccurate angle at a corner of the pyramid would throw the whole building off. They probably used a method that closely approximates the technique for bisecting a line segment given in Construction 3. The difference was that the line segment was a rope stretched taut between two markers and the compass was a cord anchored at one end.

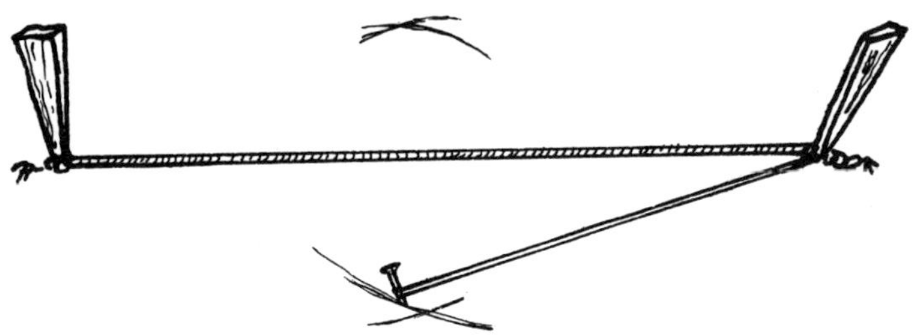

They discovered that if they cut a cord longer than half the length of the stretched rope and then swung arcs from both markers, the points where these arcs intersected would lie on a line at right angles to the original line, in this case, the stretched rope.

In order to insure that the courses of massive stone blocks, many blocks weighing as much as 2.5 tons each, were perpendicular to the ground, the builders found that if they hung a heavy weight by a piece of cord and then let it swing until coming to rest, the right angle would be insured. This device is called a plumb line and is still in use today.

The blocks themselves had to be fitted together in such a way that the course level was preserved and the corners contained the proper angles. Thus the need arose for a consistent and accurate carpenter's square. Such a tool necessitates a right angle which is most easily obtained from a right triangle. It is believed that the discovery of the right-angled triangle was made by the rope knotters of early Egypt. Their job was to tie equally spaced knots in long ropes which would be used then for measuring. The distance between two consecutive knots was the unit measure, for in large-scale building there had to be measures that were constant regardless of who did the measuring.

With these crude tools, the pyramids were built so skillfully that the joinings of the massive blocks are scarcely visible.

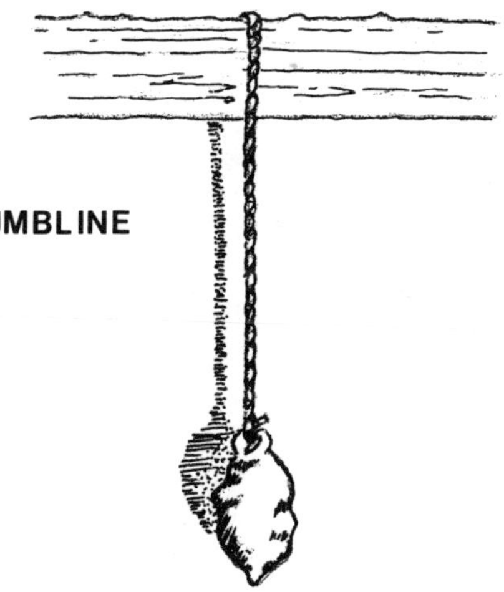

PLUMBLINE

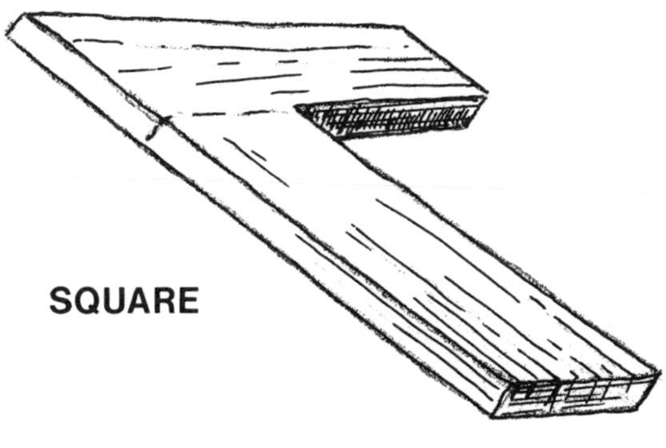

SQUARE

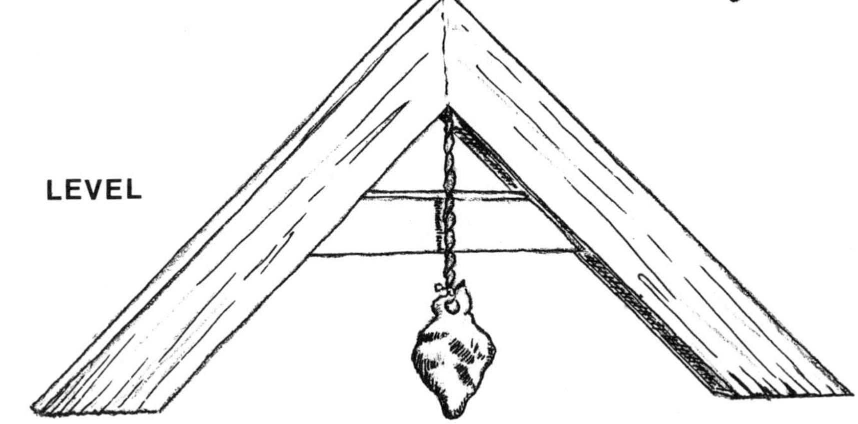

LEVEL

In stretching the ropes around three pegs, again the minimum number needed to enclose an area by straight sides, they discovered that a rope with exactly 13 knots (12 units) formed a triangle that had one right angle and whose sides could be measured in whole numbers, important for a culture (not so unlike ours!) that could not deal comfortably with fractions. The legs of this triangle measured 3 and 4 units, respectively, and the hypotenuse measured five units. It had the added property that the sum of the squares on the legs equalled the square on the hypotenuse, although it was the Greeks about 2,000 years later who were first intrigued by this relationship.

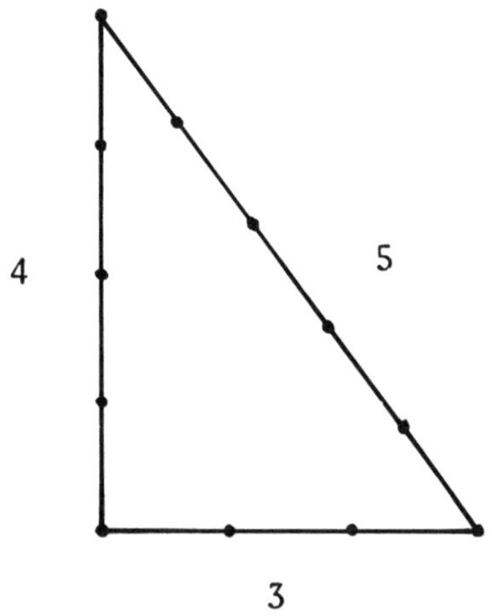

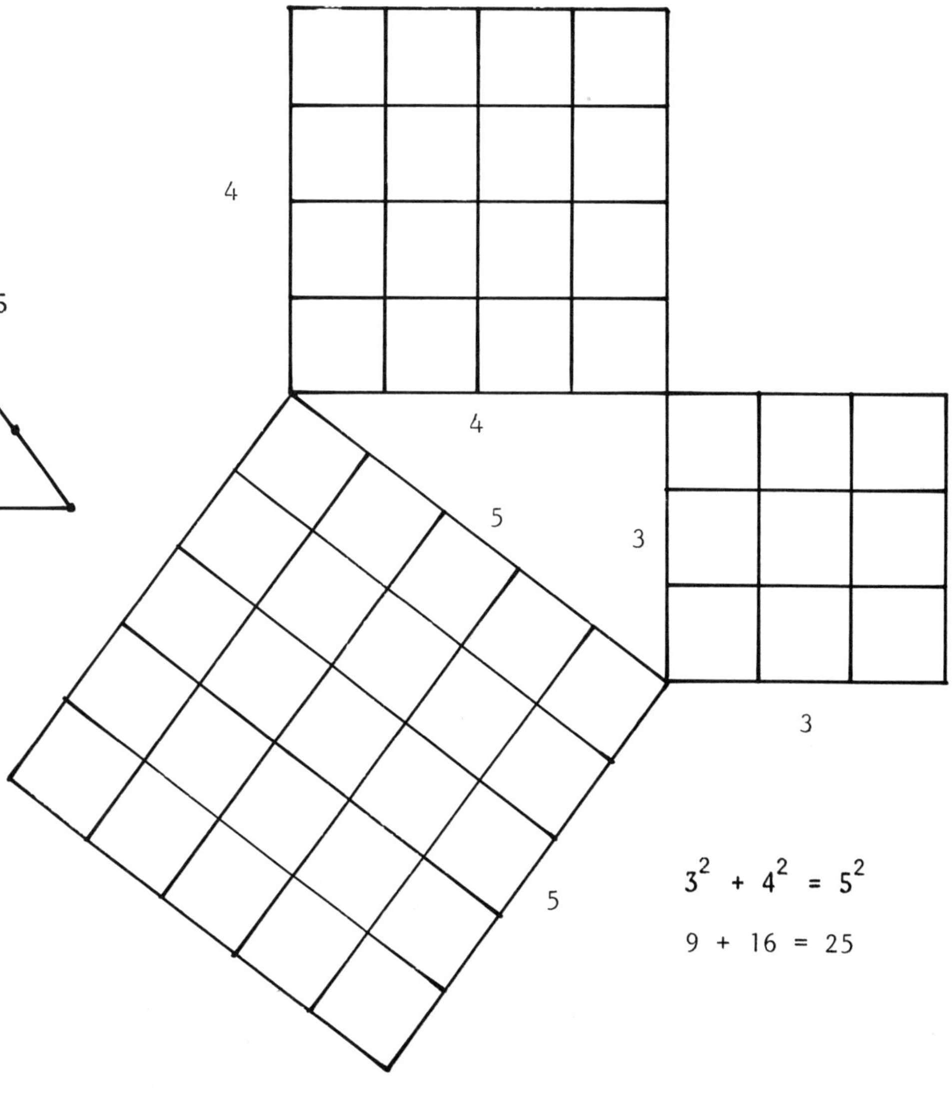

$$3^2 + 4^2 = 5^2$$

$$9 + 16 = 25$$

This triangle is sometimes still referred to as the "rope knotters" or "rope stretcher's" triangle.

This triangle also bore an interesting relationship to the Golden Rectangle; that is, if the hypotenuse of the 3-4-5 triangle were swung in such a way that it formed a right angle with the side measuring three, a rectangle would be formed with a length of five and a width of three. The ratio of 5 to 3 is 1.66666+ , a fair approximation of ϕ with an error of less than five one-hundredths. So, it appears that, possibly by accident, the rope knotters discovered a triangle that not only contained a right angle but also had harmonic proportions.

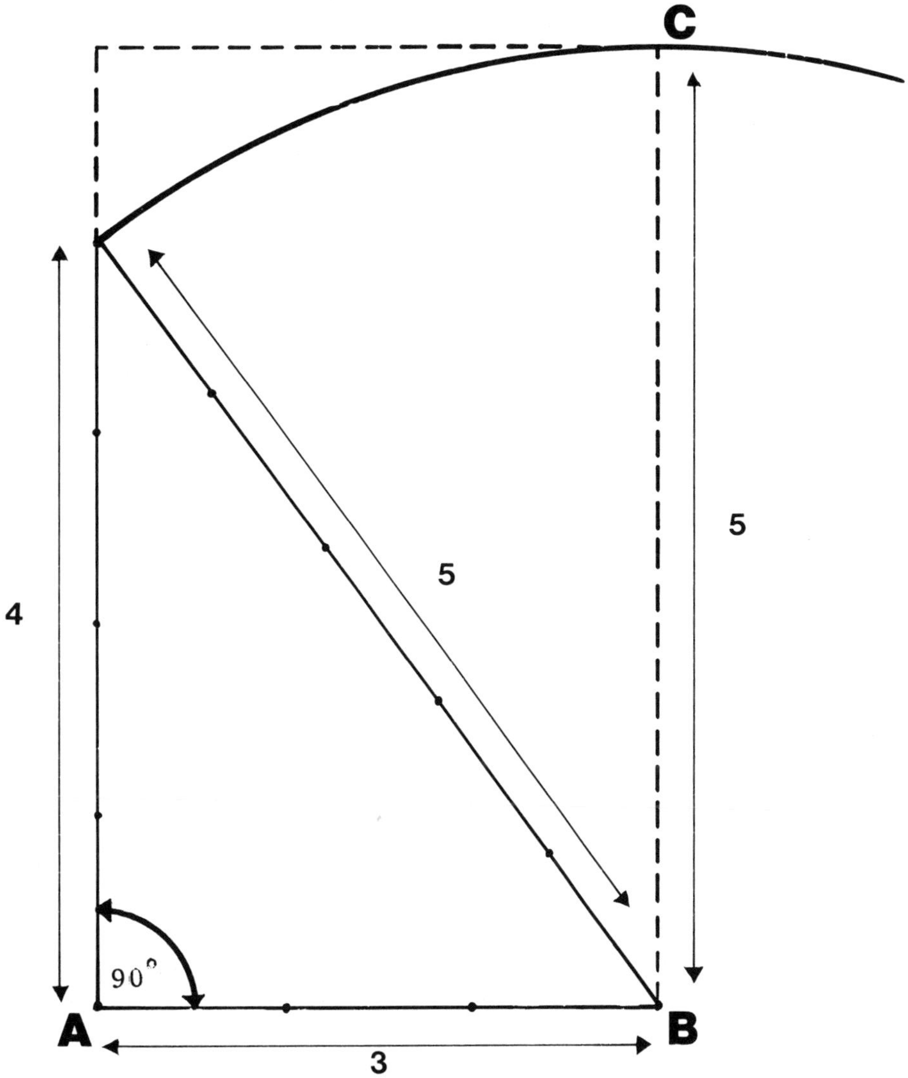

Actually, very little is known about how the mathematical discoveries of Ancient Egypt were made. Probably, the common man knew little more than how to make crude calculations that helped him cope with everyday problems. The true keepers of knowledge were the priests, and it is believed that they may well have kept the more important mathematical finds secret as was their custom with other scientific knowledge. Some scholars feel that the Rhind Papyrus, written by the scribe Ahmes in the 17th century B.C., and believed by some to contain all the mathematical knowledge of the day, was only a rough tool for the common man. Certainly, if that were not the case, some astounding "accidents," as we shall see shortly, occurred during the building of the Great Pyramid of Khufu thirteen hundred years before.

The Babylonians, contemporaries of the Egyptians, discovered fifteen other right triangles whose sides could be measured in whole numbers. These triangles also had the property that the sum of the squares on the legs was equal to the square on the hypotenuse. A few examples are given below.

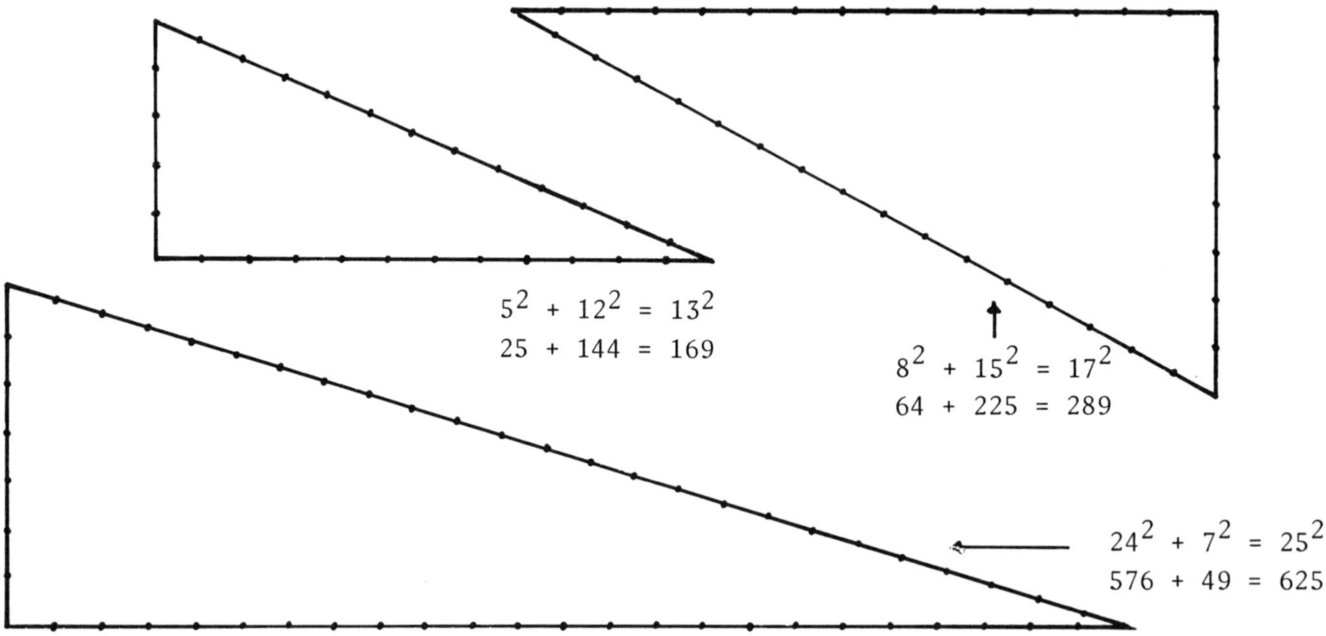

Approximately two thousand years later the Pythagorean Brotherhood of Greece devised a method for finding triangles which had the aforementioned property. Rather, it may be said, that the Pythagoreans discovered a rule for finding triples of numbers that could be applied to the sides of right triangles. They were the first to investigate mathematics as a purely abstract endeavor regardless of real world applications. Like philosophy, pure mathematics became a liberal art.

The rule they used was as follows:

> If n is any <u>odd</u> number, then $\frac{n^2 - 1}{2}$ and $\frac{n^2 + 1}{2}$ provide the other two numbers.

For example: Let n = 9, then

$$\frac{n^2 - 1}{2} = \frac{9^2 - 1}{2} = \frac{81 - 1}{2} = \frac{80}{2} = 40$$

and

$$\frac{n^2 + 1}{2} = \frac{9^2 + 1}{2} = \frac{81 + 1}{2} = \frac{82}{2} = 41$$

$9^2 + 40^2 = 41^2$ since $81 + 1600 = 1681$

This rule <u>did not</u> yield every triple of numbers with this special property. For example, $8^2 + 15^2 = 17^2$, but the three numbers 8, 15 and 17 do not fit the rule. Any three integers whose squares relate in such a manner is called a <u>Pythagorean triple</u> or a <u>Pythagorean triad</u>. The Pythagoreans later proved that the sides of <u>any right triangle</u> have the property that the square of the hypotenuse is equal to the sum of the squares of the legs. If the legs are labeled a and b respectively, and the hypotenuse is labeled c, we write $c^2 = a^2 + b^2$. This is called the <u>Pythagorean Theorem</u> and is proved below.

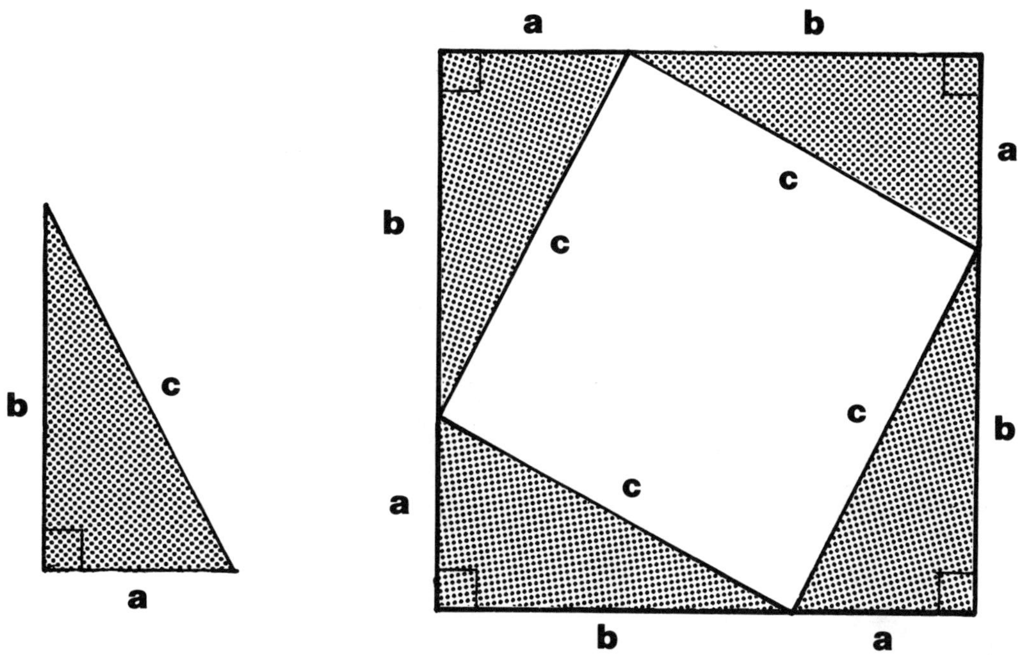

Remembering that the area of a triangle is ½ the base times the altitude and that the area of a square is the square of the side, we can find the area of the small square (c^2) which is the area of the large square, $(a + b)^2$, minus the area of the four triangles, $4(\frac{1}{2}ab)$.

$$c^2 = (a + b)^2 - 4(\tfrac{1}{2}ab)$$
$$c^2 = a^2 + 2ab + b^2 - 2ab$$
$$c^2 = a^2 + b^2$$

This important theorem was a rather distressing revelation for the Pythagoreans who felt that the harmony of the universe was explicable in terms of ratios of whole numbers. The concept of <u>irrationality</u> (in numbers) was an anathema to them, but they realized that the ratio of the diagonal of a square to its side could never be expressed by whole numbers no matter how they tried.

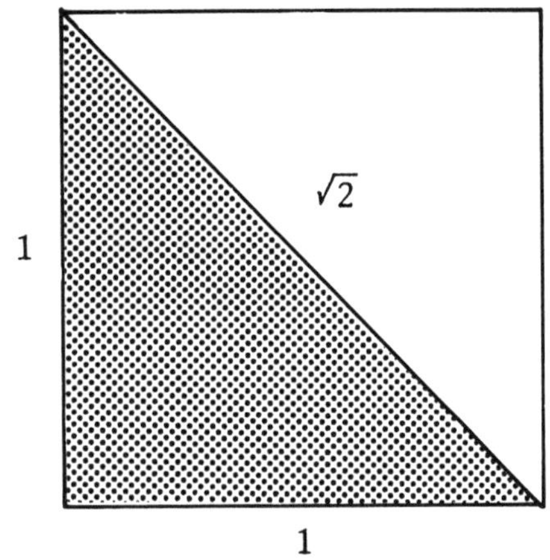

$1^2 + 1^2 = 2$ making the square of the hypotenuse 2 and, therefore, the measure of the hypotenuse $\sqrt{2}$.

Since there is no whole number or fraction that can be multiplied by itself to produce 2, $\sqrt{2}$ is irrational.

Legend has it that this discovery was made during a sea voyage and that the poor fellow who made the discovery was thrown overboard for his efforts.

The discovery of the Pythagorean Theorem altered the course of mathematical thought for the Greeks. From numerical measurement they turned their attention to the geometry of shapes, for what they could not explain with whole number ratios, they could express by geometric figures. We will see further illustrations in Chapter 4.

The 3-4-5 Triangle is probably the most familiar example of the Pythagorean relationship. Two constructions for this special triangle follow.

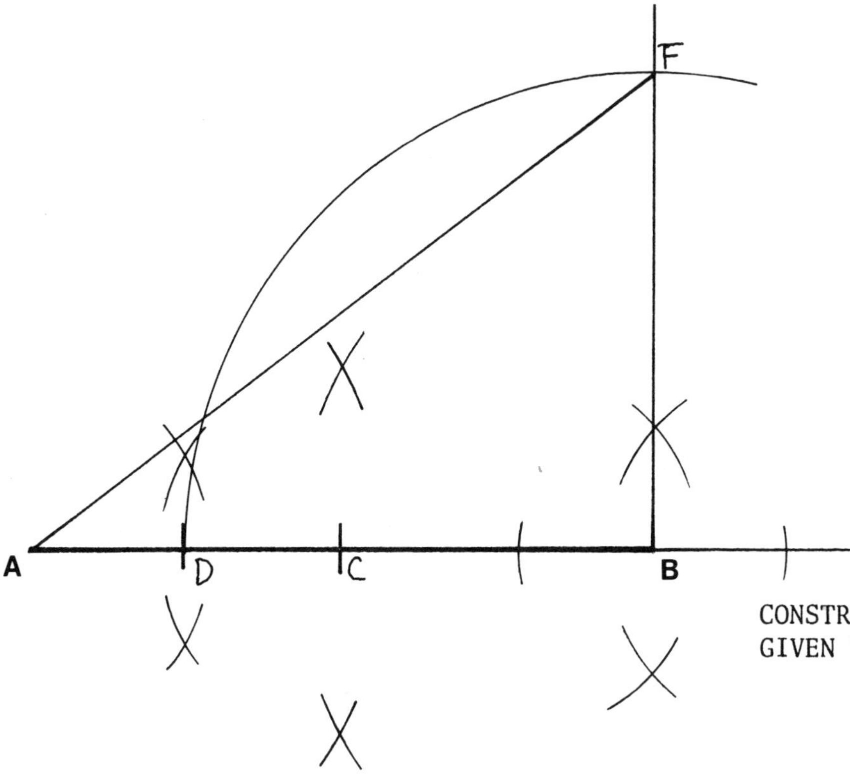

CONSTRUCT A 3-4-5 TRIANGLE GIVEN THE LONGER LEG

Given $\overline{AB}$, construct $\triangle ABF$ such that it is a 3-4-5 Triangle.

1. Bisect $\overline{AB}$ and label its midpoint C.
2. Bisect $\overline{AC}$ and label its midpoint D.
3. Extend $\overrightarrow{AB}$ through B.
4. Construct a perpendicular to $\overleftrightarrow{AB}$ at B.
5. Place the metal tip on B and the pencil tip on D and cut an arc that intersects the perpendicular drawn in Step 4.
6. Label the point of intersection F.
7. Draw $\overline{AF}$.

Now $\triangle ABF$ is a 3-4-5 Triangle with BF:AB:AF = 3:4:5 and $\angle B = 90^{\circ}$.

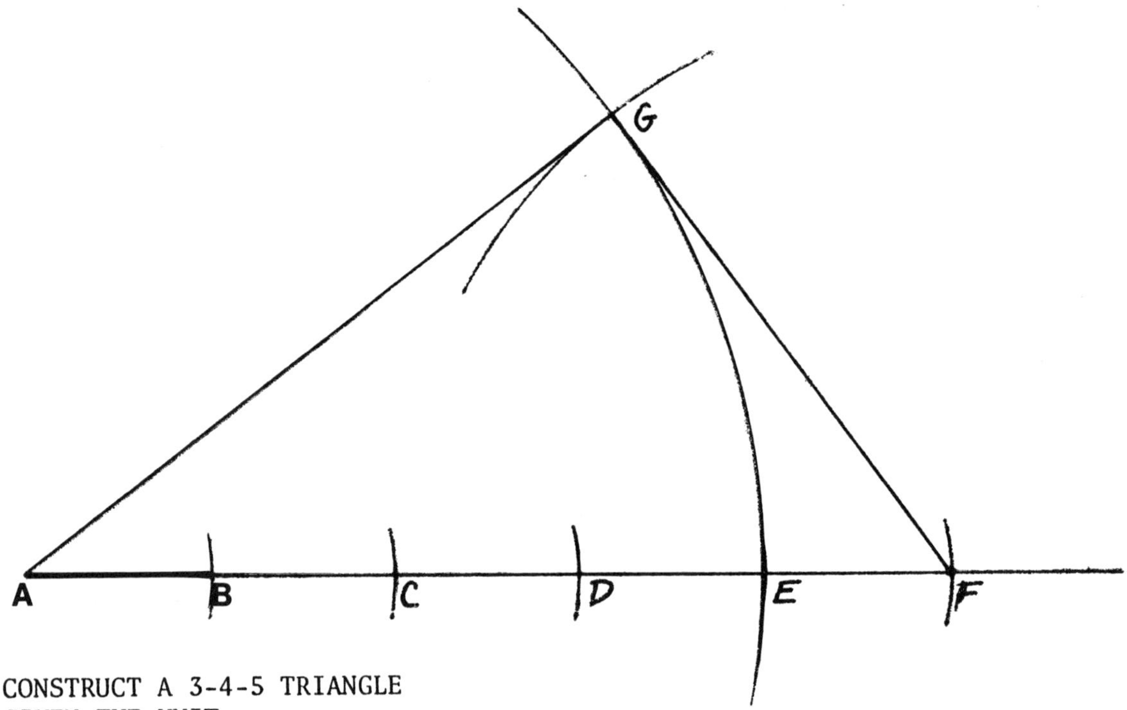

CONSTRUCT A 3-4-5 TRIANGLE
GIVEN THE UNIT

Given AB as the unit, construct $\triangle AFG$ so that its sides measure 3, 4 and 5 respectively.

1. Extend $\overrightarrow{AB}$ through B.
2. Open the compass so that it measures the same as AB.
3. Place the metal tip on B and cut an arc that intersects $\overrightarrow{AB}$ on the side of B opposite A.
4. Repeat Step 3 three more times each time placing the metal tip on the most recently constructed point of intersection on $\overrightarrow{AB}$.
5. Label the points of intersection C, D, E and F respectively.
6. Place the metal tip on A and the pencil tip on E and cut an arc on one side of $\overrightarrow{AB}$.
7. Place the metal tip on F and the pencil tip on C and cut an arc that intersects the one drawn in Step 6.
8. Label the point of intersection G.
9. Draw $\overline{AG}$ and $\overline{GF}$.

Now $\triangle AFG$ is a 3-4-5 Triangle with FG:GA:AF = 3:4:5 and $\angle G = 90^{\circ}$.

Let us take a backward jump in history once again to the time of the Egyptians and have a closer look at the Great Pyramid of Khufu. As you recall, this pyramid is set on a square base with four triangular faces. A cross section of the pyramid yields a triangle different from the faces and one which has some unique properties.

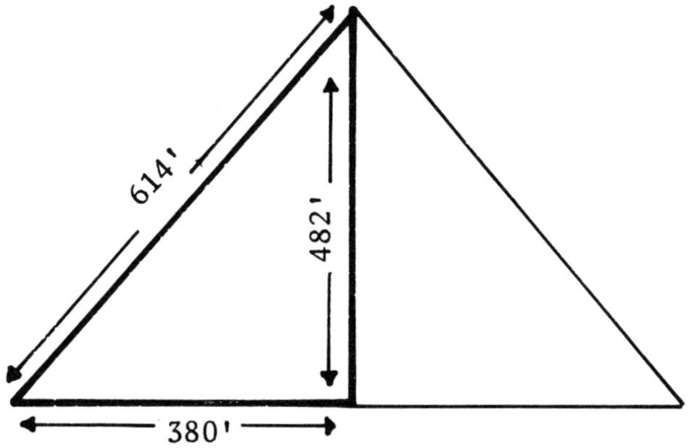

The Triangle of Price

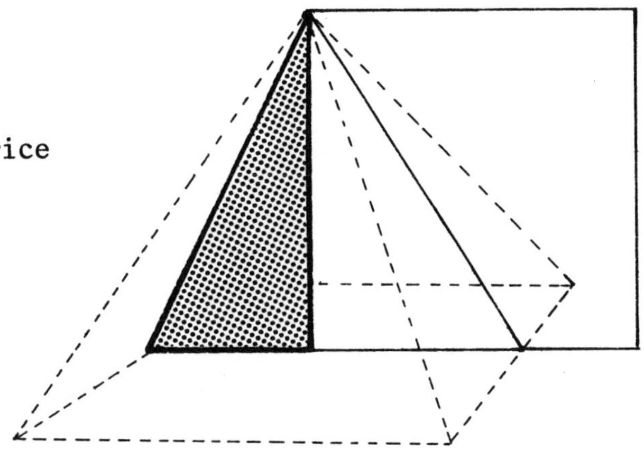

If an altitude is dropped from the vertex of the triangle, it bisects the original angle and forms the right triangle that has come to be known as the Egyptian triangle or the Triangle of Price, after its discoverer. The Price triangle has the distinction of being the *only* right triangle whose sides are in geometric progression.

Looking at the diagram, we have the three numbers 380, 482 and 614 as the lengths of the sides.

$\frac{482}{380}$ = 1.27 to two decimal places

and

$\frac{614}{482}$ = 1.27 to two decimal places

$\frac{614}{380}$ = 1.62 (ϕ rounded to two decimal places)

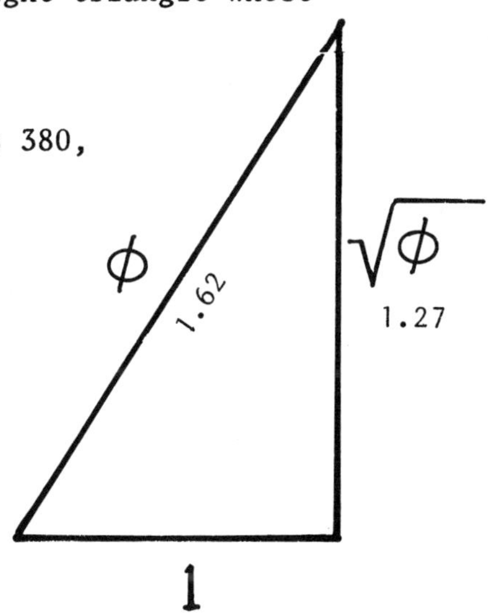

Incredibly enough, 1.27 = $\sqrt{\phi}$ to two decimal places. Therefore, if we relabel the sides of this triangle, naming the shortest side "one," or unity, the remaining sides are $\sqrt{\phi}$ and ϕ.

Therefore, the Triangle of Price, like the pentagram and pentagon mentioned in Chapter 2, contains harmonic properties, and the following construction is based on the fact that the ratio of a diagonal to a side of a regular pentagon is ϕ. The next construction demonstrates how nearly alike are the 3-4-5 Triangle and the Triangle of Price. You will recall that the 3-4-5 Triangle can be expanded into a rectangle which very closely approximates a Golden Rectangle. It is the Triangle of Price which expands to precisely a Golden Rectangle. Because they are so similar, the 3-4-5 Triangle, the Triangle of Price and the Harmonic Triangle discussed in the next section may be used interchangeably for most design purposes.

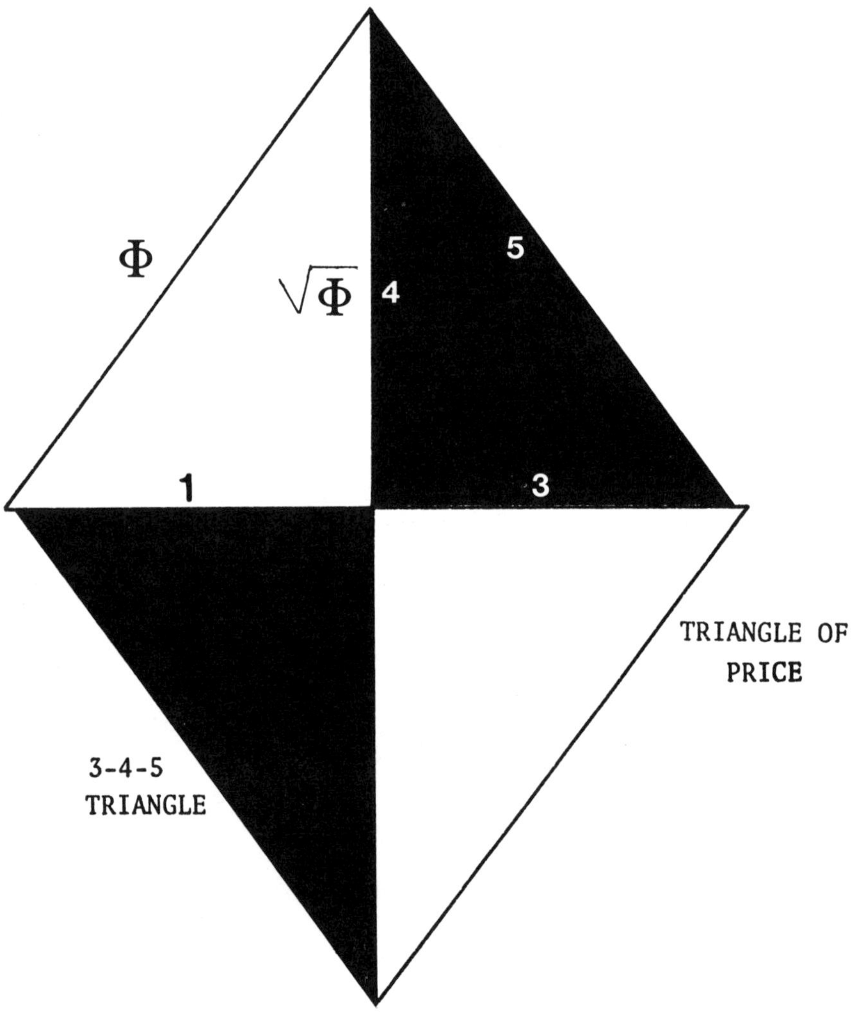

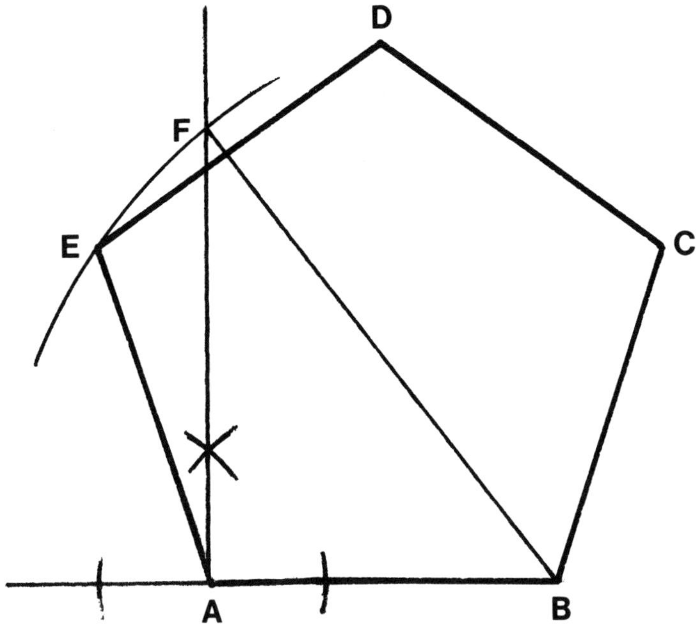

CONSTRUCT A TRIANGLE OF PRICE
GIVEN A REGULAR PENTAGON

Given regular pentagon ABCDE, construct $\triangle ABF$ such that it is a Triangle of Price.

1. Extend $\overleftrightarrow{AB}$ through A.
2. Construct a perpendicular to $\overleftrightarrow{AB}$ through A.
3. Place the metal tip on B and the pencil tip on E and cut an arc that intersects the line constructed in Step 2.
4. Label the point of intersection F.
5. Draw $\overline{BF}$.

Now $\triangle ABF$ is a Triangle of Price with AB = 1, AF = $\sqrt{\phi}$ and BF = ϕ.

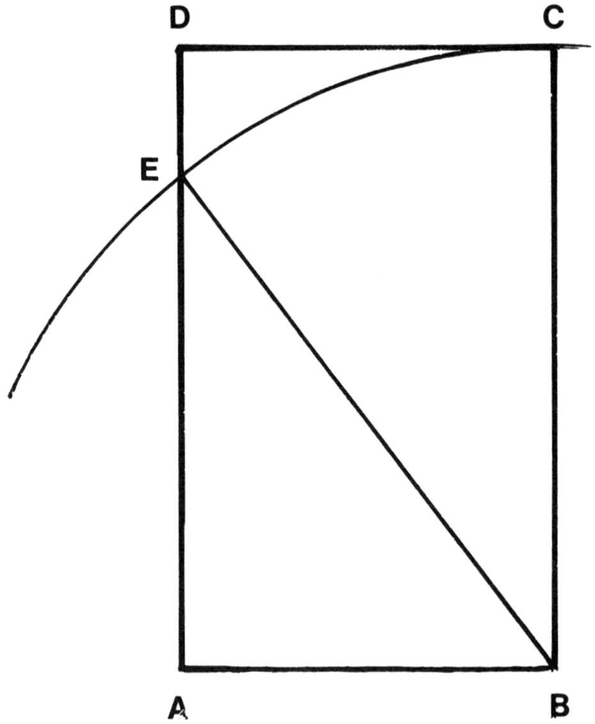

CONSTRUCT A TRIANGLE OF PRICE
GIVEN A GOLDEN RECTANGLE

18

Given Golden Rectangle ABCD, construct ΔABE such that it is a Triangle of Price.

1. Place the metal tip on B and the pencil tip on C and cut an arc that intersects $\overline{AD}$.
2. Label the point of intersection E.
3. Draw $\overline{EB}$.

Now ΔABE is a Triangle of Price with $AB = 1$, $AE = \sqrt{\phi}$ and $EB = \phi$.

As we explained earlier on, standardized measurement became a necessity for the Egyptians. In early times, when building for themselves, men could use the lengths of particular parts of their own bodies as constant units for measure. We still refer to the height of a horse in terms of "hands," or say an object is so many "heads" high, and use rulers that are a "foot" long.

The Egyptians used the foot, palm, fingers, and forearm for measuring. The diagram shows the relationship between digits, palms, and cubits. Actually, when standardized, the Egyptians used two variations for the cubit; the little cubit measured just under half a meter and the Royal Cubit just over half a meter.

See Appendix B for the relationship between the little cubit, the royal cubit and ϕ.

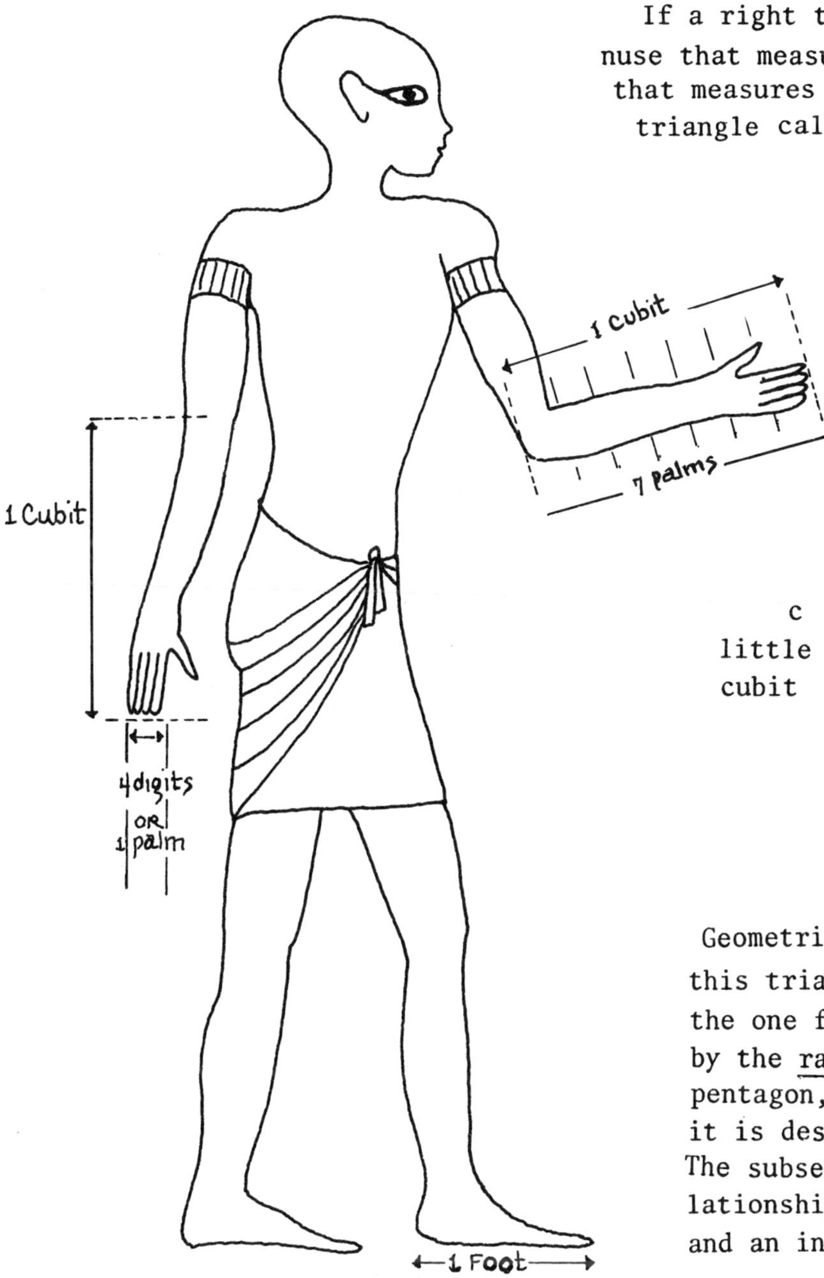

If a right triangle is formed with a hypotenuse that measures a Royal Cubit and one leg that measures a little cubit, the result is a triangle called the Harmonic Triangle. Oddly enough, the remaining side has a measure that is an average obtained from the various standard "foot" measurements used among different cultures of that time.

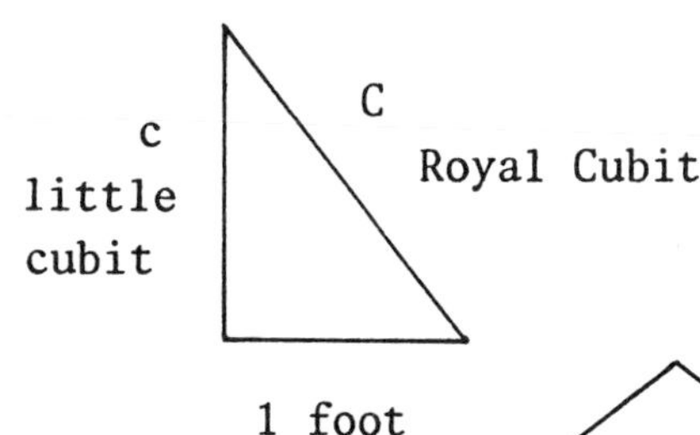

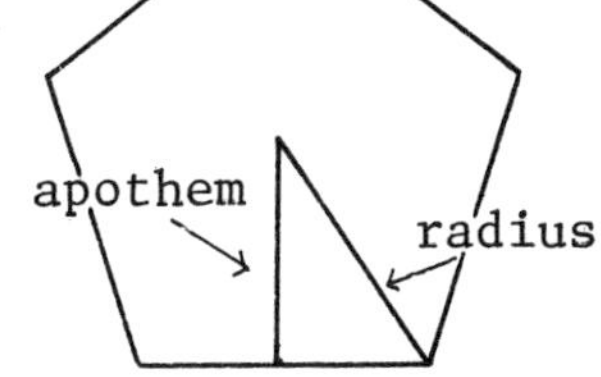

Geometrically, this triangle is the one formed by the radius and apothem of a regular pentagon, and a method for constructing it is described on the following page. The subsequent construction shows the relationship between the Harmonic Triangle and an inscribed pentagon.

CONSTRUCT AN HARMONIC TRIANGLE GIVEN A REGULAR PENTAGON

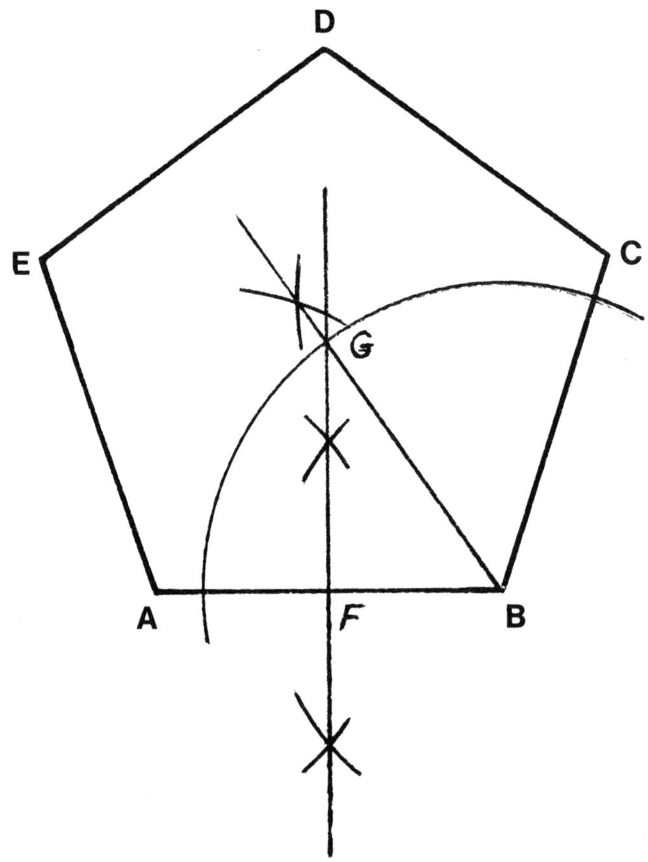

Given regular pentagon ABCDE, construct ΔFBG so that it is an Harmonic Triangle.

1. Construct the perpendicular bisector of $\overline{AB}$.
2. Label F the midpoint of $\overline{AB}$.
3. Construct the bisector of $\angle B$ and extend it so that it intersects the line drawn in Step 1.
4. Label the point of intersection G.

Now ΔFBG is an Harmonic Triangle.

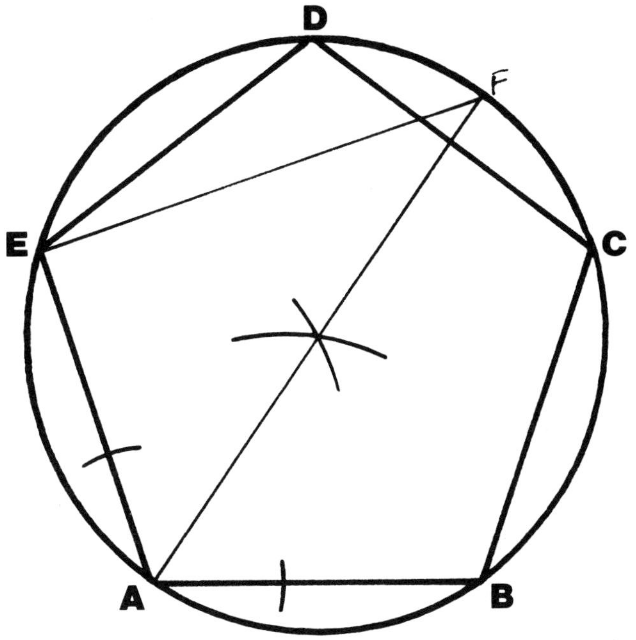

CONSTRUCT AN HARMONIC TRIANGLE
GIVEN A CIRCLE CIRCUMSCRIBED
ABOUT A REGULAR PENTAGON

Given regular pentagon ABCDE inscribed in a circle, construct Harmonic Triangle AEF.

1. Bisect ∠A and extend the bisector to intersect the circle.
2. Label the point of intersection F. ($\overline{AF}$ is a diameter of the circle.)
3. Draw $\overline{EF}$.

Now △AEF is an Harmonic Triangle.

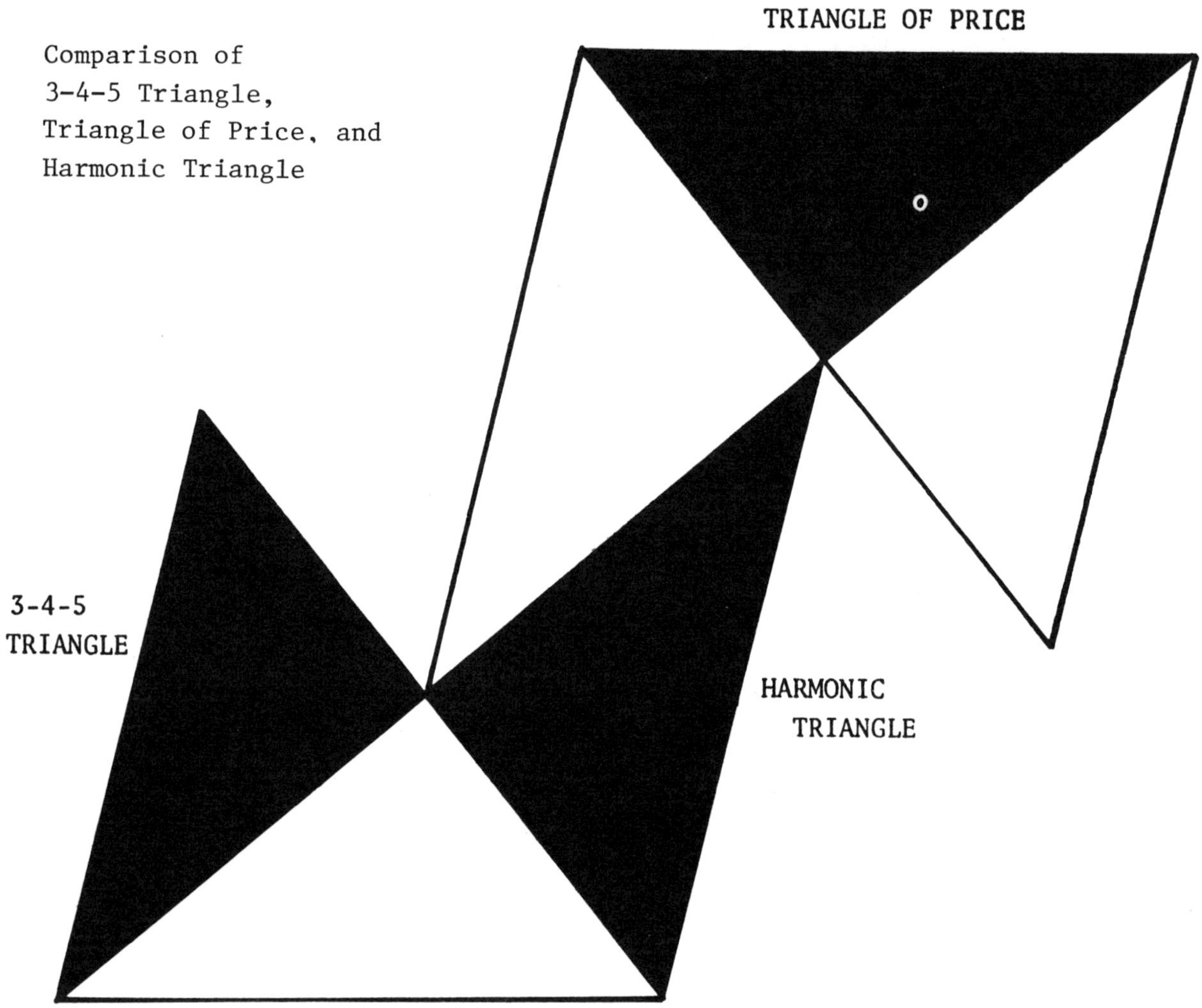

Comparison of
3-4-5 Triangle,
Triangle of Price, and
Harmonic Triangle

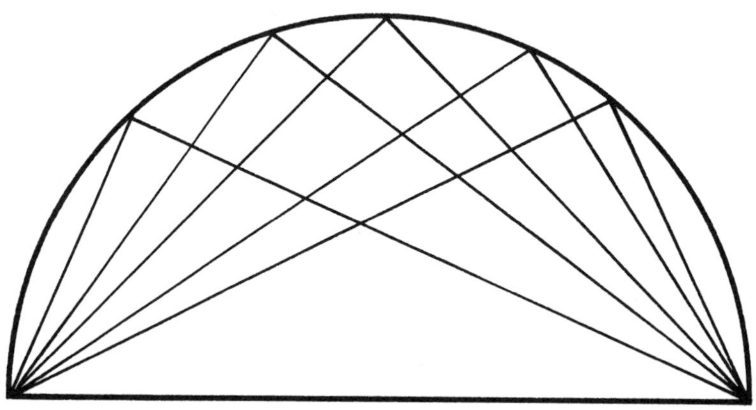

It should be noted that any triangle inscribed in a semicircle is a right triangle

The next two constructions provide methods whereby you can obtain right triangles by first defining a semicircle.

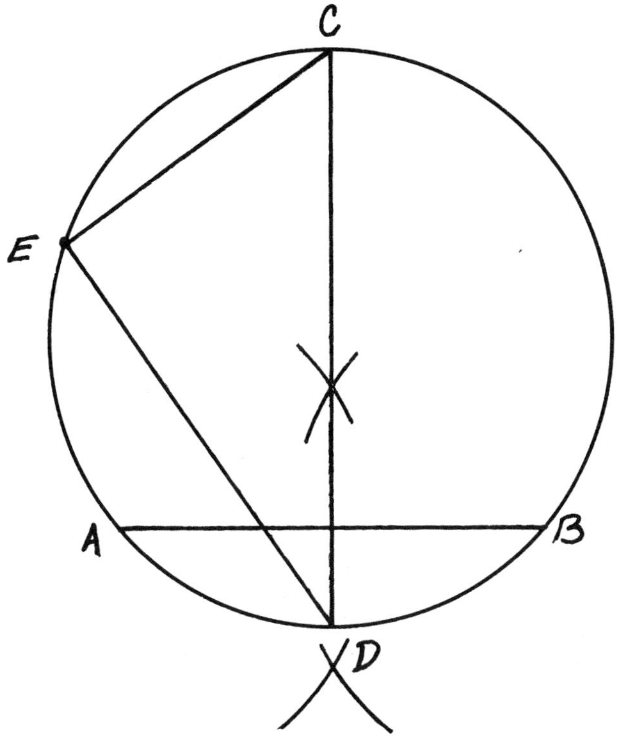

CONSTRUCT A RIGHT TRIANGLE IN A SEMICIRCLE GIVEN THE CIRCLE

Given a circle, construct a right triangle in a semicircle.

1. Draw a <u>chord</u> and label its endpoints A and B.
2. Construct the perpendicular bisector of $\overline{AB}$ and extend it so that its endpoints lie on the circle.
3. Label the endpoints C and D. ($\overline{CD}$ is a <u>diameter</u> of the circle.)
4. Choose point E anywhere on the circle.
5. Draw $\overline{EC}$ and $\overline{ED}$.

Now $\triangle$CED is a right triangle with $\angle$E the right angle.

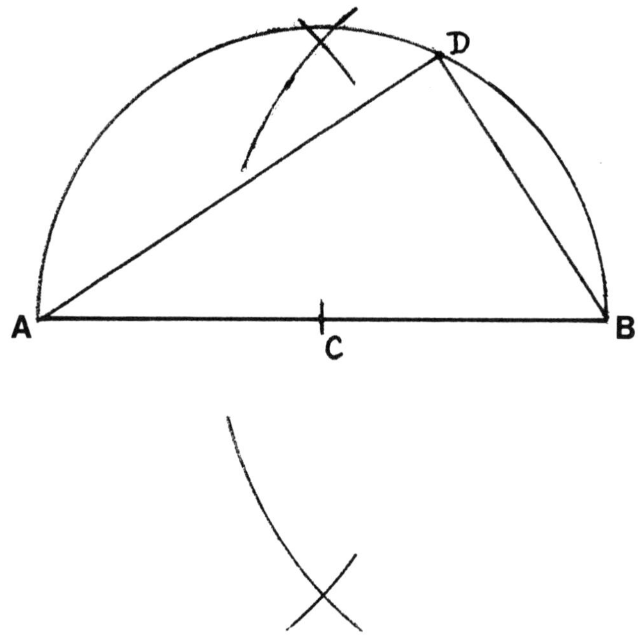

CONSTRUCT A RIGHT TRIANGLE IN A SEMICIRCLE GIVEN THE DIAMETER

22

Given $\overline{AB}$, construct a right triangle in semicircle $\widehat{AB}$.

1. Bisect $\overline{AB}$ and label its midpoint C.
2. Place the metal tip on C and cut an arc from A to B.
3. Choose <u>any</u> point D on $\widehat{AB}$.
4. Draw $\overline{DA}$ and $\overline{DB}$.

Now $\triangle ADB$ is a right triangle with $\angle D$ a right angle.

Student Work
Chapter 4, based on project 4

Suzanne Tronerud

According to some sources the Harmonic Triangle was used as the standard drafting tool of the Middle Ages. If so, it has since been replaced by another right triangle that does not have harmonic proportions but whose sides are in a much simpler ratio. This is the so-called 30-60-90 Triangle or Timeaus Triangle. It is the triangle built with the radius and apothem of the regular hexagon rather than the pentagon, and its angles measure 30°, 60° and 90° respectively.

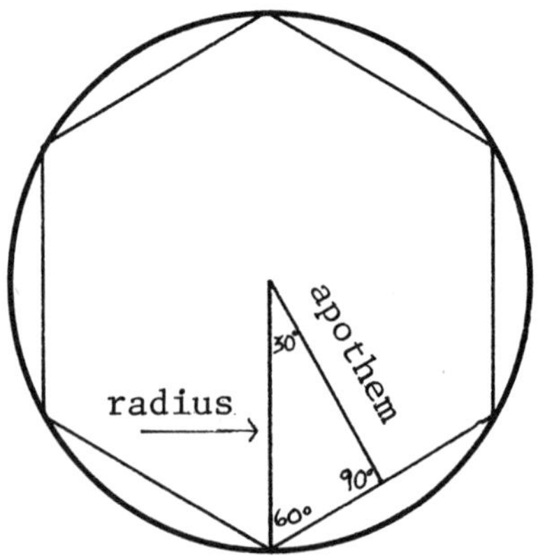

If we let the side opposite the 30° angle measure 1, then the hypotenuse measures 2 and, by the Pythagorean Theorem, the remaining side measures $\sqrt{3}$.

$$a^2 + b^2 = c^2$$
$$a^2 + 1^2 = 2^2$$
$$a^2 + 1 = 4$$
$$a^2 = 3$$
$$a = \sqrt{3}$$

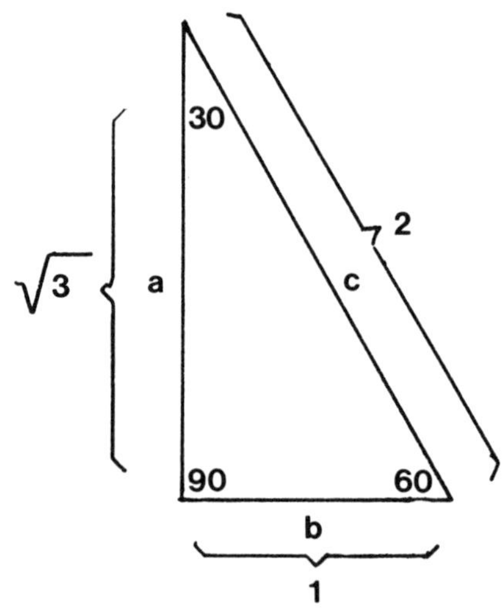

Similarly, if b is any number, say x, c is twice its length or 2x, and a is $x\sqrt{3}$ since $a^2 + b^2 = c^2$.

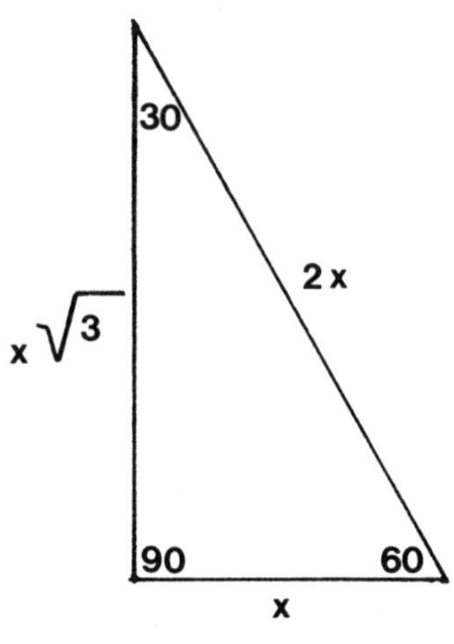

$$a^2 + x^2 = (2x)^2$$
$$a^2 + x^2 = 4x^2$$
$$a^2 = 3x^2$$
$$a = \sqrt{3x^2}$$
$$a = x\sqrt{3}$$

Some scholars feel that with the change from the use of the Harmonic Triangle to the use of the 30-60-90 Triangle much of the beauty and harmony in architectural design was lost. The Harmonic Triangle is based on the pentagon found in Nature only as a structure of life and never found in static or crystalline forms. The 30-60-90 Triangle, on the other hand, comes from the hexagon which is found in some lower order creatures but which abounds in inorganic forms. To some this relationship with the non-living suggests sterility, hence lack of good proportion, and yet Plato felt this was the most beautiful of all triangles. We feel that the crystalline and the organic are two aspects of the unity that is Nature. Moral judgments of better or worse do not apply.

The other standard draftsman's triangle is the one that is derived by dividing a square along its diagonal. This triangle then has a pair of congruent sides and a pair of congruent angles, the latter always measuring 45°.

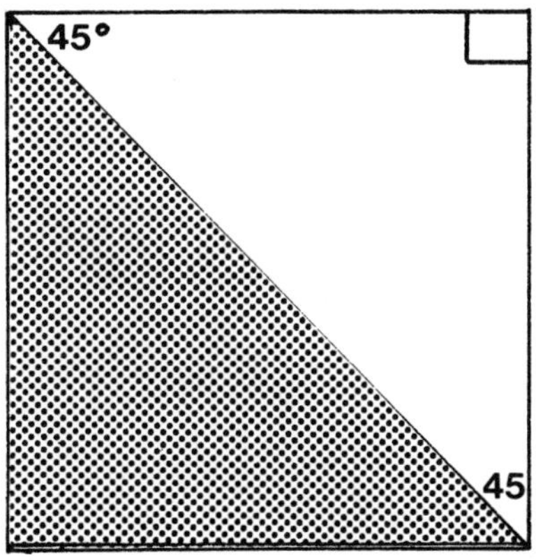

We have already seen that if the sides of the square measure 1, then the diagonal measures that irrational number $\sqrt{2}$ that caused the Pythagoreans such discomfort. Yet it is the Pythagorean Theorem which allows us to find the measure of the hypotenuse regardless of the measure of the side.

We know that $a^2 + b^2 = c^2$, but this time $a = b$, so we have

$$a^2 + a^2 = c^2$$
$$2a^2 = c^2$$
$$\sqrt{2a^2} = c$$
$$a\sqrt{2} = c$$

so if $a = 1$, $c = \sqrt{2}$
if $a = 2$, $c = 2\sqrt{2}$
if $a = 3$, $c = 3\sqrt{2}$ and so on.

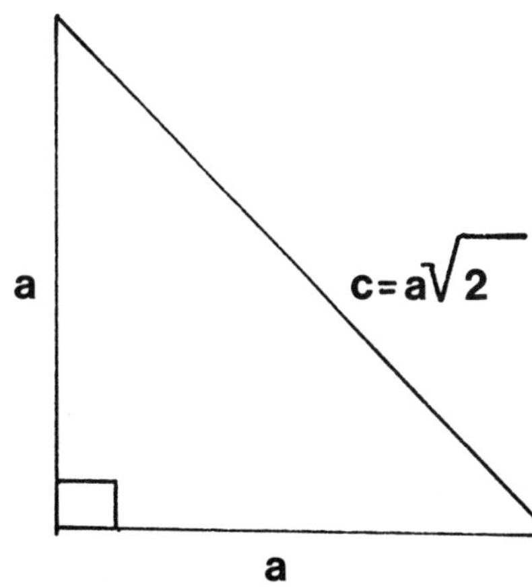

This triangle is called the 45-45-90 Triangle or the Isosceles Right Triangle, and its construction follows.

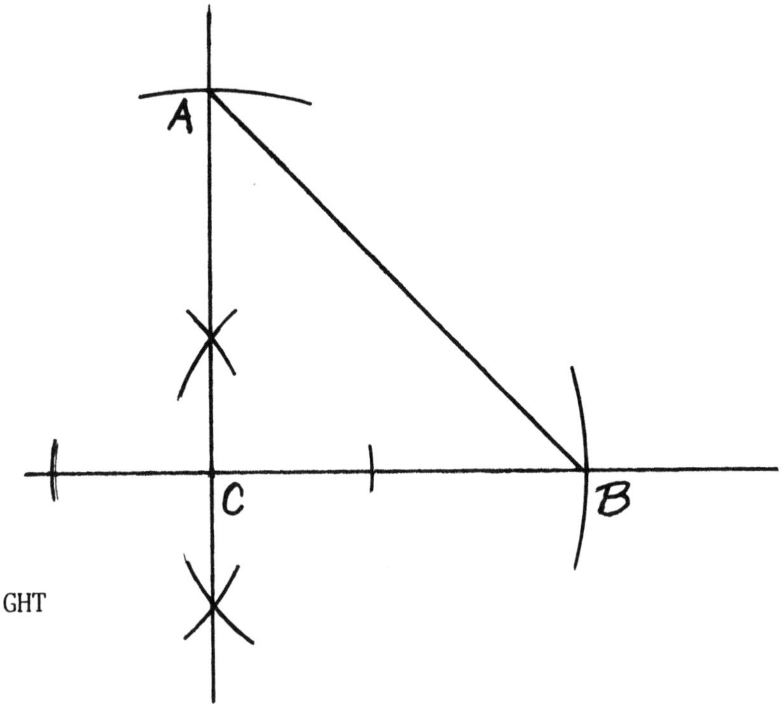

CONSTRUCT AN ISOSCELES RIGHT TRIANGLE

Construct a right triangle ABC such that $\overline{CA} \cong \overline{CB}$.

1. Construct a pair of perpendicular lines.
2. Label their point of intersection C.
3. Open the compass to measure the desired length of the legs of the triangle.
4. Place the metal tip on C and cut one arc on each line.
5. Label the points of intersection A and B.
6. Draw $\overline{AB}$.

Now $\triangle ABC$ is an Isosceles Right Triangle with $\overline{CA} \cong \overline{CB}$. Also $\angle A \cong \angle B$ and each measures 45°.

Any triangle that has two congruent sides is called Isosceles, and any isosceles triangle has a pair of congruent angles called base angles. The third angle is called the vertex angle.

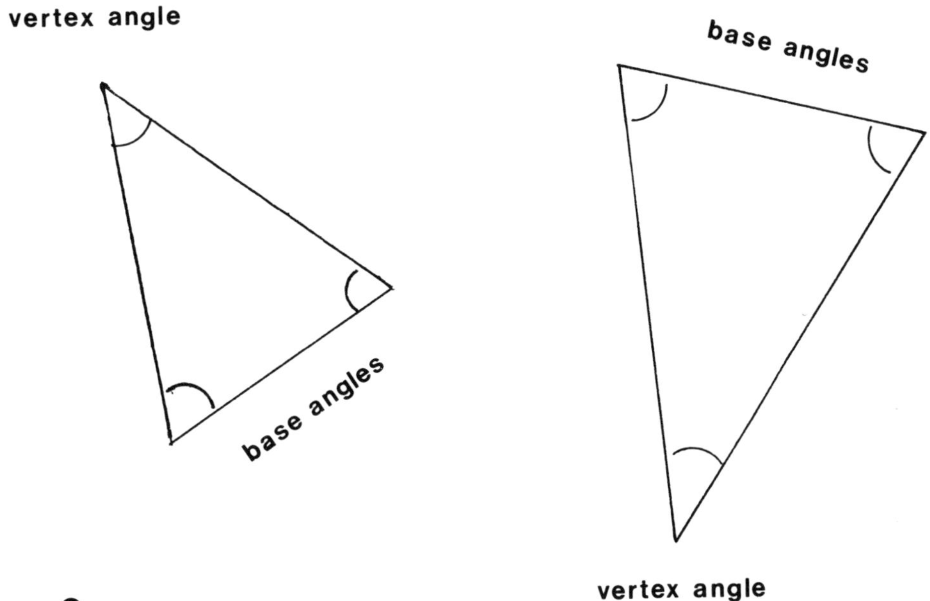

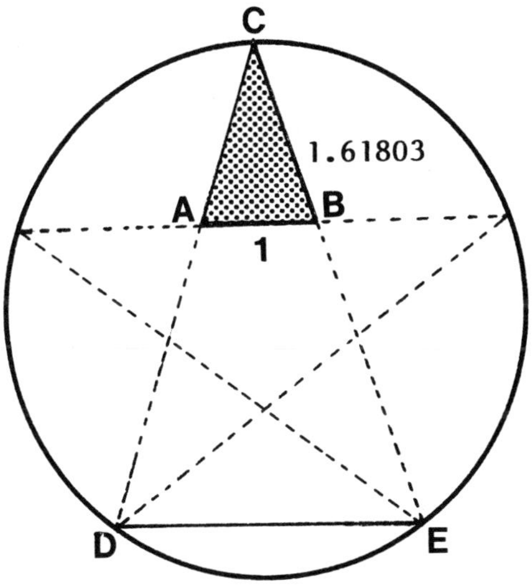

One isosceles triangle relates directly to the pentagon and its pentagram and picks up the harmonic proportions inherent in these figures.

ΔABC is isosceles with base angles that measure 72° and a vertex angle of 36°. As you recall from Chapter 2, $\frac{CB}{BA} = 1.61803 = \phi$. If we draw $\overline{DE}$, we have another triangle, ΔDEC, which is similar to ΔABC. Since similar triangles have corresponding sides in proportion, we have

$$\frac{CB}{BA} = \frac{CE}{ED} = \phi$$

This triangle is called the Sublime Triangle or the Triangle of the Pentalpha, and from the preceeding diagram we can see that it is a simple matter to construct it by drawing the two diagonals from one vertex of a regular pentagon.

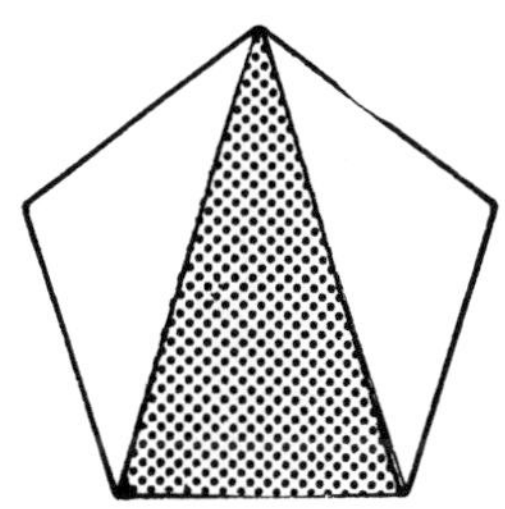

The following construction shows the relationship between the Triangle of the Pentalpha and the Golden Rectangle. Because of this relationship it has come to be known as the Golden Triangle, which is the name we shall use from this point on.

CONSTRUCT A GOLDEN TRIANGLE
GIVEN A GOLDEN RECTANGLE

24

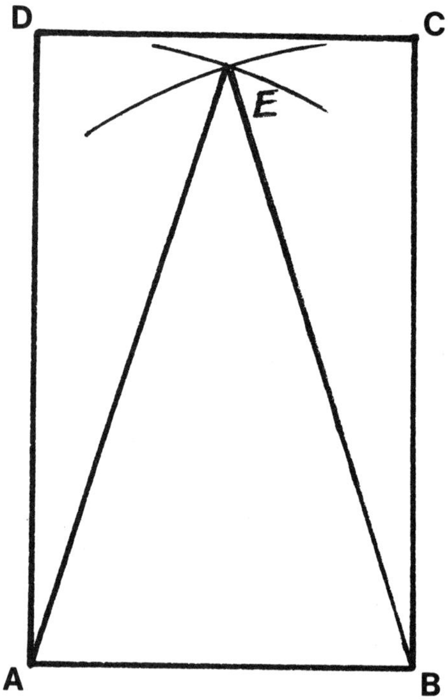

Given Golden Rectangle ABCD, construct $\triangle ABE$, a Golden Triangle

1. Place the metal tip on B and the pencil tip on C and cut an arc into the interior of ABCD.
2. Without changing the setting, place the metal tip on A and cut an arc that intersects the one drawn in Step 1.
3. Label the point of intersection E.
4. Draw $\overline{EA}$ and $\overline{EB}$.

Now ABE is a Golden Triangle with
$\frac{EB}{AB} = \frac{EA}{AB} = \phi$.

The intimate relationship between the pentagon and the Golden Triangle is demonstrated in a figure that is known as the Lute of Pythagoras.

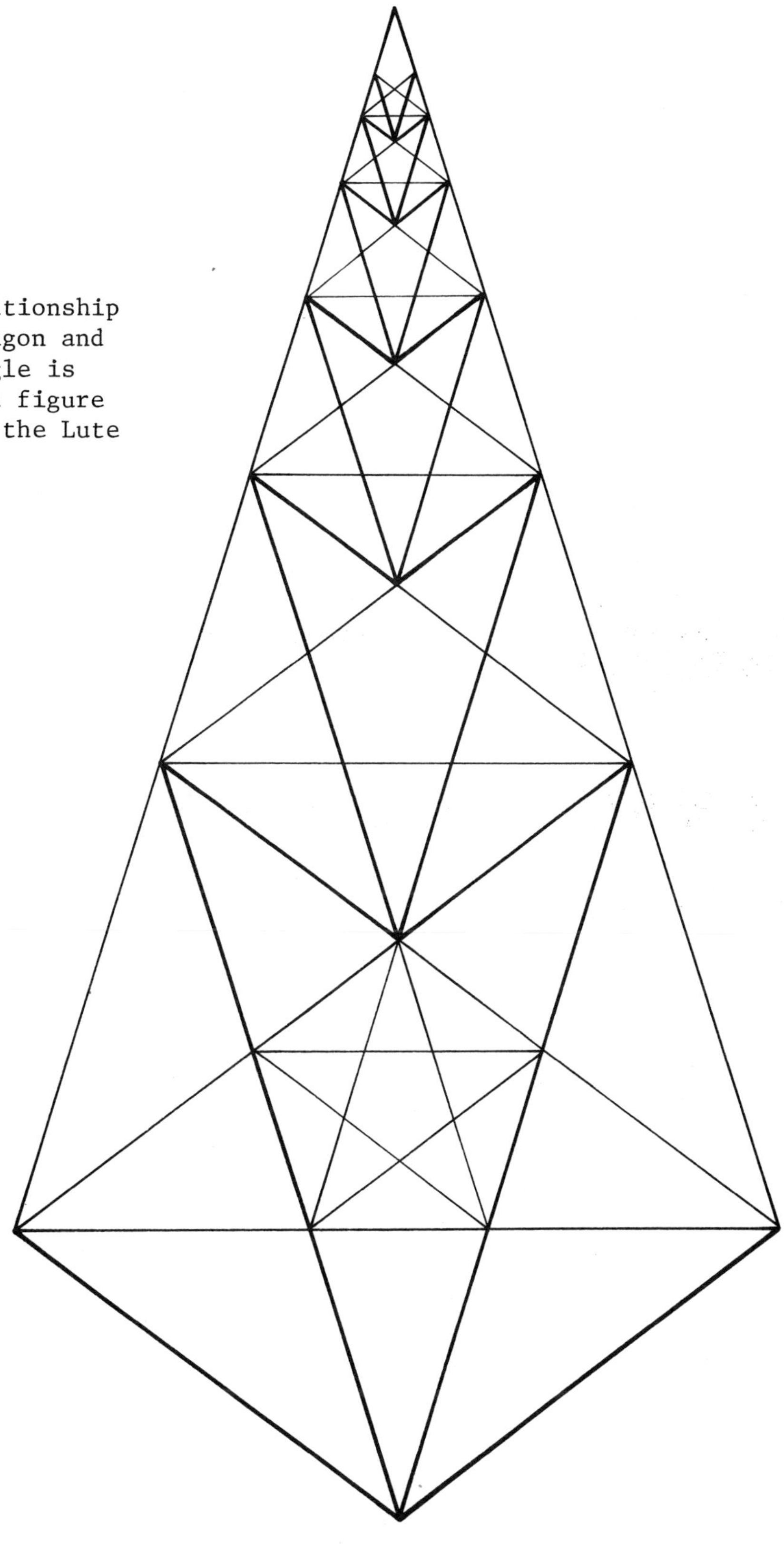

In the construction of this figure, a "ladder" is built within a Golden Triangle by first opening the compass to the length of the base of the triangle, placing the metal tip on the vertices of the base angles, and cutting arcs on the opposite sides respectively. The process is continued upward through the triangle, each time setting the compass by the previously cut pair of arcs. The points of intersection are then joined by line segments.

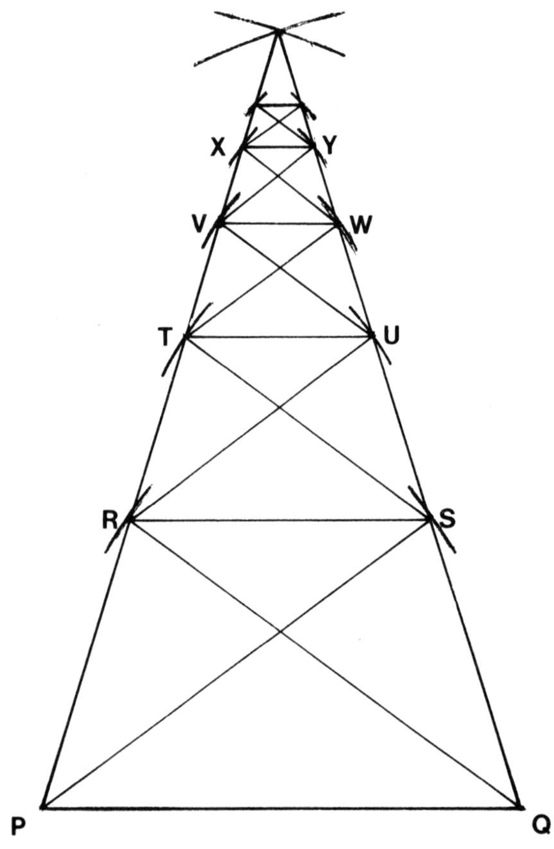

Thus we have:

PQ = PS = QR
RS = RU = ST
TU = TW = UV
VW = VY = WX

The pentagonal figures are then formed by joining corresponding points on either side of the triangle to the point where the intersecting segments meet in the previous "rung."

Finally, point Z is determined by setting the compass to measure QS and cutting arcs from points P and Q respectively. $\overline{PZ}$, $\overline{ZQ}$, $\overline{RZ}$ and $\overline{SZ}$ are drawn, and the figure is completed with the addition of the small pentagram in the interior of QSRPZ.

Not only does this figure contain all the Ø-proportions already mentioned in the pentagon, pentagram and Golden Triangle, but also the rungs of the ladder divide the sides of the large Golden Triangle into Ø-proportions.

That is: $\frac{QS}{SU} = \frac{SU}{UW} = \frac{UW}{WY} = \emptyset$

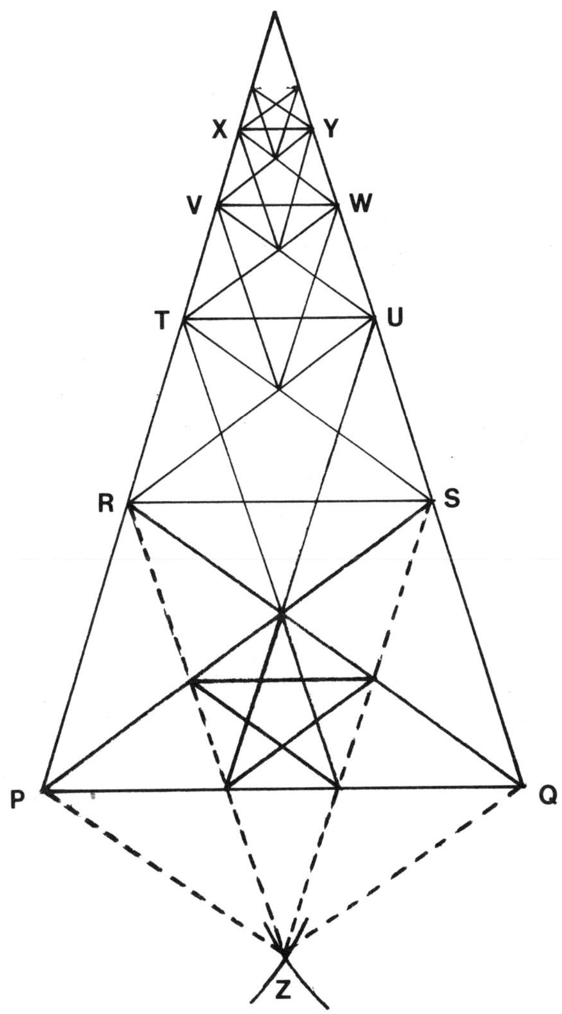

Photo by Richard Newman

Julie Conway

Student Work
Chapter 3, project 1

The following diagrams further demonstrate the use of the pentagram and the Golden Triangle as elements of design in dividing areas harmoniously.

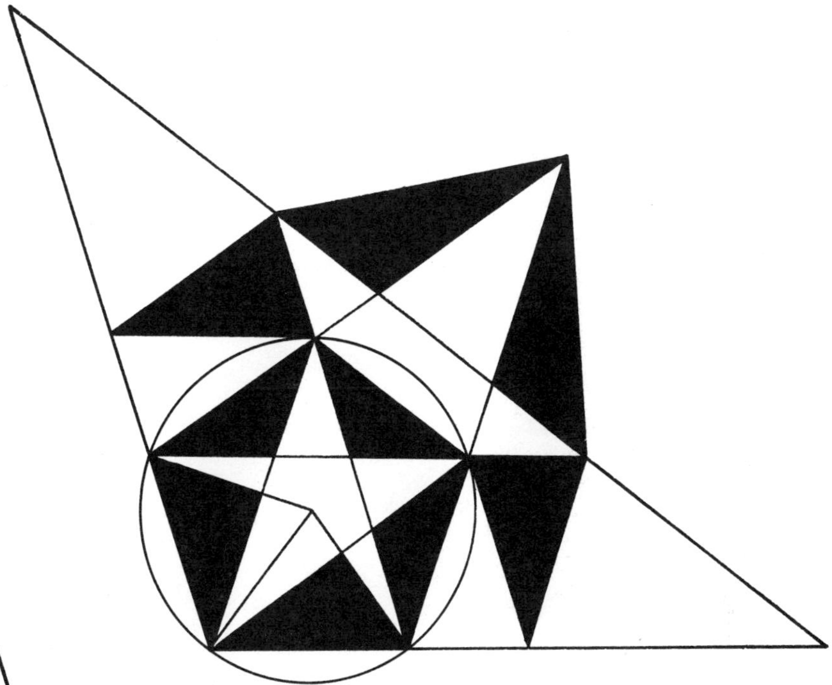

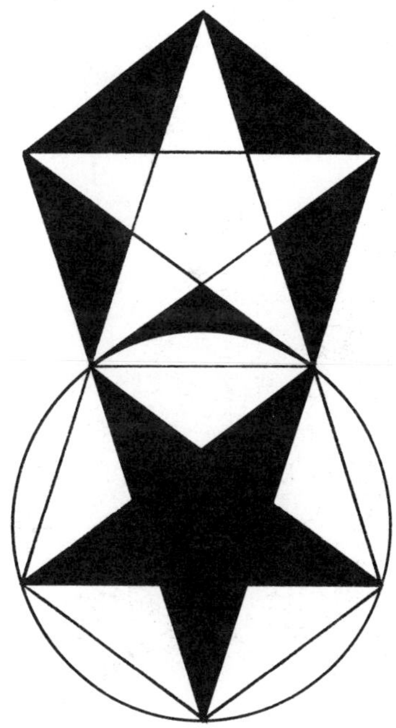

Although the Isosceles Right Triangle and the Triangle of the Pentalpha are special cases, isosceles triangles in general can be useful design tools, and they can be constructed by the following method.

CONSTRUCT AN ISOSCELES TRIANGLE

25

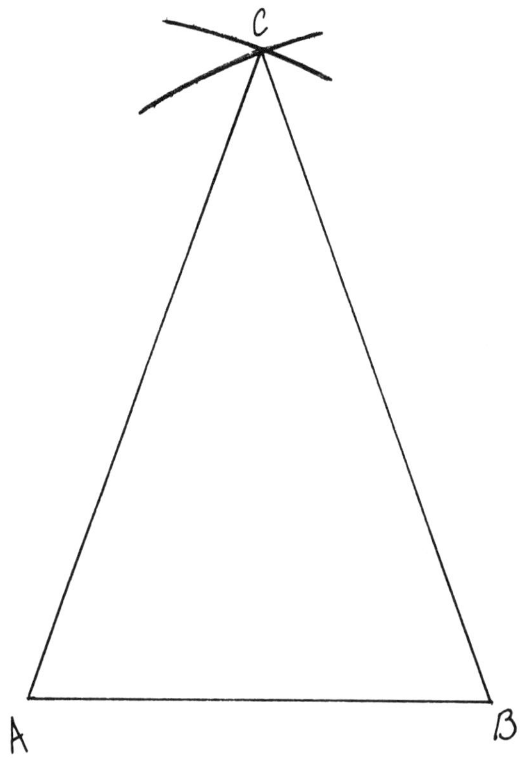

Construct $\triangle ABC$ such that $\overline{AC} \cong \overline{BC}$.

1. Draw a line segment that measures the desired length of the <u>base</u> of the triangle.
2. Label its endpoints A and B.
3. Open the compass so that it measures the desired length of the <u>legs</u> of the triangle. (Note: This measure must be more than $\frac{1}{2}AB$ or a triangle cannot be formed.)
4. Place the metal tip on A and cut an arc on one side of $\overline{AB}$.
5. Without changing the setting, place the metal tip on B and cut an arc that intersects the one drawn in Step 4.
6. Label the point of intersection C.
7. Draw $\overline{AC}$ and $\overline{BC}$.

Now $\triangle ABC$ is isosceles with $\overline{AC} \cong \overline{BC}$ and $\angle A \cong \angle B$.

The last triangle we wish to consider is the Equilateral Triangle. It is the triangle whose sides and angles are all congruent. The sides may measure any length but the angles will always be 60^0.

The equilateral triangle has been used in the design of symbols throughout history. It is the Christian symbol for the Trinity. For both the Ancient Chinese and the Japanese, the triangle is the symbol for Heaven, Earth and Man.

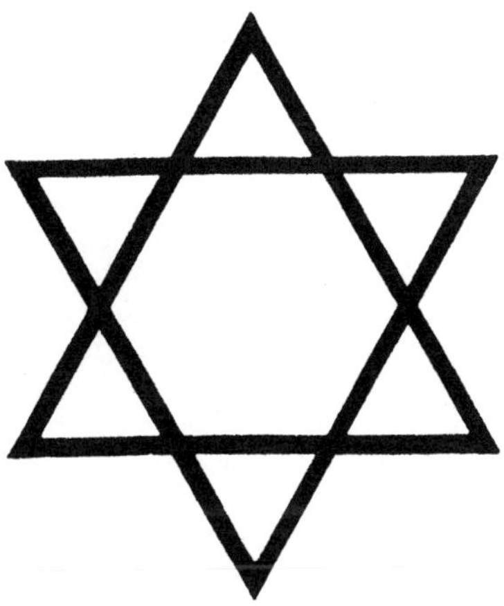

When known as the Jewish Star of David, the intersecting pair of Equilateral Triangles symbolizes the power of Jehovah moving outward from the center, as well as being seen as the sixfold symbol of creation and perfection.

As Solomon's Seal it symbolizes the synthesis of the four elements -- earth, air, fire and water. It also symbolizes the union of complements. The triangle pointing upwards represents Man striving for heaven while the downward pointing triangle represents Man with earthly concerns.

In India the interlocking of the two triangles is that of Shiva and Shaki forming the Wheel of Vishnu.

For the ancient Egyptians, the symbol represented fertility by suggesting the sexual union of the pagan goddess Ashtaroth and Adonis.

Nature uses the equilateral triangle as a design element, too.

Some other examples of the use of equilateral trianles are pictured here.

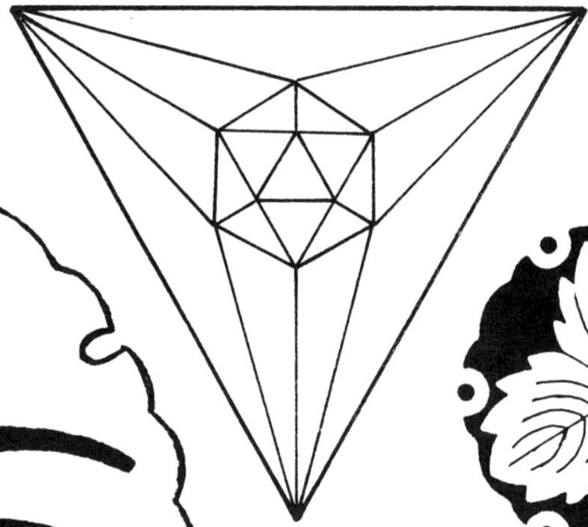

If the equilateral triangle also appeals to you as a design element, its construction is a very simple matter.

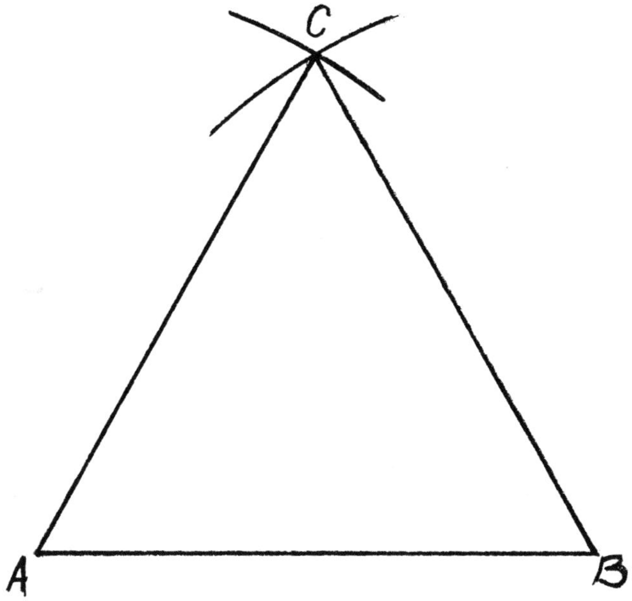

CONSTRUCT AN EQUILATERAL TRIANGLE

Construct $\triangle ABC$ such that $\overline{AB} \cong \overline{BC} \cong \overline{AC}$.

1. Draw a line segment that measures the desired length of a side of the triangle.
2. Label its endpoints A and B.
3. Place the metal tip on A and the pencil tip on B and cut an arc on one side of $\overleftrightarrow{AB}$.
4. Without changing the setting, place the metal tip on B and cut an arc that intersects the one drawn in Step 3.
5. Label the point of intersection C.
6. Draw $\overline{AC}$ and $\overline{BC}$.

Now $\triangle ABC$ is equilateral with $\overline{AB} \cong \overline{BC} \cong \overline{AC}$. Also $\angle A \cong \angle B \cong \angle C$ and each measures 60°.

The following diagram illustrates the harmonic subdivision of the equilateral triangle.

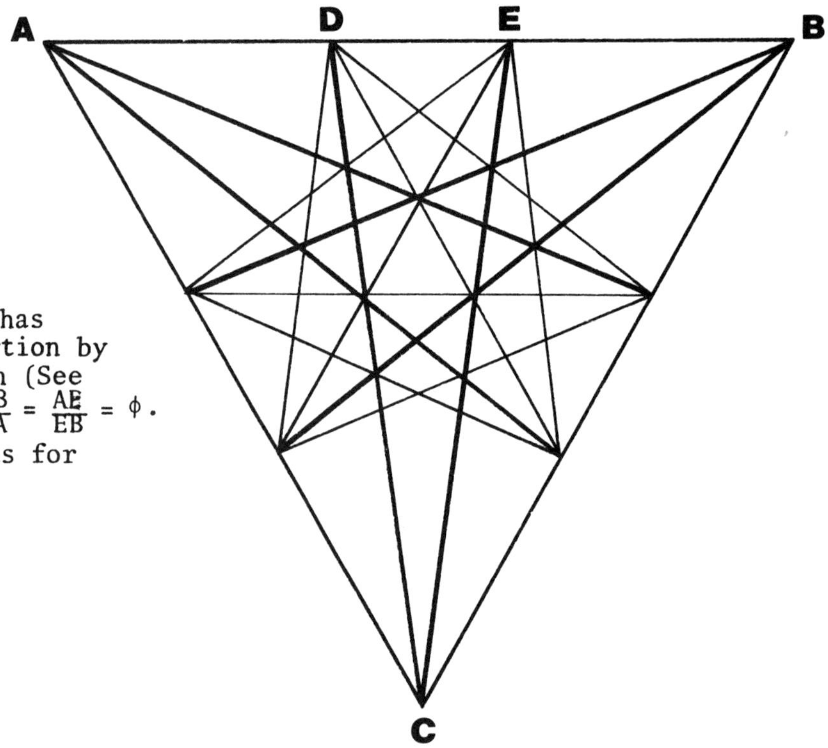

Each of the sides of ΔABC has been divided into ϕ proportion by finding the Golden Section (See Construction 9). That is $\frac{DB}{DA} = \frac{AE}{EB} = \phi$. The same relationship holds for the other two sides.

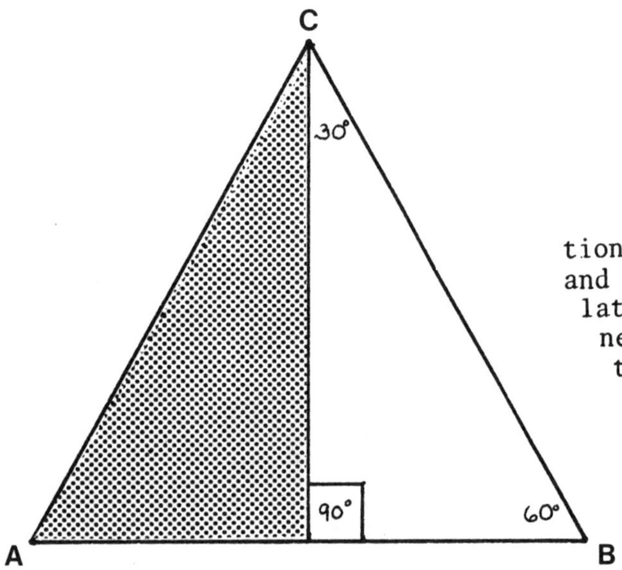

Let us briefly examine the relationship between the equilateral triangle and the 30-60-90 Triangle. Since the equilateral triangle has angles of 60°, we need only bisect one of these to divide the angle into two 30° angles and the triangle into two congruent 30-60-90 Triangles.

Because of the symmetric properties of the equilateral triangle, any angle bisector is also the perpendicular bisector of the opposite side, providing the requisite right angle for the 30-60-90 Triangle.

Photo by Richard Newman

Student Work
Chapter 3, project 5

Ruth DesRoches

SPECIAL TRIANGLES

NAME(S)	TYPE	ANGLES
3-4-5 Rope Stretcher's Rope Knotter's Pythagorean	Right Triangle	36°52'12" 53° 7'48" 90°
Triangle of Price Egyptian Triangle	Right Triangle	38°30', 51°30', 90°
Harmonic Triangle	Right Triangle	36°, 54°, 90°

The above three triangles may be interchanged for most design purposes.

NAME(S)	TYPE	ANGLES
30-60-90 Triangle Timaeus Triangle	Right Triangle	30°, 60°, 90°
Isosceles Right 45-45-90	Right Triangle	45°, 45°, 90°
Golden Triangle Sublime Triangle	Isosceles Triangle	36°, 72°, 72°
Equilateral Triangle	Equilateral Triangle	60°, 60°, 60°

	SIDES	DIAGRAM
________	Only right triangle whose sides are in arithmetic progression, in the ratio 3:4:5	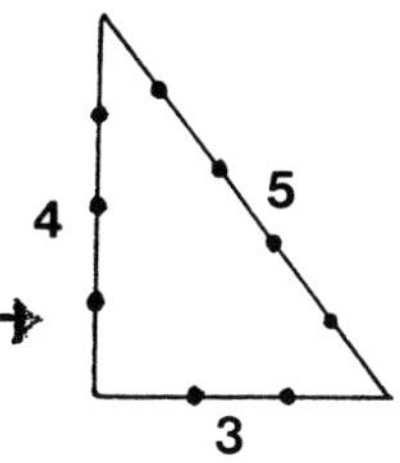
________	Only right triangle whose sides are in geometric progression, in the ratio $1:\sqrt{\phi}:\phi$	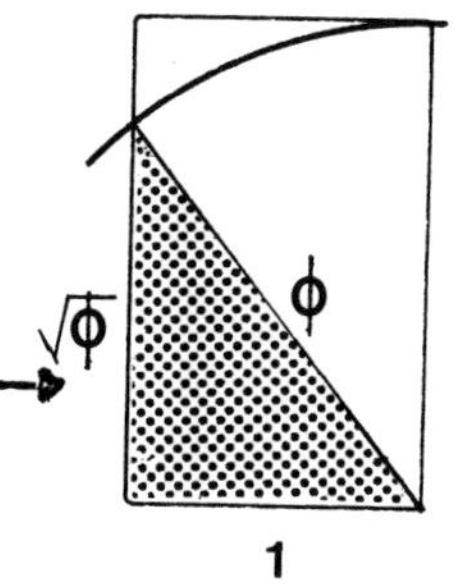
________	In the ratio of 1 foot : a little cubit:a royal cubit	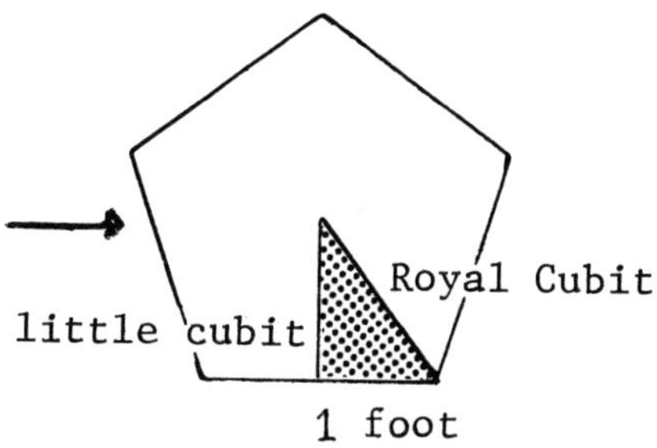
________	In the ratio $1:\sqrt{3}:2$	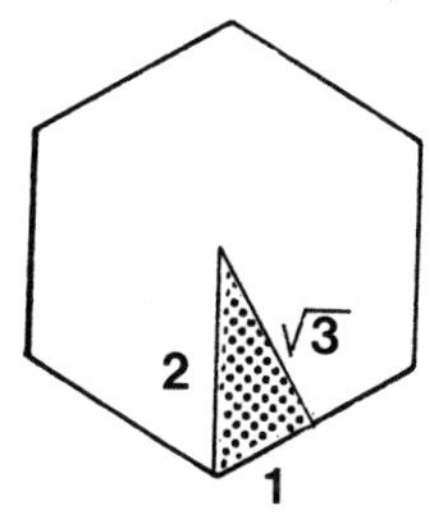
________	In the ratio $1:1:\sqrt{2}$	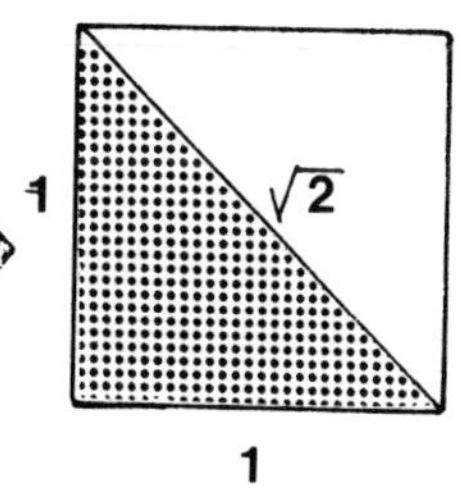
________	Two congruent sides Ratio of leg to base is ϕ	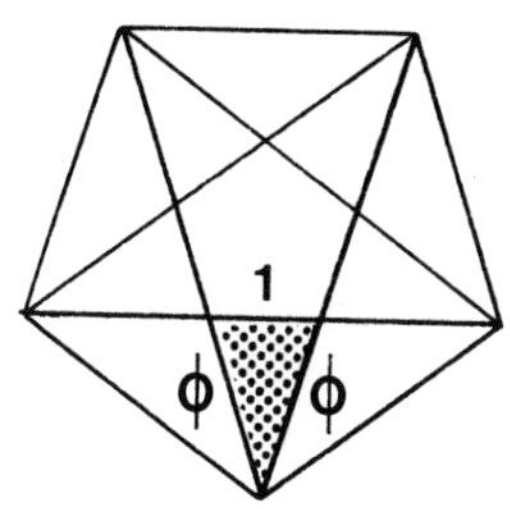
________	Three congruent sides	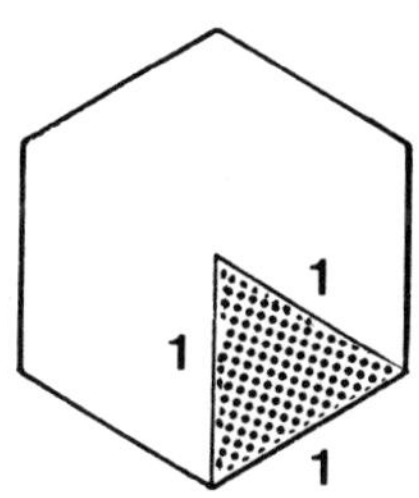

PROBLEM 1

Go out into the landscape with a friend(s), some pegs and some cord. Using these materials, bound an area of landscape. Now, using the process of triangulation, mark the area into triangles, measure the individual ones and add together to obtain the total of the bounded area. Document this experience with photographs, drawings or diagrams.

PROBLEM 2

Construct one of each of the special triangles that appear on the chart on the previous page.

PROBLEM 3

Construct a Golden Triangle with leg length of at least five inches. Use the information given on pp. 107 - 109 to construct the Lute of Pythagoras within the triangle.

PROBLEM **4**

Given the illustrations of butterflies and moths following, escribe a triangle around each. What kinds of special triangles occur? (You'll need to use a <u>protractor</u> to measure the angles of the triangles.) Write a summary of your findings.

PROBLEM **5**

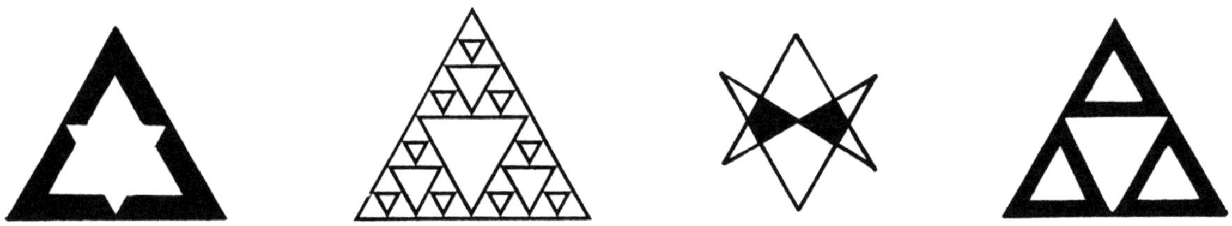

Each of the above symbols in some way utilizes the concept of intersecting equilateral triangles. Reconstruct each one on a separate sheet of paper. For ease of construction let the side of the largest triangle in each symbol measure a minimum of four inches.

PROBLEM **6**

Construct an equilateral triangle with sides at least four inches long. Subdivide the sides harmonically and join points as in the illustration on page 118. Strengthen the design by alternately coloring in areas with black.

PROJECT **1**

Given the information in problem 3, create two imaginary butterflies that look different but will fit into the same (any one of your choice) special triangle. Use whatever materials catch your fancy.

PROJECT **2**

Now that you have learned to construct several different types of triangles, create an interesting pattern, which may be either two dimensional or three dimensional, using:

a. The same shape triangle, but altering the sizes and using a monochromatic coloring scheme.

b. Two different types or triangles and a complementary color scheme.

c. A variety of triangular shapes and an analogous color scheme.

PROJECT **3**

Using any materials whatsoever, construct a pyramid on a square base that is at least 12" high.

PROJECT **4**

On an illustration board, construct a Golden Rectangle in dimensions of your own choosing. Using the process of triangulation divide the area enclosed by the rectangle. Color by using one of the following:

a. Black and one other color;

b. Black and white;

c. Five values of grey;

d. A pair of complementary colors;

e. A triad of colors.

PROJECT **5** Sculptural Form

Twenty equilateral triangles join to form a figure called the icosahedron. The pattern below shows how to lay out the 20 triangles in the plane. By bending, folding, scoring and pulling together, the icosahedron is formed.

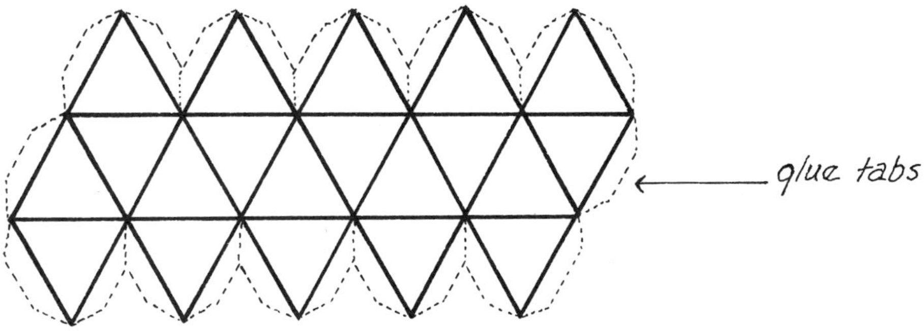

Create your own variation on this Platonic Solid by using one of the following:

a. Making a template out of heavy paper or acetate following the directions for Construction 26. Use this template to make all 20 units.

b. Divide the triangle itself into harmonic divisions following the directions in Construction .

c. Join all 20 units, or multiples of 20, to create a sculpture.

d. Use materials of your choice to make the finished form.

e. Consider the possibilities of using:

1. felt material
2. patchwork units } these would need to be stuffed to fill out the forms
3. knitted units }
4. crochet units }
5. illustration board
6. bristol paper
7. plexiglas
8. sheet copper, brass or aluminum
9. colored acetate
10. stained glass pieces
11. balsa wood in sheets or sticks
12. clay units
13. any other materials you might fancy

PROJECT 6

Now that you have learned of other useful triangles for the artist-designer, construct a set of heavy cardboard templates of the triangles that you would find useful. These could be then manufactured in a more permanent material.

PROJECT 7

Assume you own your own business. Use the harmonic subdivisions of an equilateral triangle to design a logo for your company. The choice of black and white or color is yours, but the finished work should be on illustration board and the triangle should have sides that measure ten inches. Use acrylic, casein, or gouache paints to obtain good surface consistency.

PROJECT 8

Make a kite using the Lute of Pythagoras as a pattern. Construct the Lute out of tissue paper, acetate, cellophane or fabric in any color scheme of your choice. Support your kite with balsa wood strips, string around the edges, add a tail and make a windy day more beautiful.

Further Reading

Eves, Howard. An Introduction to the History of Mathematics. 3d ed. New York: Holt, Rinehard and Winston, 1969.

Fleming, William. Arts and Ideas. New York: Holt, Rinehart and Winston, 1980.

Gillings, Richard J. Mathematics in the Time of the Pharaohs. New York: Dover Publications, Inc., 1972.

Grillo, Paul Jacques. Form, Function and Design. New York: Dover Publications, Inc., 1960.

Hogben, Lancelot. Mathematics in the Making. Garden City, New York: Doubleday and Company, Inc. 1960.

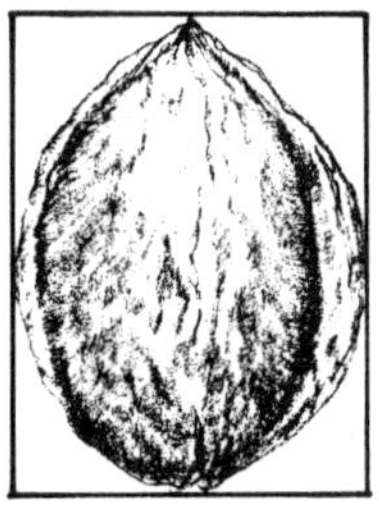

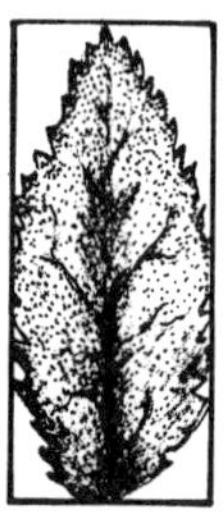

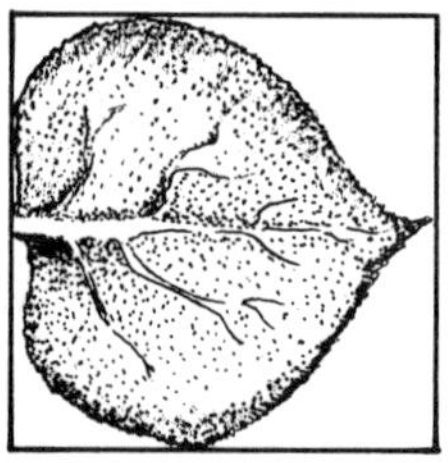

Dynamic Rectangles

4

Composition is to the artist what formulas are to the mathematician. Each is a skeletal understructure that gives support to the flesh of an idea.

For the visual artist, the idea of composition is rooted in the concept of spatial division. Composition in drawing or painting or designing involves bounding a portion of the plane and then sub-dividing the enclosed area into meaningful divisions to be used with either non-objective or representational images.

Prior to the time of the Ancient Egyptians, all spatial directions were considered equally valid for composition. With the Egyptians, the rectangle became the supreme compositional format because of the relationship between the vertical and the horizontal at a 90° angle.

The first concern of the artist is what size shall this rectangle be? When painting was more closely allied with the mural form, it was the intended architectural space-surface that determined this. With the advent of easel painting, the artist was left to his own fancies as to what size this portable rectangle would be. Naturally, if the work was meant for the palace of kings or the parlors of burghers, the ultimate size of the finished work was influenced by this factor. But if the artist was no longer tied to a particular audience or a particular place, then the rectangle became an arbitrary choice. There should always be a connection between the scale of an idea and its particular presentation, but within these confines, how does an artist go about choosing the "right" proportional relation between the vertical and the horizontal? And after that decision is determined, the artist must decide how to break up the spaces within the rectangle so as to provide both structure and meaning. For space, beyond its physical limitations, also acts symbolically as well.

The artist could choose to use rectangles whose sides are in whole number ratios such as those pictured below.

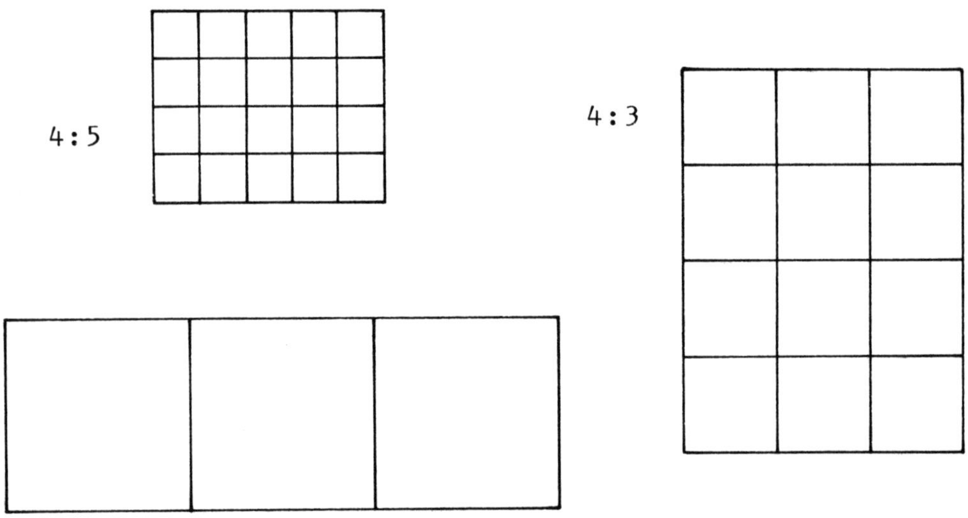

1:3

This type of rectangle leads to incompatibility within the same composition when used in combination, even though individually each can be evenly sub-divided by the <u>unit square</u>. The unit length may vary, as illustrated above, and is dictated by the particular instrument of measure. For example, a rectangle may measure 3-7/8" x 4-1/2" and still its sides are in whole number ratio if the unit square measures 1/8 inch on a side.

There is, however, a group of rectangles, all derived from the square, that gives us a family that relates proportionally and gives harmony. Members of this family are called the "Dynamic Rectangles," sometimes called the "Euclidean Series," because they give harmony within each rectangle and between several rectangles. As we shall see when we examine the structure of the rectangles, there are sub-groups of this family that work well together, especially those that relate to the Golden Rectangle. Since the square is an integral part of each member of this family, it can be used harmoniously with any of these rectangles within a given composition.

According to authors Jay Hambidge and Edward B. Edwards, "Dynamic Rectangles" have "the quality of form in motion" because the subdivision of the parent rectangle produces proportionately related units which allow for continued similarity of shape which is suggestive of the principle of growth found in Nature. The rectangles provide "structural skeletal forms" which are essential for any good composition and design.

Except in certain cases, these rectangles do not have sides in whole number ratios. One side serves as the unit measure and the other side is named by the square root of one of the natural numbers other than one so that the sides of the Dynamic Rectangles are in such ratios as $1:\sqrt{2}$; $1:\sqrt{3}$; $1:\sqrt{4}$; etc. Except in the cases where the radicand is a perfect square, i.e., 4, 9, 16, etc., the length of the rectangle indicates an irrational number. Because of the appearance of these irrational numbers, the Dynamic Rectangles are sometimes called "root rectangles." A Root-two Rectangle is the one whose sides measure one and $\sqrt{2}$.

Although the lengths of sides of the Dynamic Rectangles are incommensurable, except in the cases mentioned above, the areas of the squares on the sides are. In the following diagram, $AB^2=1$ and $CB^2=3$, and the ratio of the areas is 1:3. ABCD is a $\sqrt{3}$ Rectangle, but this principle will hold for all the Dynamic Rectangles since the width is always considered one.

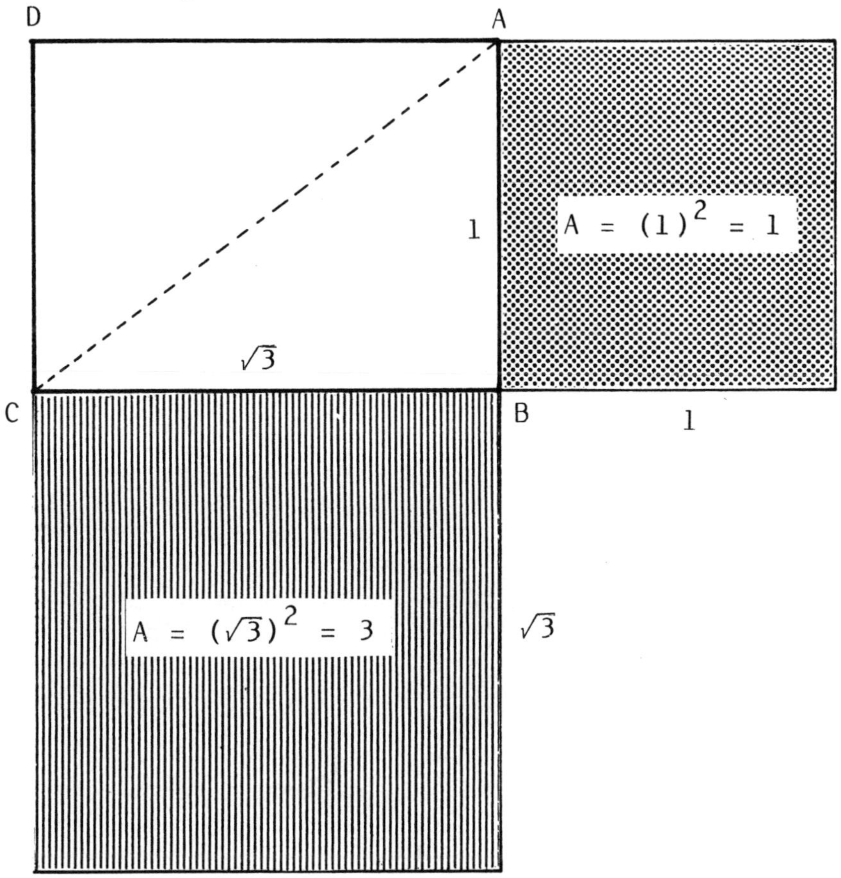

In both the construction of the Dynamic Rectangles and their subsequent subdivisions, the diagonal plays an important role. It is used repeatedly in the following construction for setting the compass, although it is never actually drawn.

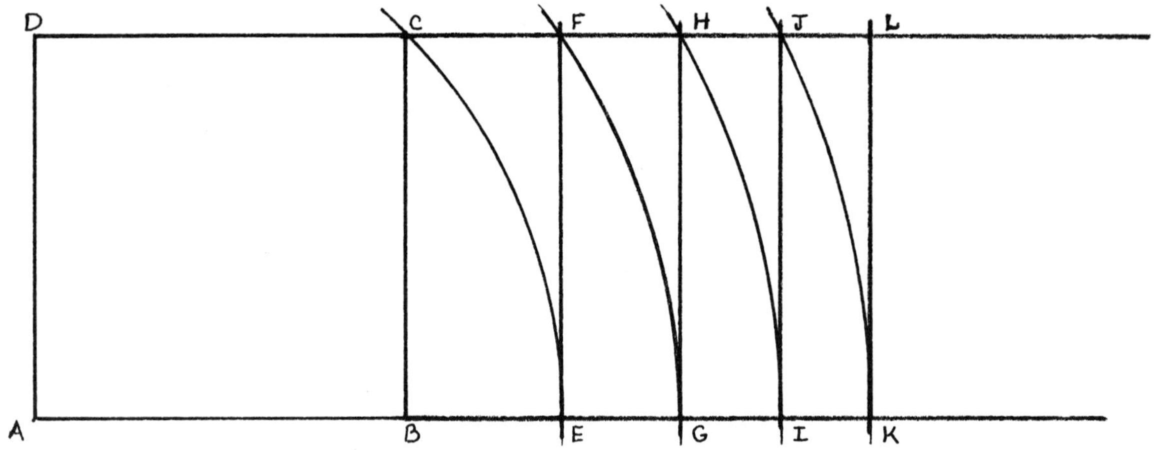

Given square ABCD, construct the dynamic rectangles such that each has side $\overline{AD}$.

GENERATE THE DYNAMIC RECTANGLES FROM THE UNIT SQUARE

27

1. Extend $\overrightarrow{AB}$ and $\overrightarrow{DC}$.
2. Place the metal tip on A and the pencil tip on C and and cut an arc that intersects $\overrightarrow{AB}$.
3. Label the point of intersection E.
4. At E construct a perpendicular to $\overleftrightarrow{AB}$ (refer to Construction 5) and extend it to intersect $\overrightarrow{DC}$.
5. Label the point of intersection F.
6. Draw $\overline{EF}$.

Now AEFD is a $\sqrt{2}$ rectangle.

7. Place the metal tip on A and the pencil tip on F and cut an arc that intersects $\overrightarrow{AB}$.
8. Label the point of intersection G.
9. At G construct a perpendicular to $\overleftrightarrow{AB}$ and extend it to intersect $\overrightarrow{DC}$.
10. Label the point of intersection H.
11. Draw $\overline{GH}$.

Now AGHD is a $\sqrt{3}$ rectangle.

12. Place the metal tip on A and the pencil tip on H and cut an arc that intersects $\overrightarrow{AB}$.
13. Label the point of intersection I.
14. At I construct a perpendicular to $\overleftrightarrow{AB}$ and extend it to intersect $\overrightarrow{DC}$.
15. Label the point of intersection J.
16. Draw $\overline{IJ}$.

Now AIJD is a $\sqrt{4}$ rectangle.

17. Place the metal tip on A and the pencil tip on J and cut an arc that intersects $\overrightarrow{AB}$.
18. Label the point of intersection K.
19. At K construct a perpendicular to $\overleftrightarrow{AB}$ and extend it to intersect $\overrightarrow{DC}$.
20. Label the point of intersection L.
21. Draw $\overline{KL}$.

Now AKLD is a $\sqrt{5}$ rectangle.

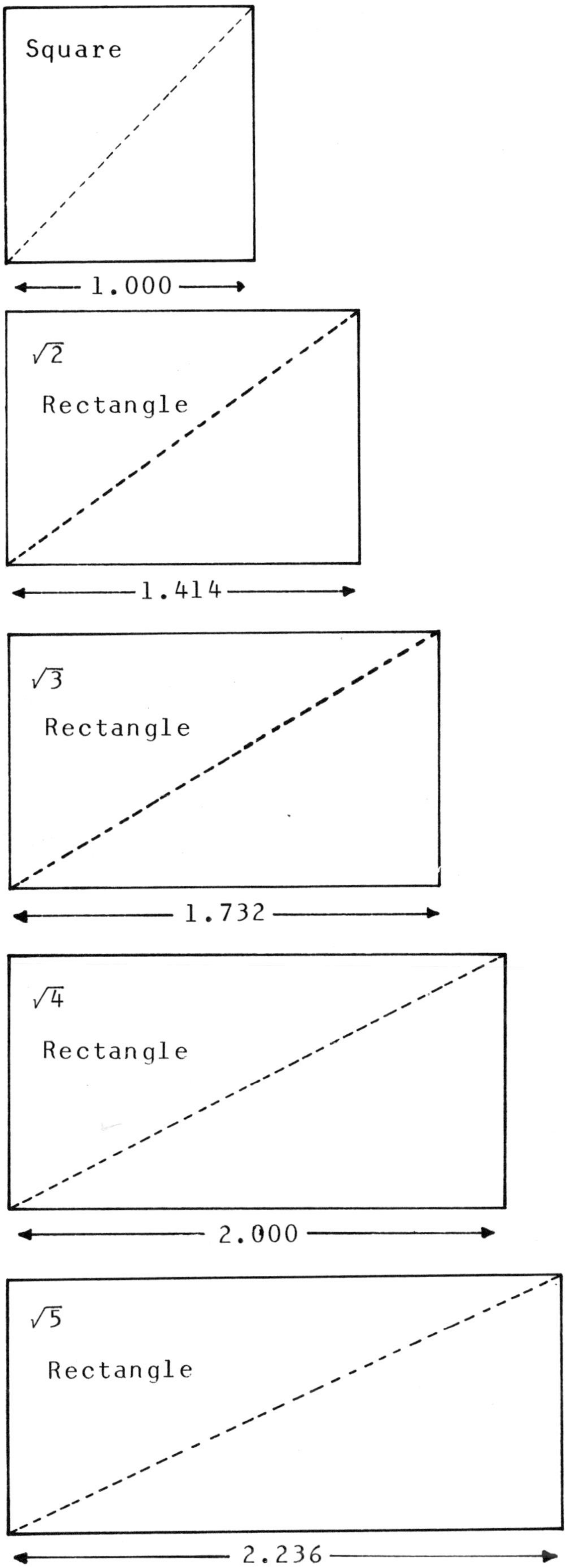

The previous construction could be carried on indefinitely, but it has been found that for practical purposes there is little use for Dynamic Rectangles beyond the √5 Rectangle. The illustration to the left shows the Dynamic Rectangles individually, together with the square that generated them.

The dotted line indicates a diagonal, important for the construction of the subsequent rectangle. Later on in the chapter we shall examine the importance of the diagonal as a design tool and design element.

Although the Dynamic Rectangles can always be constructed by the previous method, it may not be the most efficient one, since to construct a $\sqrt{5}$ Rectangle, you must first construct the $\sqrt{2}$, $\sqrt{3}$, and $\sqrt{4}$, respectively. Therefore, the next series of constructions provide alternate methods for obtaining the Dynamic Rectangles whereby each rectangle may be constructed separately. You are encouraged to note the relationship between the given geometric figure and the resulting Dynamic Rectangle.

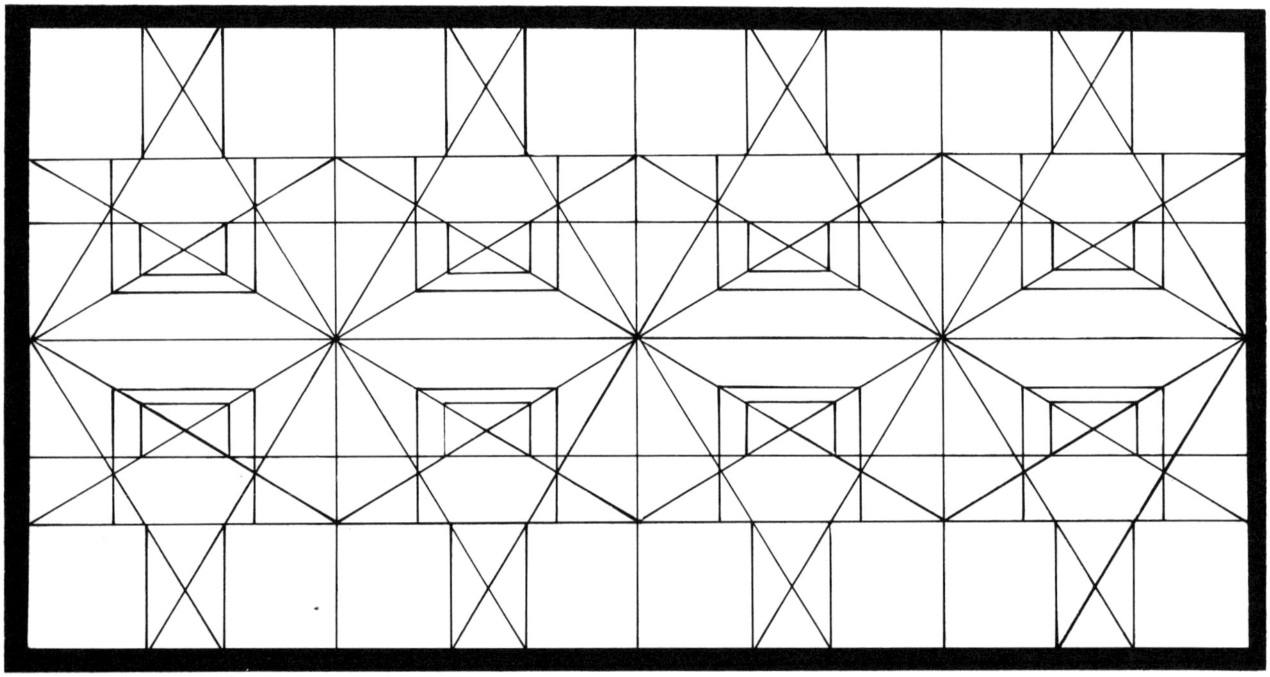

Angela Hsiung

Student Work
Chapter 4, project 5

The Unit Square

When we speak of "the unit square," we mean the square whose side will serve as the unit measure (or the number one) for all measurements taken subsequently within a given situation. This unit does not mean a particular length within any system of measure, but rather that once the unit is determined it becomes the basis for all measurements that follow. As the choice of the unit is arbitrary, determined by specific demands, it may vary from one situation to another. For example, the unit used to measure the length of a field will differ drastically from the one used to measure a section of galactic space or the diameter of a single cell.

Construction 13 in Chapter 2 provides the method for constructing a square. The choice of the unit is left to the individual.

In design the square is often perceived as a static form with few dynamic properties. However, the square itself may be subdivided harmoniously into the Dynamic Rectangles by the following method.

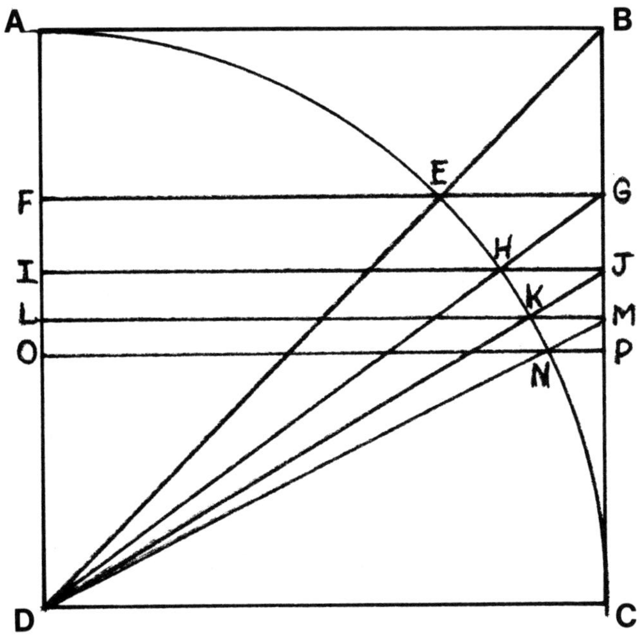

CONSTRUCT THE DYNAMIC RECTANGLES WITHIN A SQUARE

28

Given square ABCD, construct the Dynamic Rectangles within its boundaries.

1. Place the metal tip on D and the pencil tip on A and cut $\overset{\frown}{AC}$.
2. Draw $\overline{DB}$.
3. Label E the intersection of $\overset{\frown}{AC}$ and $\overline{DB}$.
4. From E construct a perpendicular to $\overline{BC}$ (refer to Construction 6) and extend it so that it intersects $\overline{AD}$.
5. Label the points of intersection F and G.

Now FDCG is a $\sqrt{2}$ Rectangle.

6. Draw $\overline{DG}$.
7. Label H the intersection of $\overline{DG}$ and $\overset{\frown}{AC}$.
8. From H construct a perpendicular to $\overline{BC}$ and extend it to intersect $\overline{AD}$.
9. Label the points of intersection I and J.

Now IDCJ is a $\sqrt{3}$ Rectangle.

10. Draw $\overline{DJ}$.
11. Label K the intersection of $\overset{\frown}{AC}$ and $\overline{DJ}$.
12. From K construct a perpendicular to $\overline{BC}$ and extend it to intersect $\overline{AD}$.
13. Label the points of intersection L and M.

Now LDCM is a $\sqrt{4}$ Rectangle.

14. Draw $\overline{DM}$.
15. Label N the point of intersection of $\overset{\frown}{AC}$ and $\overline{DM}$.
16. From N construct a perpendicular to $\overline{BC}$ and extend it to intersect $\overline{AD}$.
17. Label the points of intersection O and P.

Now ODCP is a $\sqrt{5}$ Rectangle.

The square may also be sub-divided harmoniously in ϕ proportions by use of the Golden Section (see Construction 9 in Chapter 2) and various diagonals. Several examples follow with information concerning their construction. You will find that points of intersection of diagonals and segments parallel to the sides play an important role in these designs. You may assume in each case that the segments that appear to be either parallel or perpendicular to each other in fact are.

Harmonic Sub-divisions of the Square Using ϕ

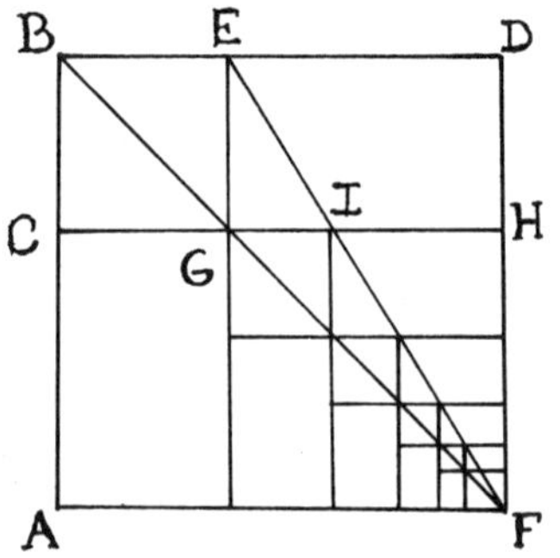

fig. 1

C and E are the Golden Sections of $\overline{AB}$ and $\overline{BD}$, respectively. Diagonals $\overline{BF}$ and $\overline{EF}$ are drawn. Square BCGE is drawn, and the process continues with the next square being determined by the intersection, point I, of $\overline{EF}$ and $\overline{CH}$.

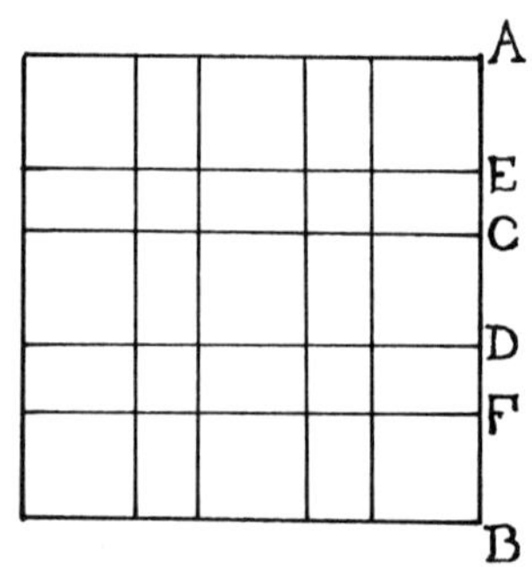

fig. 2

C and D are the Golden Sections of $\overline{AB}$. (Remember you can make the cut from either endpoint!) E and F are the Golden Sections of $\overline{AC}$ and $\overline{DB}$, respectively.

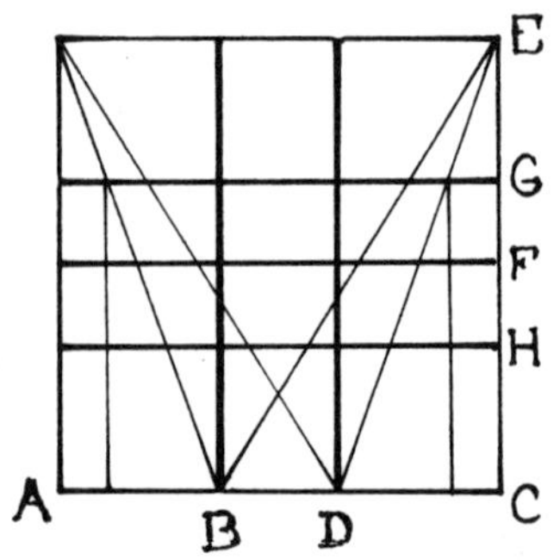

fig. 3

B and D are the Golden Sections of $\overline{AC}$. F is the midpoint of $\overline{EC}$. G and H are the Golden Sections of $\overline{EF}$ and $\overline{FC}$, respectively.

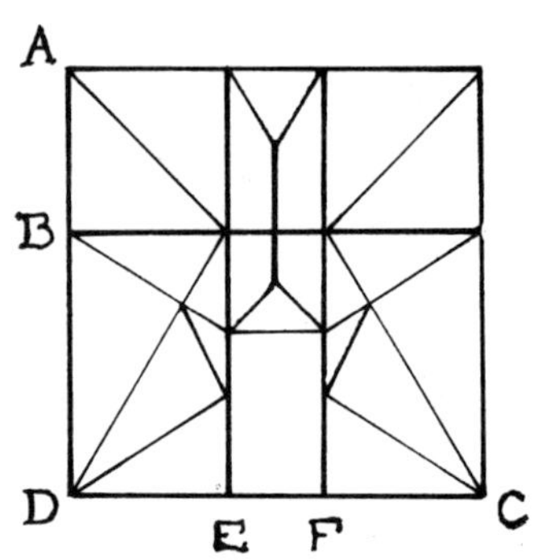

fig. 4

B is the Golden Section of $\overline{AD}$. E and F are the Golden Sections of $\overline{CD}$.

THE $\sqrt{2}$ RECTANGLE

The simplest way to construct the $\sqrt{2}$ Rectangle is to begin with the square as in Construction **27**; therefore, no other construction is needed or offered.

Some natural objects may be enclosed neatly by a $\sqrt{2}$ Rectangle, such as those illustrated below.

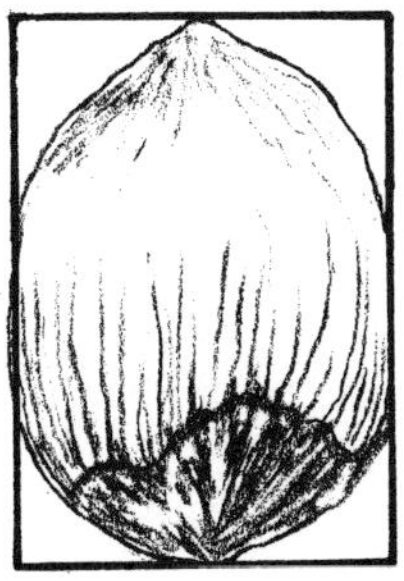

Filbert: front

Peanut
2 $\sqrt{2}$ Rectangles

Oak

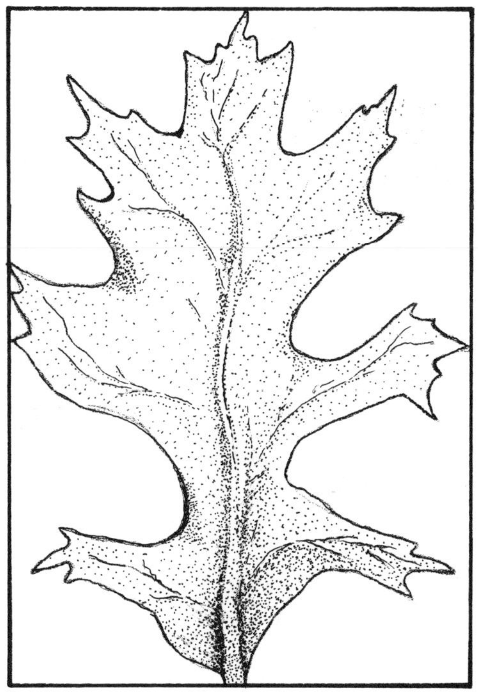

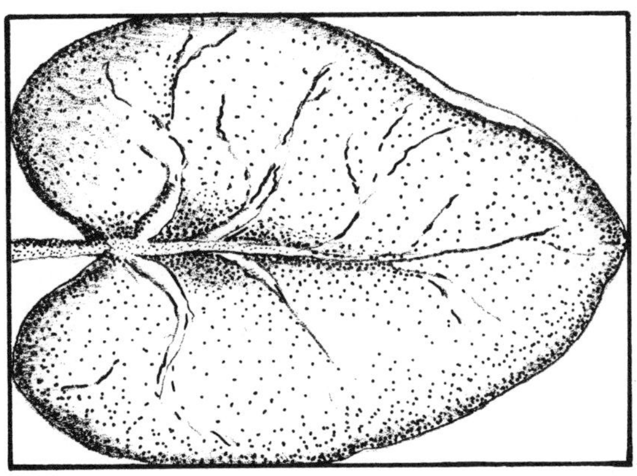

Yellow Water Lily

However, $\sqrt{2}$ Rectangles are found less frequently in nature than some of the other Dynamic Rectangles.

The $\sqrt{2}$ Rectangle has a unique property that makes it extremely practical for the graphic artist. It is the <u>only</u> rectangle in which one half of the figure is similar to the whole. To the book designer this rectangle is ideal. For if the sides of a book are in the ratio $\sqrt{2}$:1 this means that when the book is open it has the same shape as when it is closed.

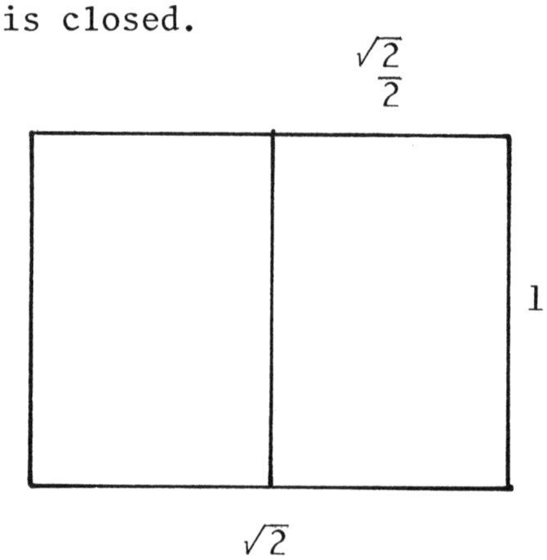

$\frac{\sqrt{2}}{1}$ is the ratio of the length to the width of the open book.

$\frac{1}{\sqrt{2}/2}$ is the ratio of the length to the width of a page (or the closed book). The following equations show these to be the same ratio.

$$\frac{1}{\sqrt{2}/2} = \frac{2}{\sqrt{2}} \cdot \frac{\sqrt{2}}{\sqrt{2}} = \frac{2\sqrt{2}}{2} = \sqrt{2}$$

The $\sqrt{2}$ Rectangle is found in an unexpected place - a wooden blanket box of a Northwest Indian tribe, the Tsimshians. It reiterates the tendency toward breaking a line segment into Ø proportions, as C is very close to the Golden Section of $\overline{AB}$. This tendency may be more pronounced in such a culture so closely attuned to the natural world.

According to Paul Grillo, in his book <u>Form, Function and Design</u>, the $\sqrt{2}$ Rectangle has become the standard for most of the windows of the villages and towns of Europe, as well as the window pane of standard proportion.

It is also used in the design of office furniture and file cabinets as it makes it possible to divide any drawer in half without wasted space.

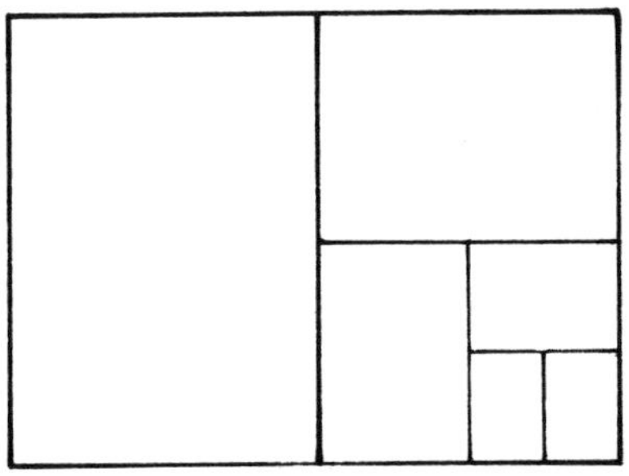

The diagram illustrates a subdivision of the $\sqrt{2}$ Rectangle which is gotten by dividing the rectangle in half through the midpoint of the longer side and repeating the process.

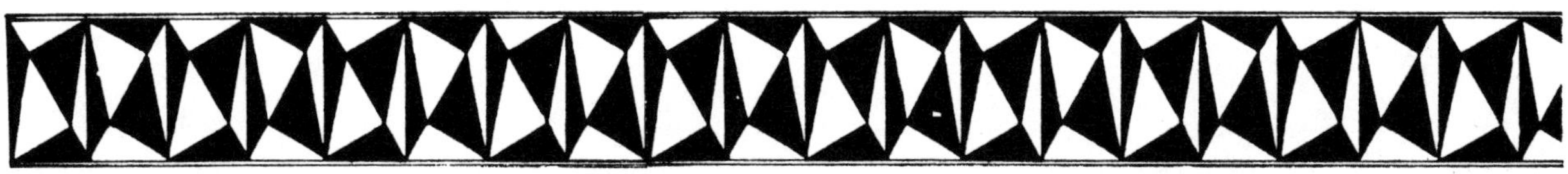

THE $\sqrt{3}$ RECTANGLE

The $\sqrt{3}$ Rectangle may be generated in several ways, each being useful for the artist-designer.

If two equilateral triangles are placed side by side and enclosed in a rectangle, that will be a $\sqrt{3}$ Rectangle.

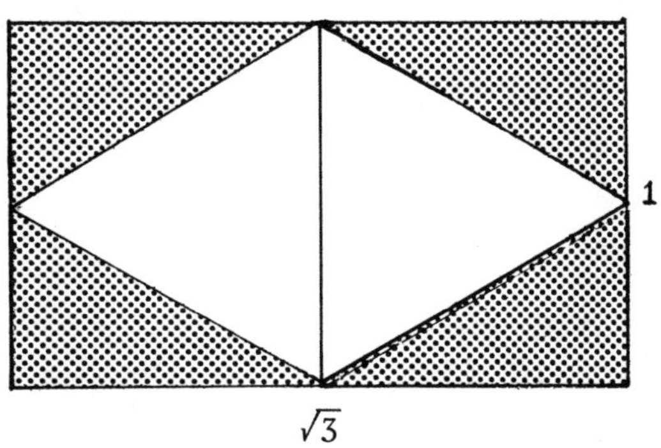

This property is the basis for the following construction, even though the two triangles are not actually drawn.

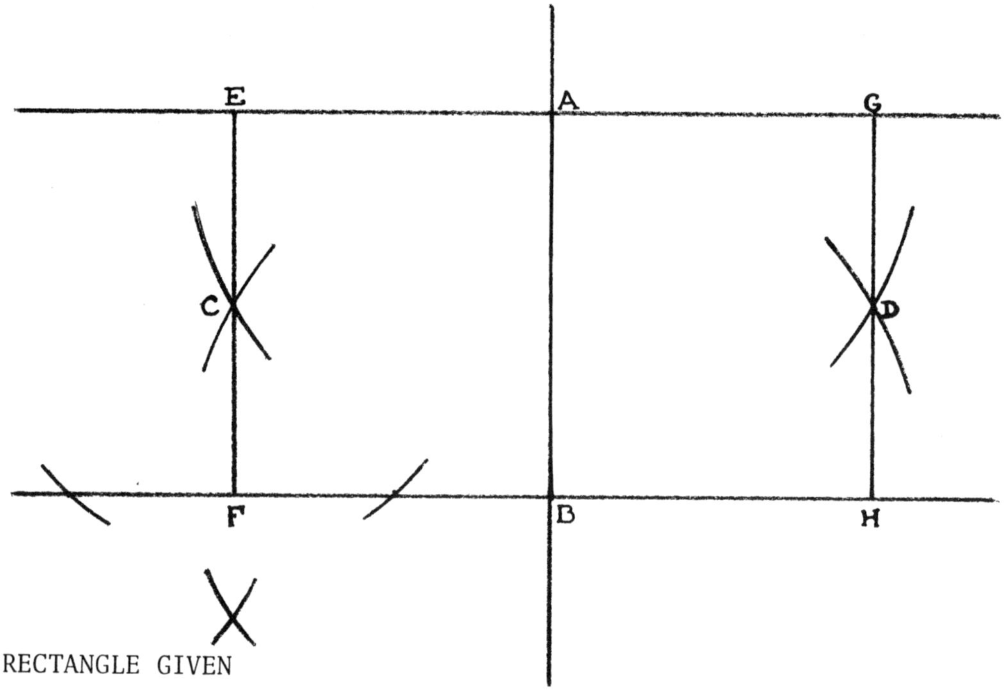

CONSTRUCT A $\sqrt{3}$ RECTANGLE GIVEN THE WIDTH

29

Given $\overline{AB}$, construct a $\sqrt{3}$ Rectangle such that the width measures AB.

1. Extend $\overleftrightarrow{AB}$.
2. Construct perpendiculars to $\overleftrightarrow{AB}$ through A and B, respectively.
3. Place the metal tip on A and the pencil tip on B and cut arcs on either side of $\overleftrightarrow{AB}$.
4. Without changing the setting, place the metal tip on B and cut arcs that intersect the ones drawn in step 3.
5. Label the points of intersection C and D, respectively.
6. From C construct a perpendicular to the line through A and extend it so that it intersects the line through B.
7. Label the points of intersection E and F, respectively.
8. Repeat step 6 from D.
9. Label the points of intersection G and H, respectively.

Now EFHG is a $\sqrt{3}$ Rectangle.

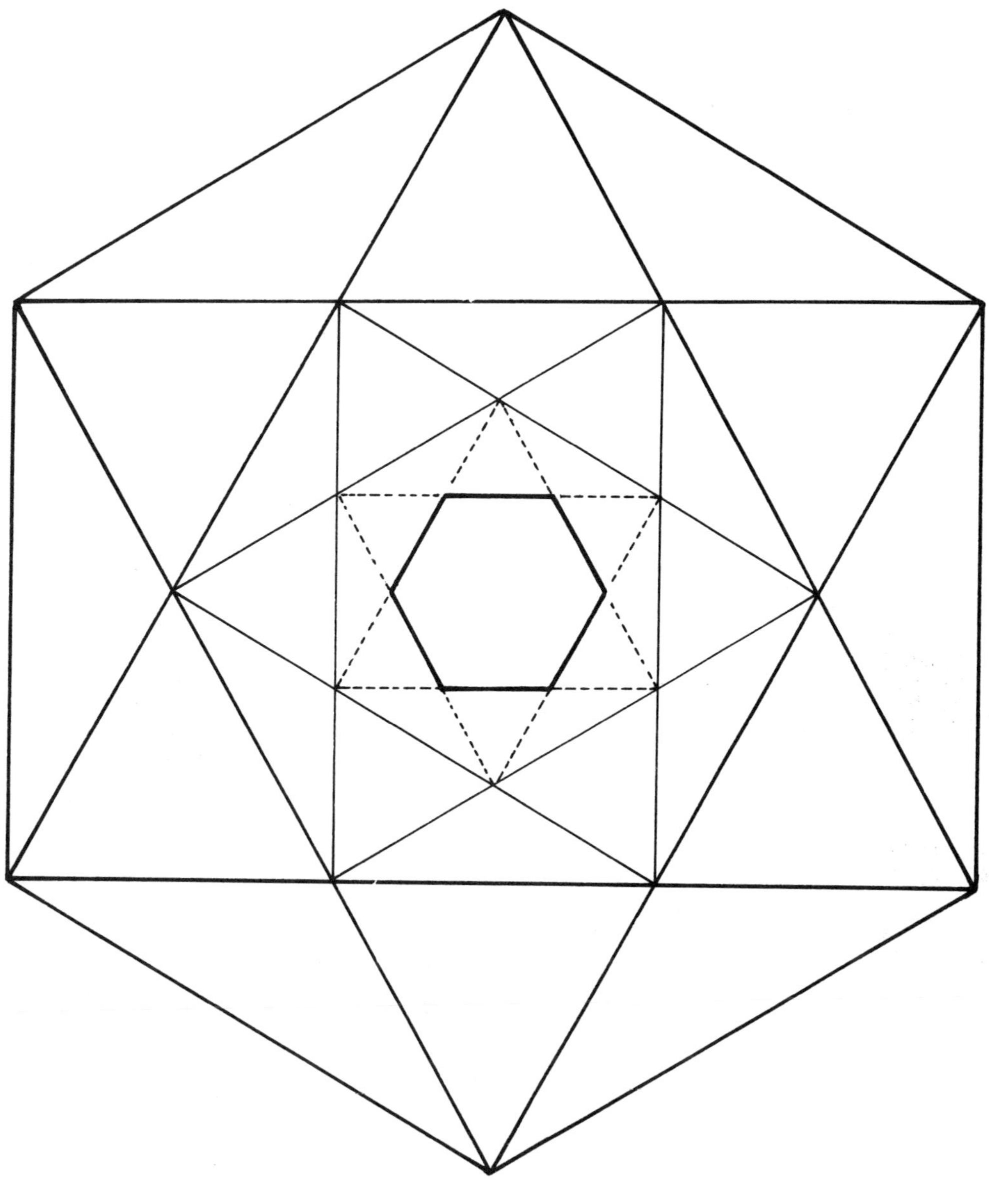

If opposite sides of a regular hexagon are joined by parallel line segments from the vertices, three intersecting $\sqrt{3}$ Rectangles are formed, producing a smaller regular hexagon in the center. The above diagram illustrates this process repeated three times. It may be carried out indefinitely (or at least until eyes, fingers and tools no longer function!)

If, in the following construction, points A, F, D, B, C and E were all connected sequentially by line segments, the resulting figure would be a regular hexagon. Therefore, $\overline{FD}$ and $\overline{EC}$ would be opposite sides as described above, and FDCE must be a $\sqrt{3}$ Rectangle.

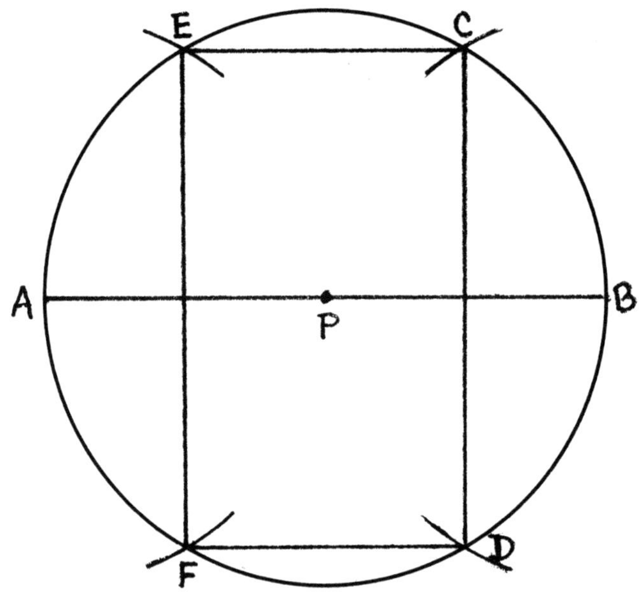

CONSTRUCT A $\sqrt{3}$ RECTANGLE GIVEN A CIRCLE

Given ⊙P, construct a $\sqrt{3}$ Rectangle.

1. Draw a diameter and label its endpoints A and B respectively.
2. Without changing the compass setting, place the metal tip on B and the pencil tip on P and cut arcs on either side of $\overleftrightarrow{AB}$.
3. Label the points of intersection C and D, respectively.
4. Without changing the setting, place the metal tip on A and cut arcs on either side of $\overleftrightarrow{AB}$.
5. Label the points of intersection E and F, respectively.
6. Draw $\overline{EF}$, $\overline{FD}$, $\overline{DC}$ and $\overline{EC}$.

Now FDCE is a $\sqrt{3}$ Rectangle.

Note: The position of the diameter is arbitrary. For the rectangle to lie horizontally the diameter should be drawn vertically. A diameter drawn diagonally will yield a rectangle that lies diagonally.

Like the $\sqrt{2}$ Rectangle, the $\sqrt{3}$ Rectangle is less often connected with organic forms but abounds in crystalline forms (as in a snowflake) since it is inextricably bound to the hexagon.

It is interesting to note that the $\sqrt{3}$ Rectangle is the one that Plato found to be the most beautifully proportioned. However, since it is rarely found in the artwork of that period, it seems his peers did not agree with his choice.

In plane design (design in two dimensions), it is considered discordant to mix the themes of $\sqrt{2}$ and $\sqrt{3}$ together within the same design; however, each may be used with the square to create harmonious composition.

In three dimensions, though, the artist may use different root rectangle themes on different faces, or planes, provided that the design theme is consistent within each plane. The illustration below shows one example in which Nature uses this principle of design.

$\sqrt{2}$ Rectangle

$\sqrt{3}$ Rectangle

Filbert: front & side

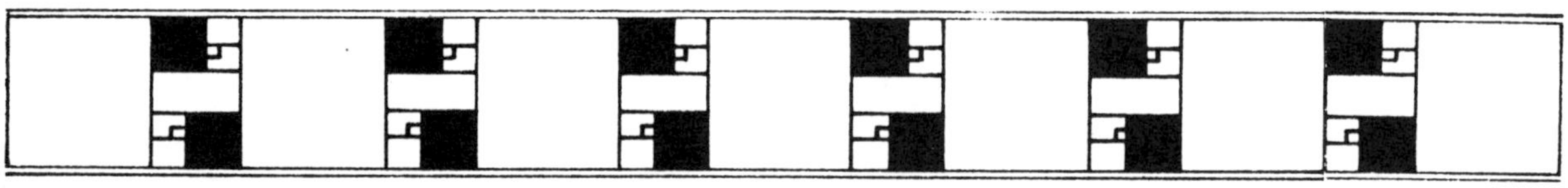

THE PHI-FAMILY RECTANGLES

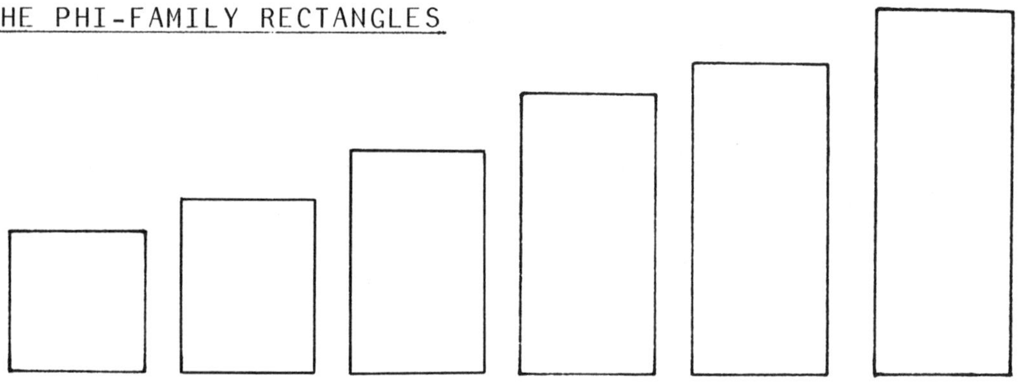

The Golden Rectangle with which you are so familiar by now, gives us a family of rectangles that works together with great compatibility for use in plane design. If the width of each of these rectangles measures 1, then the lengths measure 1 (square); $\sqrt{\phi}$ (1.272); ϕ (1.618); $\sqrt{4}$ (2); $\sqrt{5}$ (2.236); and $\phi + 1$ (2.618), respectively.

We will be discussing each of these rectangles individually later on in this chapter, but we have provided the diagram of the family tree to point out the relationship among the members of the family.

Two members of this family are root rectangles and therefore can be generated from the square. They are the $\sqrt{4}$ Rectangle and the $\sqrt{5}$ Rectangle. It should be noted that the $\sqrt{4}$ Rectangle, being a double square would also be compatible with either the $\sqrt{2}$ Rectangle or the $\sqrt{3}$ Rectangle in visual composition.

PHI-FAMILY TREE

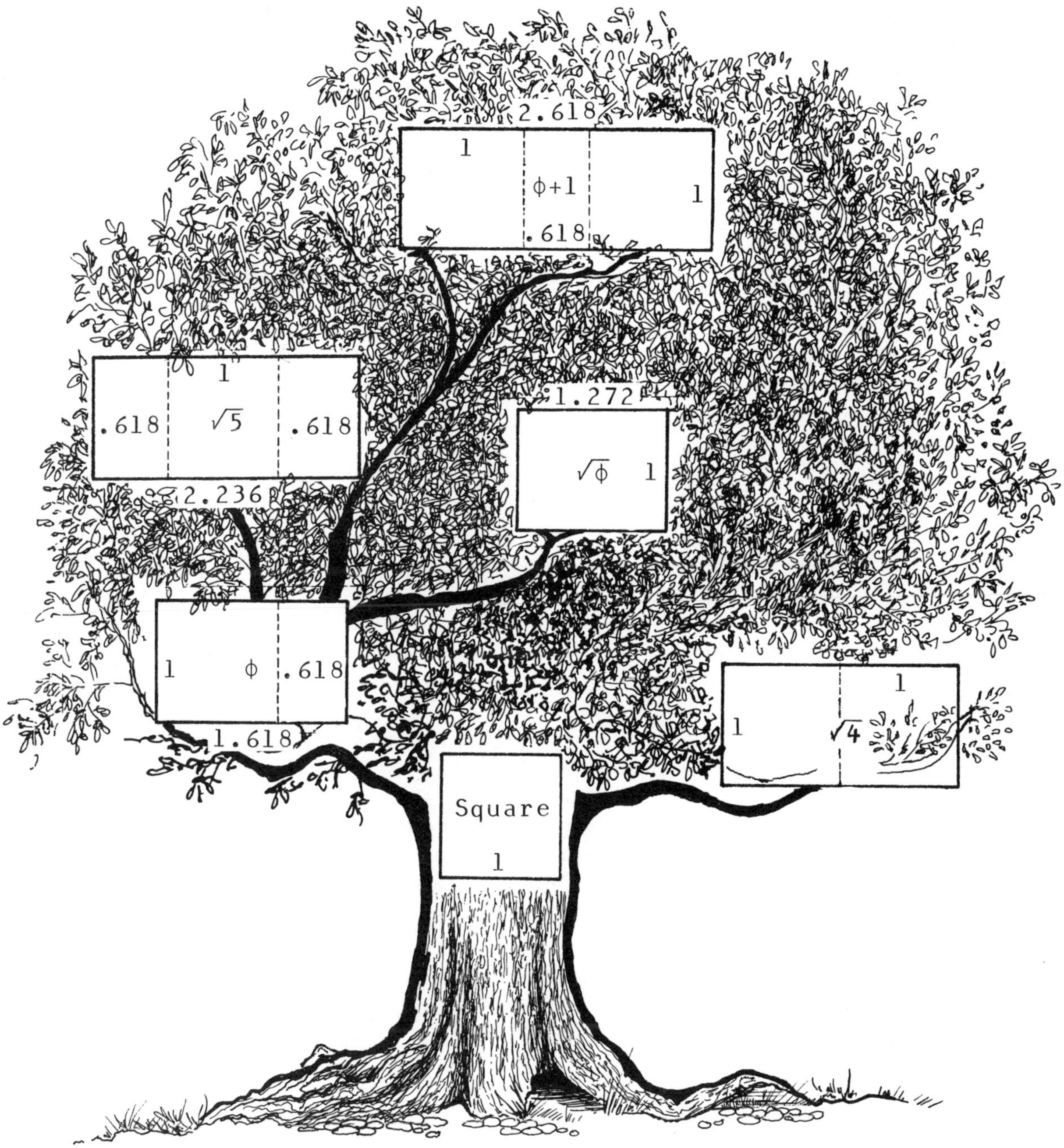

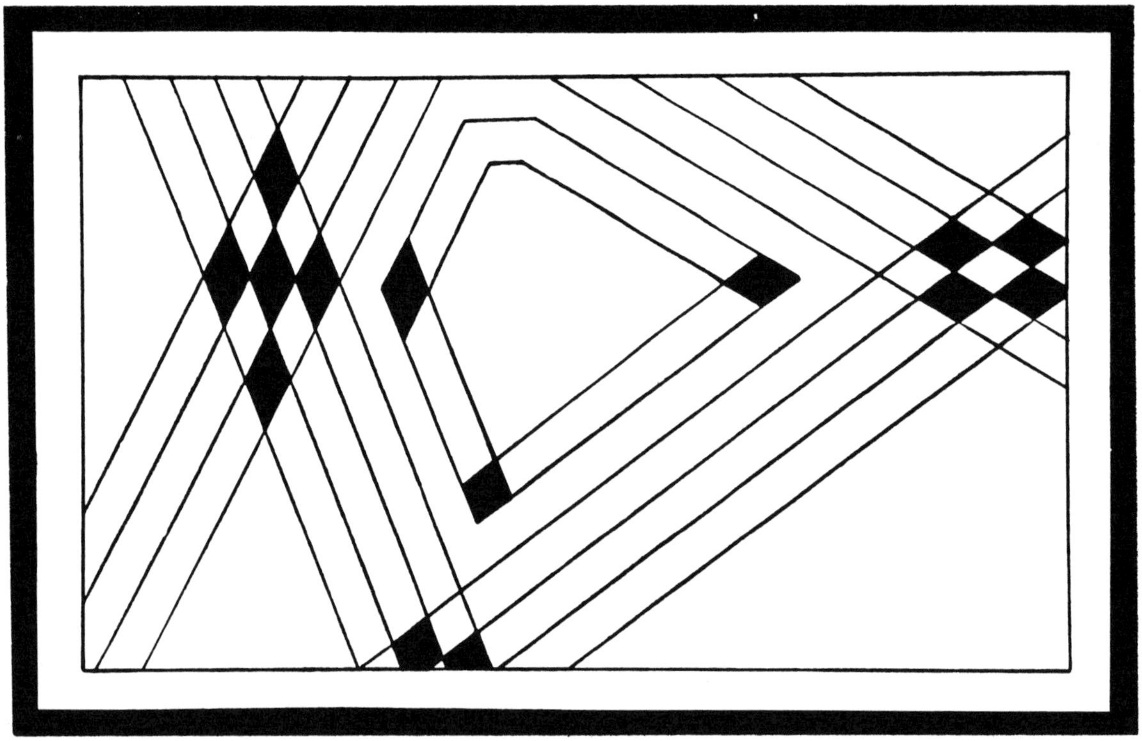

Photo by Richard Newman

Stephen Reynolds

Student Work
Chapter 4, variation on project 4

ϕ-FAMILY: $\sqrt{\phi}$ RECTANGLE

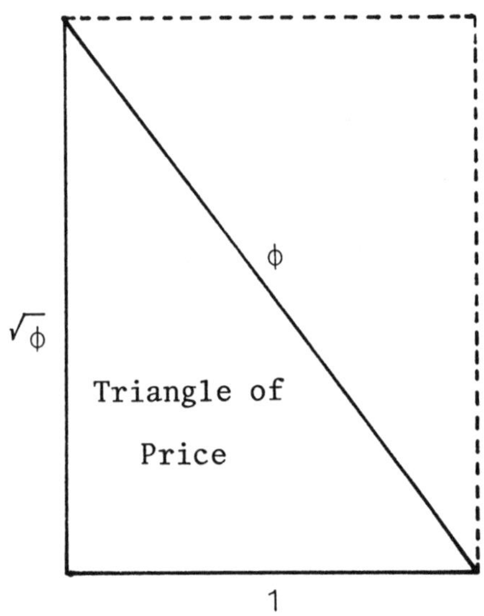

$\sqrt{\phi}$ Rectangle

The second rectangle in progression, according to the length of the sides, is the $\sqrt{\phi}$ Rectangle; the first being the unit square discussed previously. The $\sqrt{\phi}$ Rectangle is sometimes called the Rectangle of Price since it is derived from the Triangle of Price. This diagram illustrates the relationship.

You will recall that a Triangle of Price can be constructed from a Golden Rectangle (see Construction 18, Chapter 3). Therefore, it is logical to construct a $\sqrt{\phi}$ Rectangle from a Golden Rectangle in the following way.

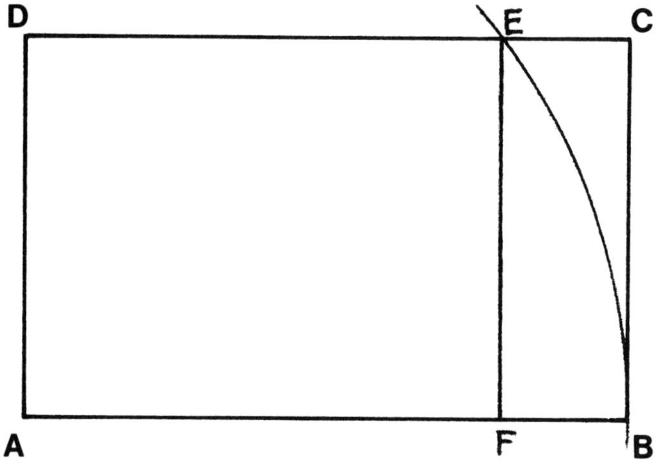

CONSTRUCT A $\sqrt{\phi}$ RECTANGLE GIVEN A GOLDEN RECTANGLE

Given Golden Rectangle ABCD, construct a $\sqrt{\phi}$ Rectangle.

1. Place the metal tip on A and the pencil tip on B and cut an arc that intersects $\overline{DC}$.
2. Label the point of intersection E.
3. Construct a perpendicular from E to $\overline{AB}$.
4. Label the point of intersection F.
5. Draw $\overline{EF}$.

Now AFED is a $\sqrt{\phi}$ Rectangle.

Most of us handle examples of $\sqrt{\phi}$ Rectangles each day without being aware of the golden connections. The standard size of notebook and typing paper (8½" x 11") very closely approximates the dimensions of the $\sqrt{\phi}$ Rectangle. Other larger standard sizes of paper also have dimensions in the same ratio as illustrated.

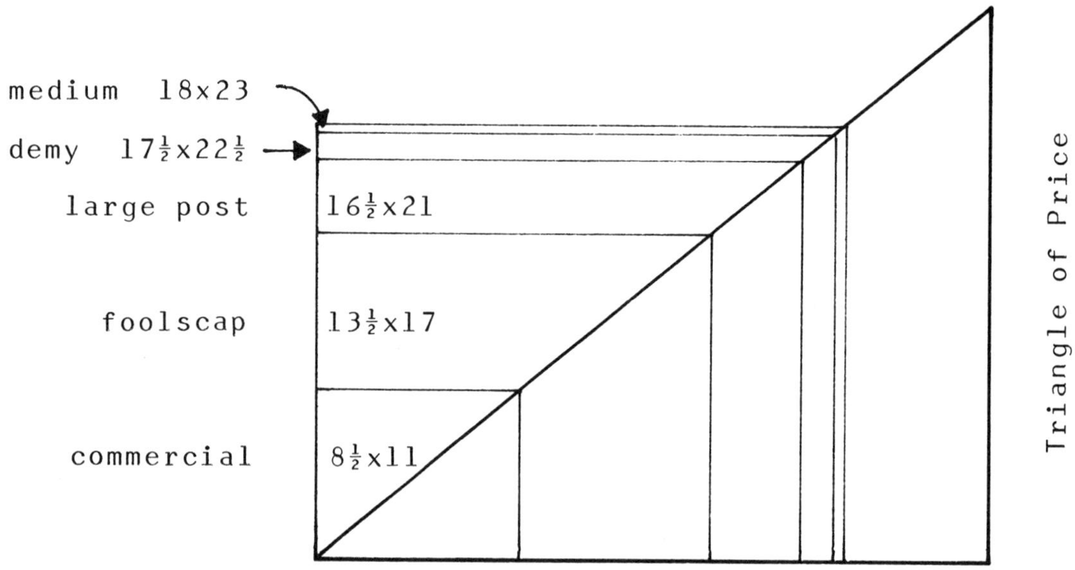

ϕ-FAMILY: $\sqrt{4}$ RECTANGLE

Although the Golden Rectangle is the next in progression, it has been discussed already in Chapter 2. Therefore, we will go on to the $\sqrt{4}$ Rectangle.

Since $\sqrt{4} = 2$, any rectangle that is twice as long as it is wide is a $\sqrt{4}$ Rectangle. The fact that the length is double the width makes this rectangle a double square, and constructing it is an easy affair.

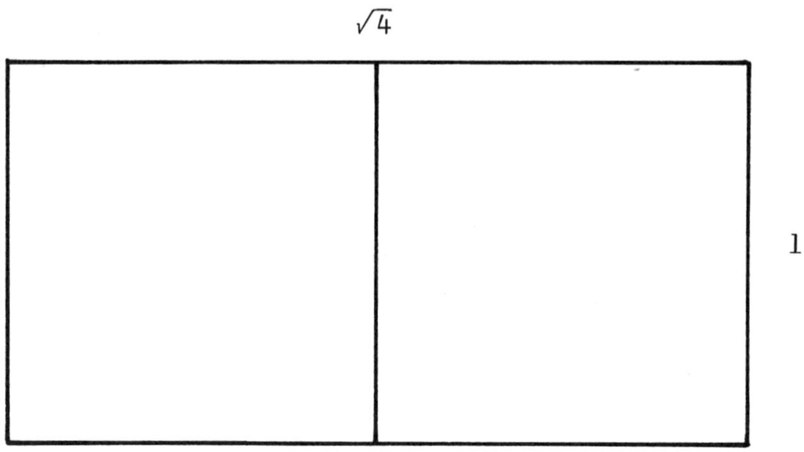

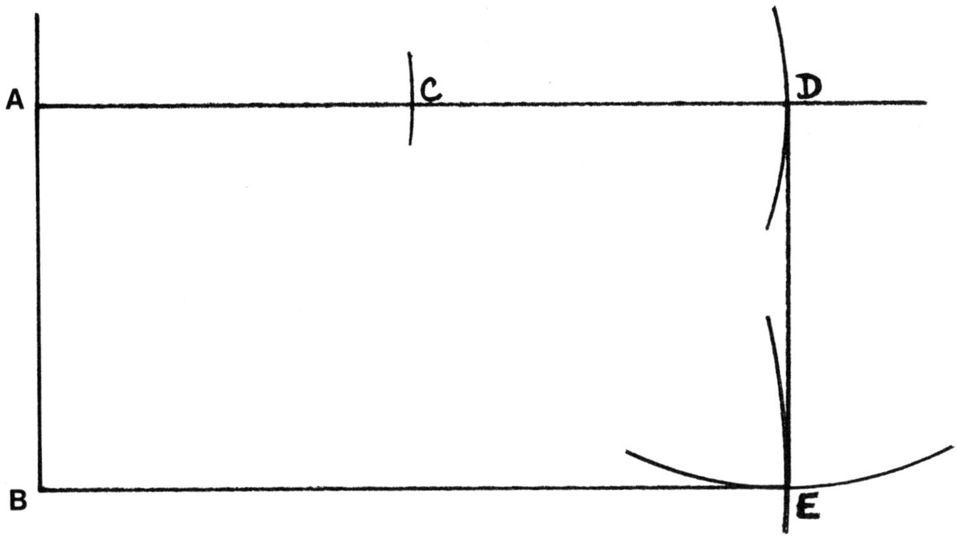

CONSTRUCT A $\sqrt{4}$ RECTANGLE GIVEN THE WIDTH

Given $\overline{AB}$, construct a $\sqrt{4}$ Rectangle such that its width equals AB.

1. Extend $\overrightarrow{BA}$.
2. Construct a perpendicular to $\overleftrightarrow{AB}$ through A.
3. Place the metal tip on A and the pencil tip on B and cut an arc that intersects the line drawn in step 2.
4. Label the point of intersection C.
5. Without changing the setting, place the metal tip on C and cut a second arc on $\overrightarrow{AC}$.
6. Label the point of intersection D.
7. Without changing the setting, place the metal tip on D and cut an arc in the interior of $\angle DAB$.
8. Open the compass so that it measures AD and place the metal tip on B. Cut an arc that intersects the one drawn in step 7.
9. Label the point of intersection E.
10. Draw $\overline{BE}$ and $\overline{DE}$.

Now ABED is a $\sqrt{4}$ Rectangle.

The $\sqrt{4}$ Rectangle is the Japanese modular unit, the tatami, from which all proportion in Japanese architecture derives. A room is built around a rectangle which is formed by a particular number of tatami mats.

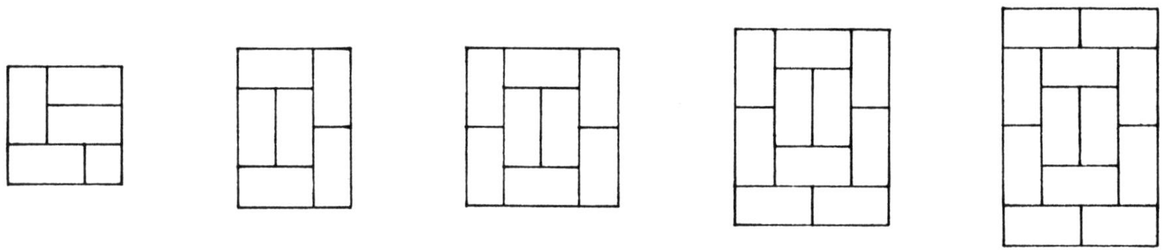

The shape and size of the house itself depends on the distribution of the room modules.

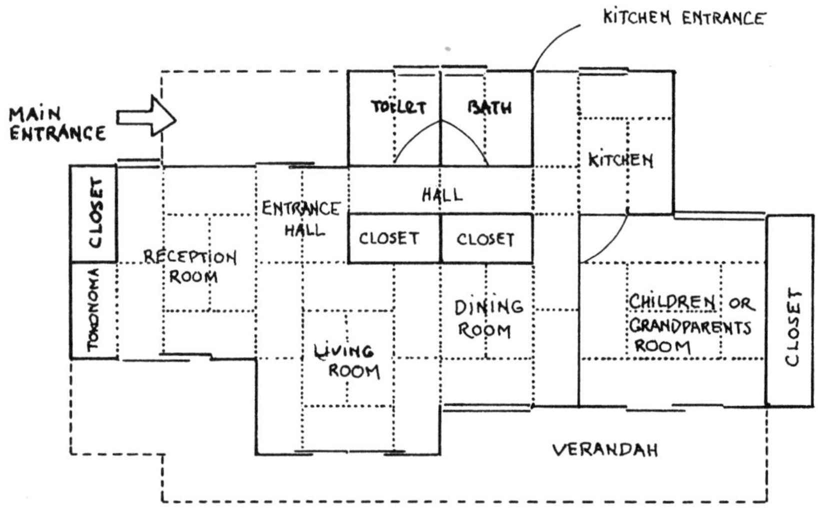

ϕ-FAMILY: $\sqrt{5}$ RECTANGLE

Next to the Golden Rectangle the $\sqrt{5}$ Rectangle is perhaps the most widely used both by Man and Nature as a structure for design.

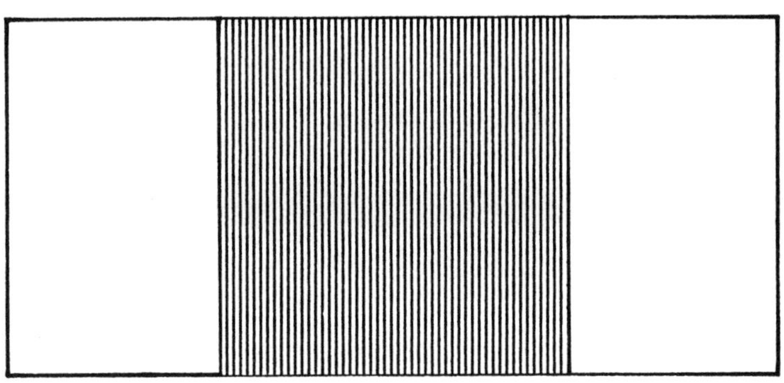

$\sqrt{5}$

The $\sqrt{5}$ Rectangle is derived by overlapping two Golden Rectangles so that their common area is a square.

This property allows for easy construction of a $\sqrt{5}$ Rectangle simply by extending the construction for a Golden Rectangle.

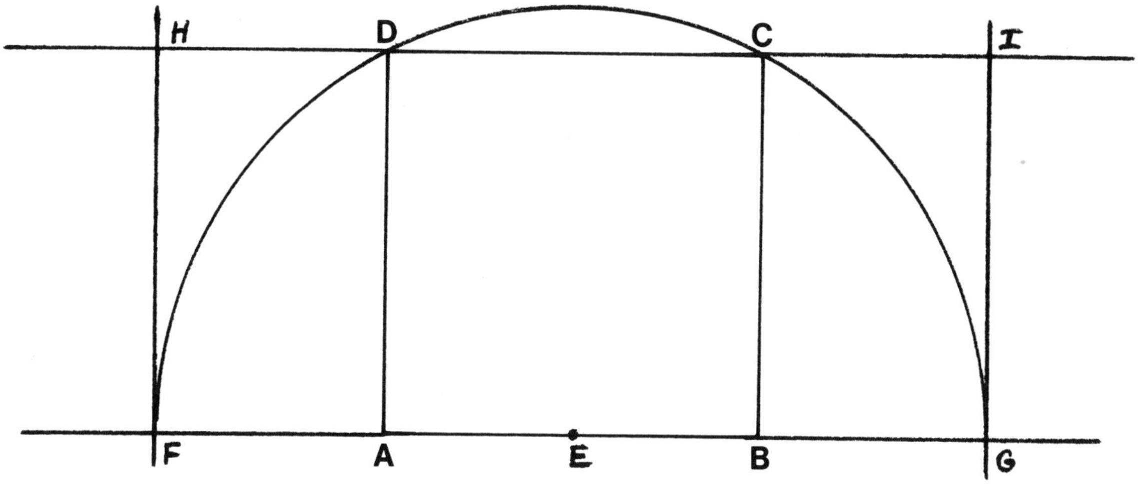

CONSTRUCT A $\sqrt{5}$ RECTANGLE GIVEN A SQUARE

Given square ABCD, construct a $\sqrt{5}$ Rectangle.

1. Extend $\overleftrightarrow{AB}$ and $\overleftrightarrow{CD}$.
2. Bisect $\overline{AB}$ and label the midpoint E.
3. Place the metal tip on E and the pencil tip on C and cut an arc that intersects $\overleftrightarrow{AB}$ twice.
4. Label the points of intersection F and G, respectively.
5. Construct perpendiculars to $\overleftrightarrow{AB}$ at F and G and extend them so they intersect $\overleftrightarrow{DC}$.
6. Label the points of intersection H and I, respectively.
7. Draw $\overline{HF}$ and $\overline{IG}$.

Now FGIH is a $\sqrt{5}$ Rectangle.

Note: AGID and FBCH are both Golden Rectangles.

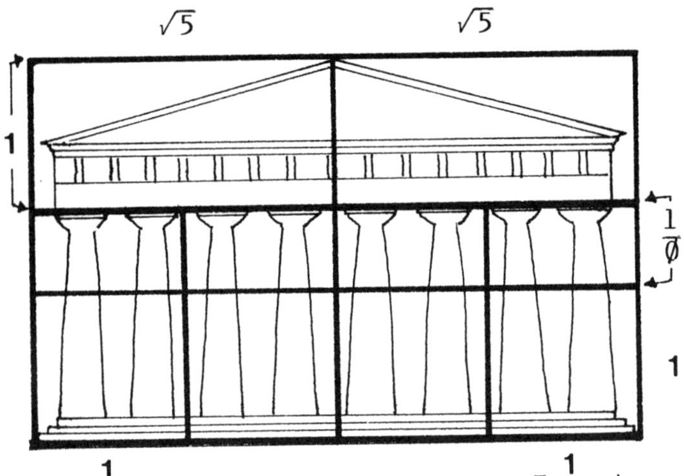

Facade

Looking once again at the Parthenon, we can see the use of the √5 Rectangle together with the square in the floor plan and within the enclosing Golden Rectangle of the facade.

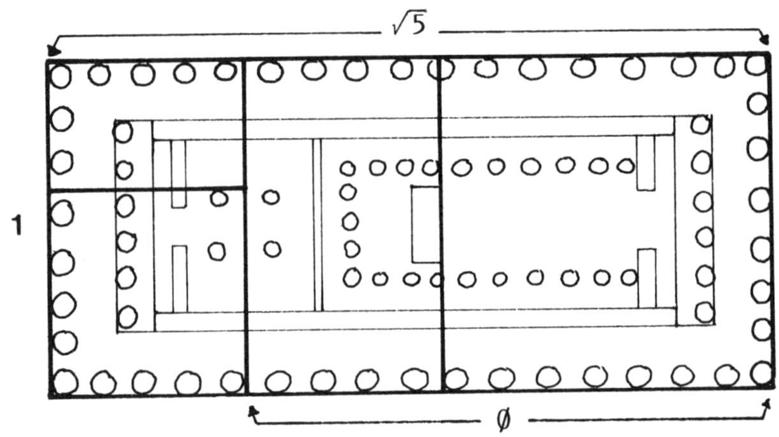

Floor Plan

In a very different art form, by a very different cultural group, we find the √5 Rectangle once again.

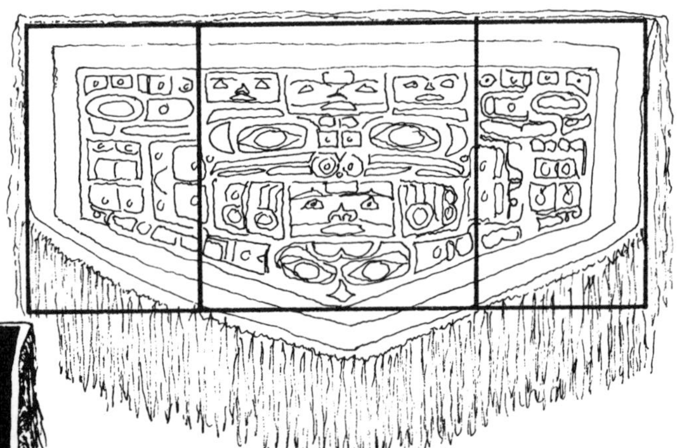

Whether a conscious choice or an intuitive one, the √5 Rectangle is used extensively in the blanket designs of the Northwest Coast Indian tribes with aesthetically sastisfying results.

Drawings after Tlingit Chilkat blanket

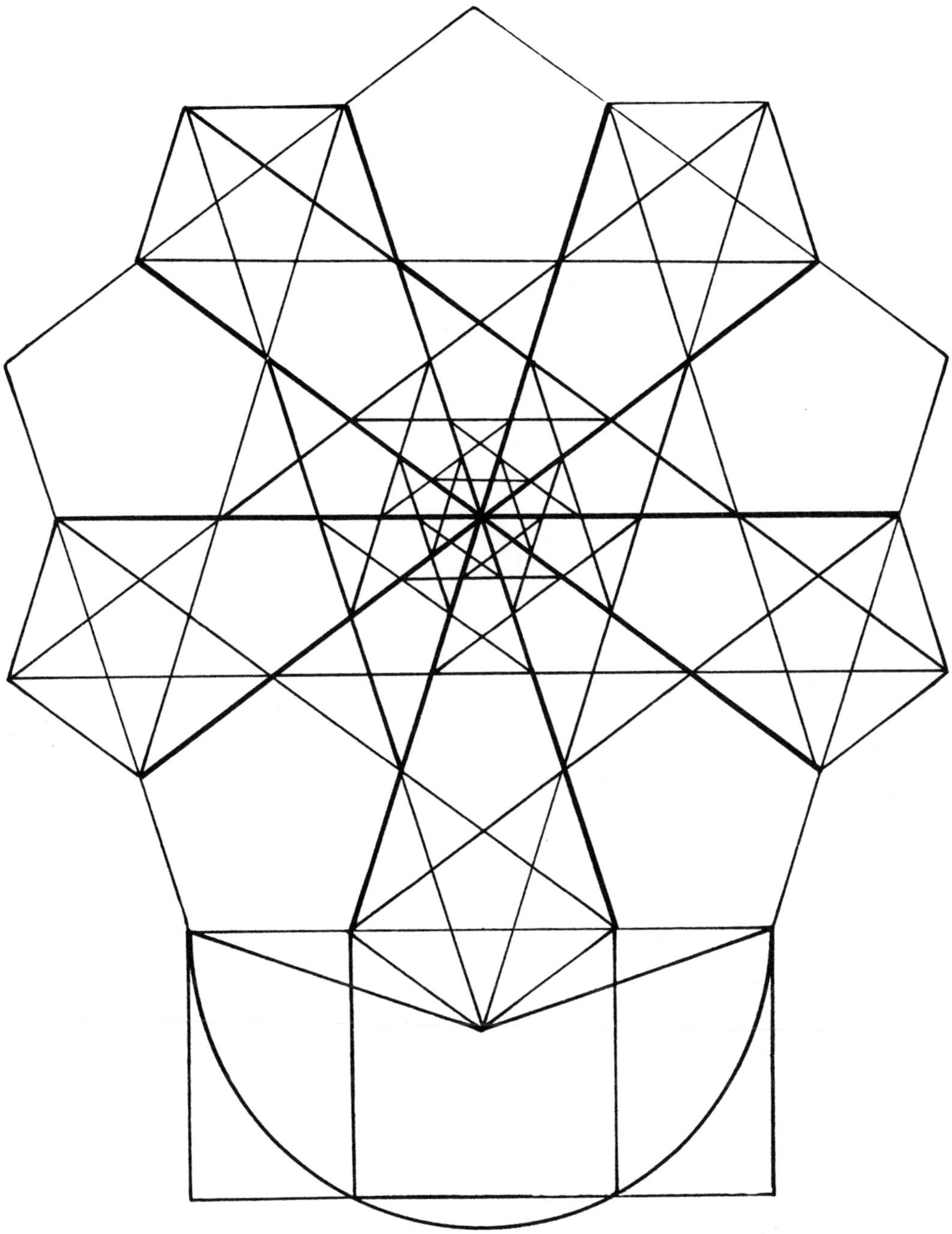

This diagram illustrates the connection made by Ø between several golden figures:

a. two intersecting pentagons giving the illusion of a decagon,

b. the Golden Triangle,

c. the Lute of Pythagoras with its corresponding pentagrams, and

d. the $\sqrt{5}$ Rectangle composed of two overlapping Golden Rectangles

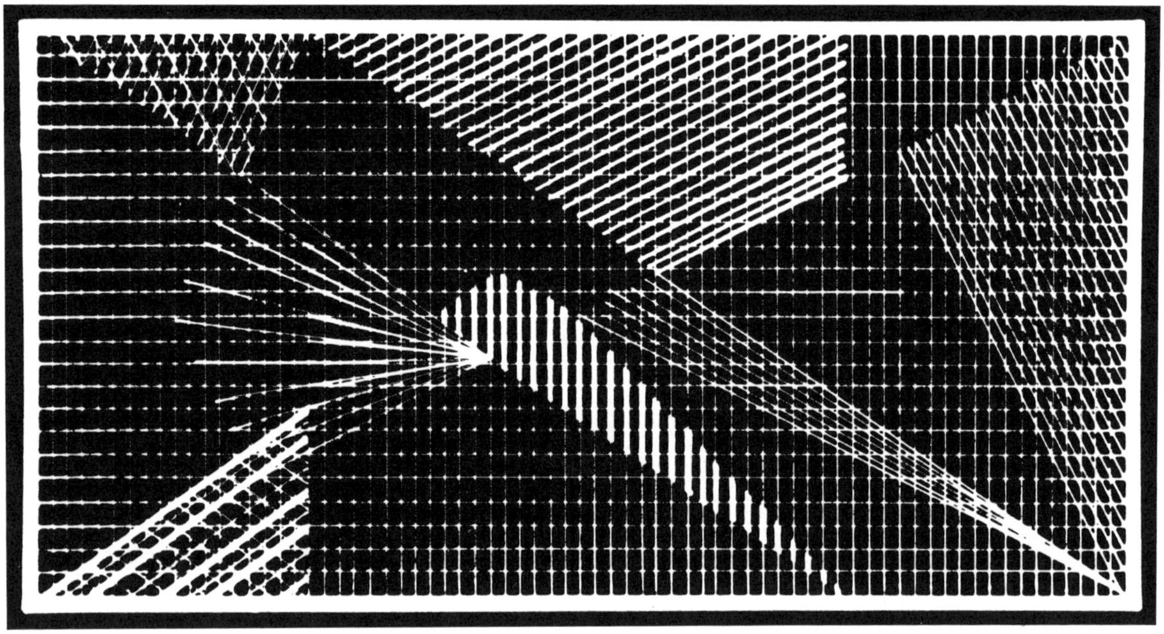

Photo by Richard Newman

Student Work
Chapter 4, variation on project 4

ϕ-FAMILY: ϕ-PLUS-ONE RECTANGLE

The other rectangle in the Ø-Family is the one the authors have christened the $\phi + 1$ Rectangle. It is obtained by adding a square to the width of an existing Golden Rectangle. Like the $\sqrt{5}$ Rectangle, the ϕ-Plus-One Rectangle contains two overlapping Golden Rectangles, this time sharing the small Golden Rectangle rather than the square.

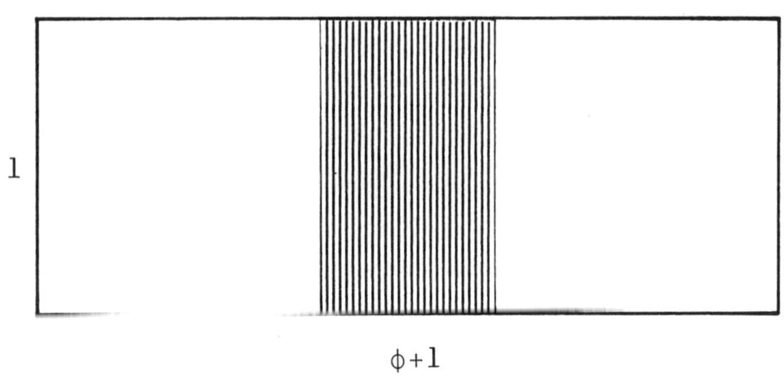

Again the construction of the Ø + 1 Rectangle is an easy extension of the construction of the Golden Rectangle.

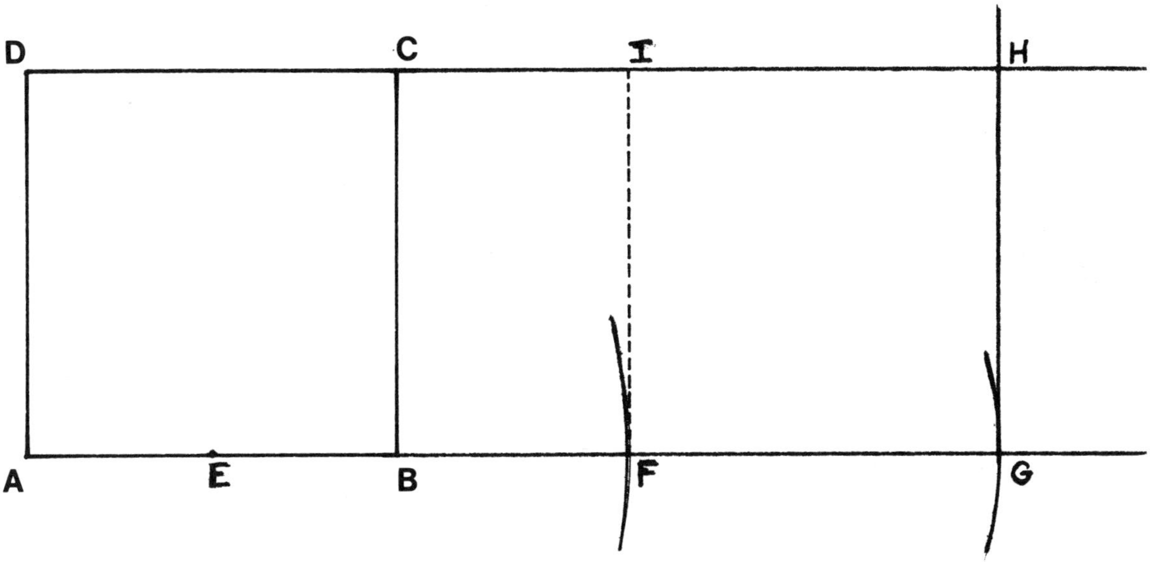

CONSTRUCT A ϕ-PLUS-ONE RECTANGLE GIVEN A SQUARE

34

Given square ABCD, construct a ϕ-Plus-One Rectangle.

1. Extend $\overrightarrow{AB}$ and $\overrightarrow{DC}$.
2. Bisect $\overline{AB}$ and label its midpoint E.
3. Place the metal tip on E and the pencil tip on C and cut an arc that intersects $\overrightarrow{AB}$.
4. Label the point of intersection F.
5. Open the compass to measure AB. Then with the metal tip on F cut an arc on $\overrightarrow{AF}$ on the side of F opposite A.
6. Label the point of intersection G.
7. Open the compass to measure AG. Then with the metal tip on D cut an arc on $\overrightarrow{DC}$.
8. Label the point of intersection H.
9. Draw $\overleftrightarrow{GH}$.

Now AGHD is a ϕ-Plus-One Rectangle

Note: If a perpendicular were drawn from F to I on $\overleftrightarrow{DH}$, AFID and BGHC are both Golden Rectangles.

ϕ-FAMILY RECTANGLES FOUND IN NATURE

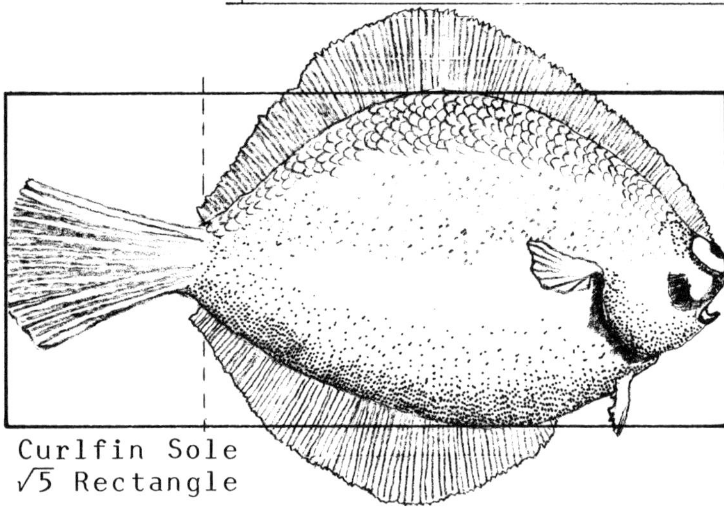

Curlfin Sole
$\sqrt{5}$ Rectangle

Almond
$\sqrt{4}$ Rectangle

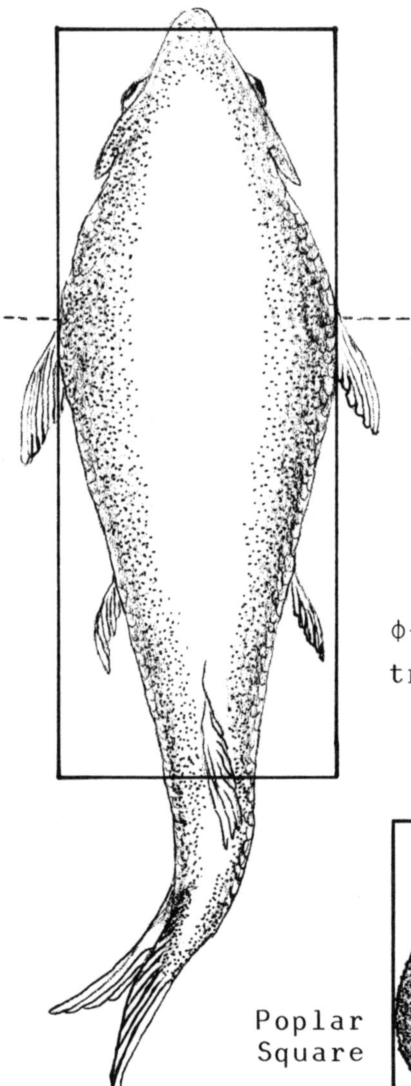

Herring
$\phi+1$ Rectangle
transverse view

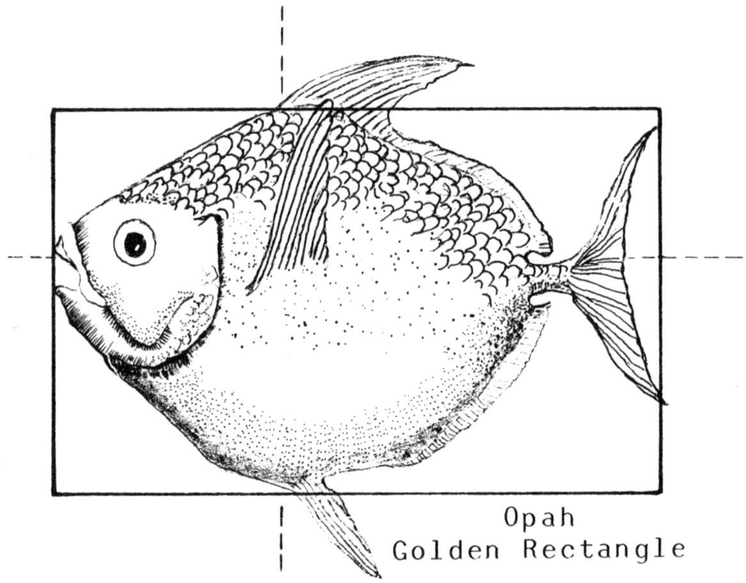

Opah
Golden Rectangle

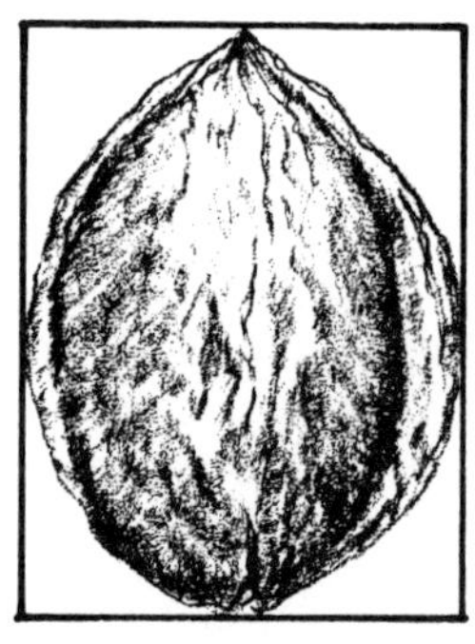

Walnut
$\sqrt{\phi}$ Rectangle

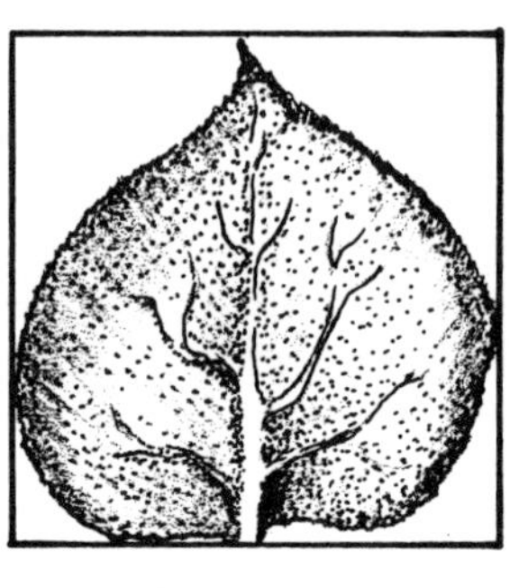

Poplar
Square

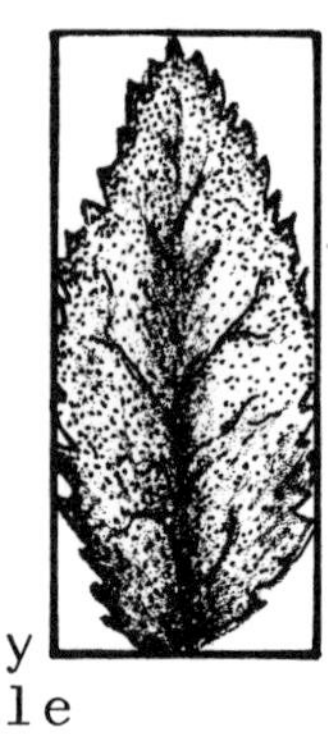

Raspberry
$\sqrt{5}$ Rectangle

THE RECIPROCAL

All the rectangles we have just discussed provide a form for designing harmoniously. However, the designer needs more than the outer proportions. He needs to be able to subdivide the rectangle into smaller units and still preserve the overall harmony.

One method of subdividing a rectangle is by use of the reciprocal. The reciprocal is a similar rectangle cut from the parent rectangle whose length is the width of the parent rectangle.

In the above diagram the shaded rectangle is the reciprocal of ABCD. Since the two rectangles are similar, they have the same shape (the position of course is altered) and their sides are in proportion. That is $\frac{AB}{BC} = \frac{BC}{EC}$.

The construction for the reciprocal of a rectangle follows.

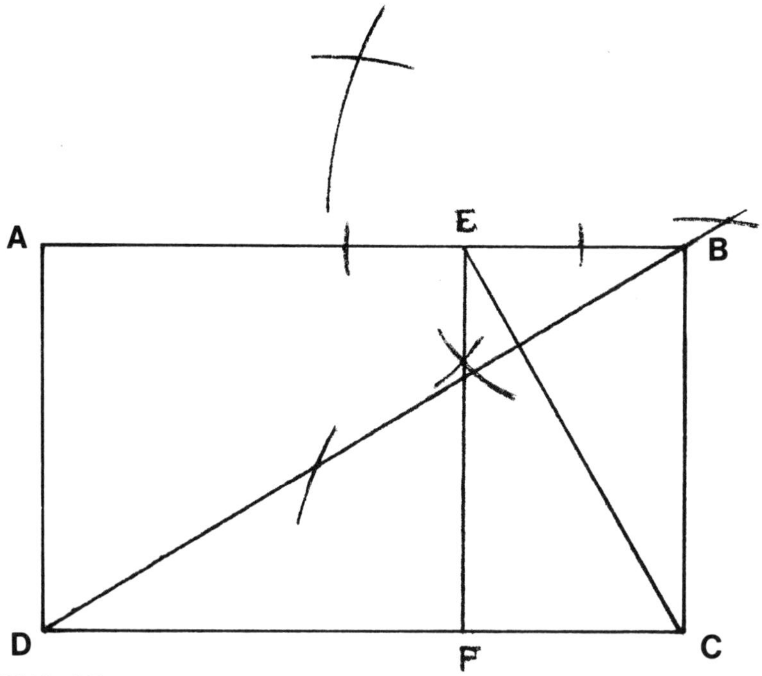

CONSTRUCT THE RECIPROCAL OF
A RECTANGLE

35

Given rectangle ABCD, construct the reciprocal.

1. Draw $\overline{DB}$.
2. From C construct a perpendicular to $\overleftrightarrow{DB}$ (refer to Construction 6) and extend it to intersect $\overline{AB}$.
3. Label the point of intersection E.
4. From E construct a perpendicular to $\overleftrightarrow{DC}$.
5. Label the point of intersection F.
6. Draw $\overline{EF}$.

Now rectangle EFCB is the reciprocal of rectangle ABCD.

$ABCD \sim FEBC$

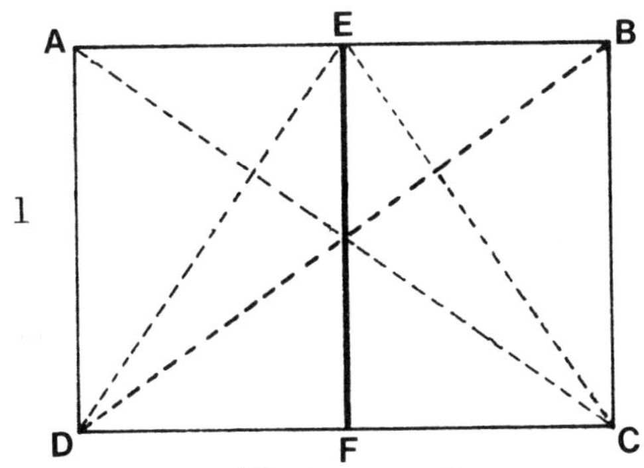

$\sqrt{2}$ Rectangle

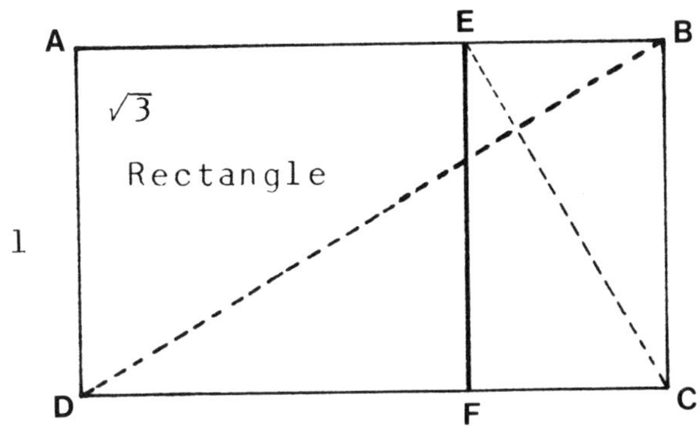

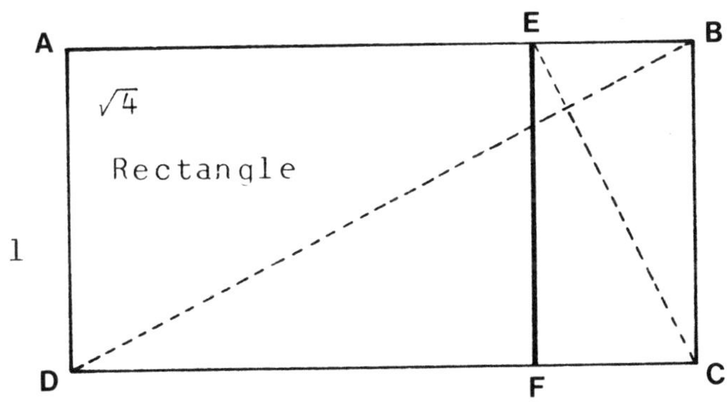

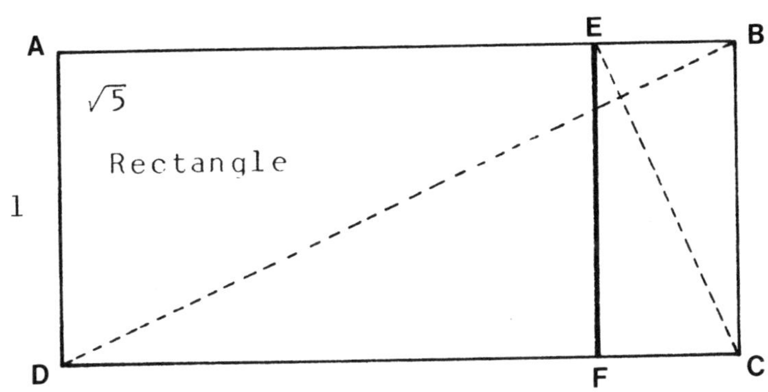

The previous construction can be used to find the reciprocals of the Dynamic Rectangles. When used together with the parent rectangle, an underlying structure for harmonious design is achieved. You will notice that as the length of the parent rectangle increases, the relative size of the reciprocal decreases. The $\sqrt{2}$ Rectangle is the *only* rectangle in which one half of the figure is similar to the whole. In each case $ABCD \sim FEBC$.

You will notice the importance of the diagonals of both the parent and reciprocal rectangles in Construction 35. In order of importance the diagonals of the parent and reciprocal are paramount. The two diagonals cut each other at right angles to form line segments and rectangles in continued proportion.

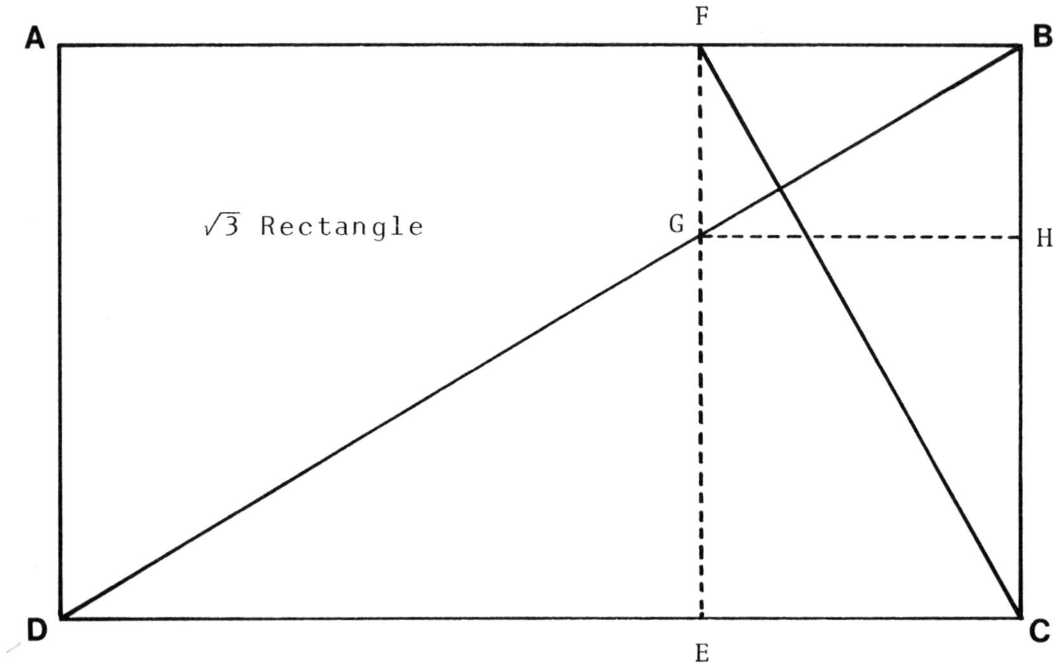

By Construction 35 we know that FECB is the reciprocal of ABCD. Then by drawing in $\overline{GH}$, we have that GHBF is the reciprocal of ECBF. So ABCD $\sim$ EFBC $\sim$ FBHG.

This property is true of all rectangles, but when coupled with a special property of Dynamic Rectangles it is what relates the Dynamic Rectangles to the phenomenon of growth in Nature.

In <u>all</u> the Dynamic Rectangles the length of the parent rectangle is a multiple of the width of the reciprocal.

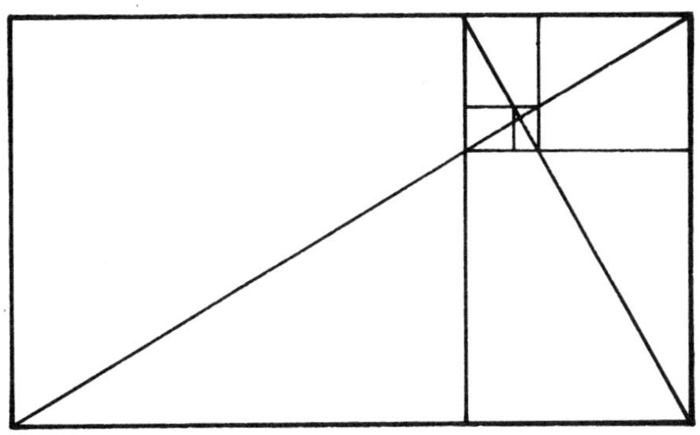

This means that as the aforementioned procedure is continued, the rectangle is subdivided into similar shapes with no leftover space. Natural growth patterns also have to do with the property of similarity and conservation of space using logarithmic spirals, both angular and curved, which can be constructed with these rectangles and their intersecting diagonals. We shall discuss this further in Chapter 6.

The fact that in the Dynamic Rectangles the reciprocal can be fitted neatly into the parent allows for many interesting and varied design possibilities. Some examples follow.

For demonstration purposes we have used only the $\sqrt{3}$ Rectangle and its reciprocal, together with various diagonals. However, the same principles of division would hold for the other Dynamic Rectangles as well.

THE GNOMON

Within any given rectangle, there exists the reciprocal together with its complement, the gnomon (pronounced no-men). This term was first used by the Babylonians to name an upright stick in the ground used as a crude sundial. For the Greeks, the term came to represent a carpenter's square, a shape similar to that in fig. 1, below. The concept was further expanded and is now taken to mean any shape that can be added to a given shape to produce a similar shape. In the figures below, the gnomon is the shaded portion. When this is added to the unshaded portion, a figure similar to the shaded portion is obtained.

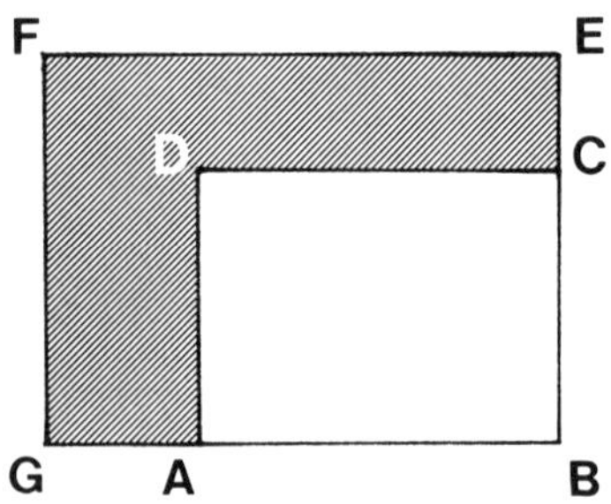

$GBEF \sim ABCD$

fig. 1

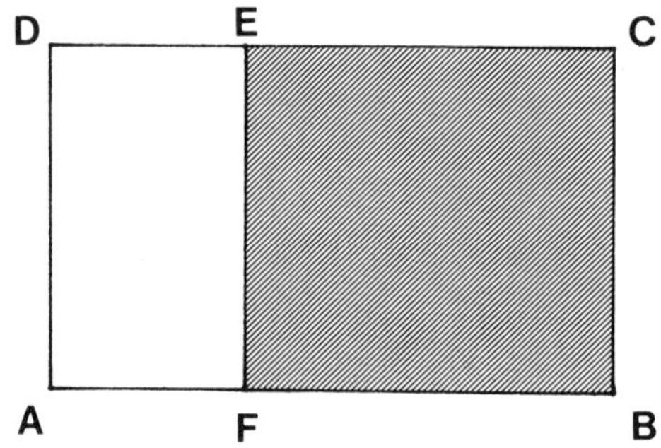

$ABCD \sim DAFE$

fig. 2

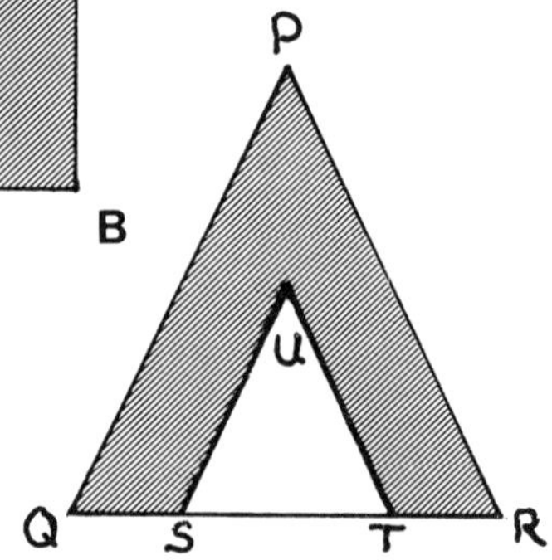

$\triangle QPR \sim \triangle SUT$

fig. 3

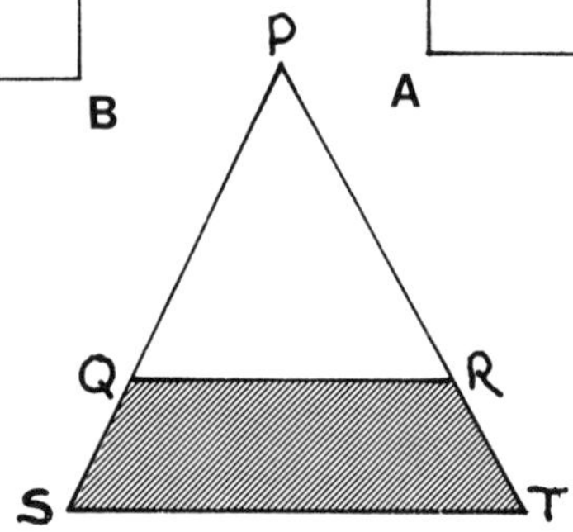

$\triangle PST \sim \triangle PQR$

fig. 4

As you can see from these examples, a single figure may have gnomons of many different sizes and shapes, and they need not be related to the shape of the original figure. We see a rectangle whose gnomon is in the shape of a carpenter's square and a rectangle whose gnomon is another rectangle. One triangle has a V-shaped gnomon, the other a gnomon in the shape of a trapezoid. In fact, a triangle may also have a gnomon that is triangular in shape and an interesting case occurs in right triangles. If an altitude is dropped from the right angle, two triangles are formed, each of which is the gnomon of the other (fig. 5).

In the above diagram we see that $\triangle ADC \sim \triangle CDB \sim \triangle ACB$. Since all three triangles are similar, either of the two smaller triangles when added to the other produces a similar triangle.

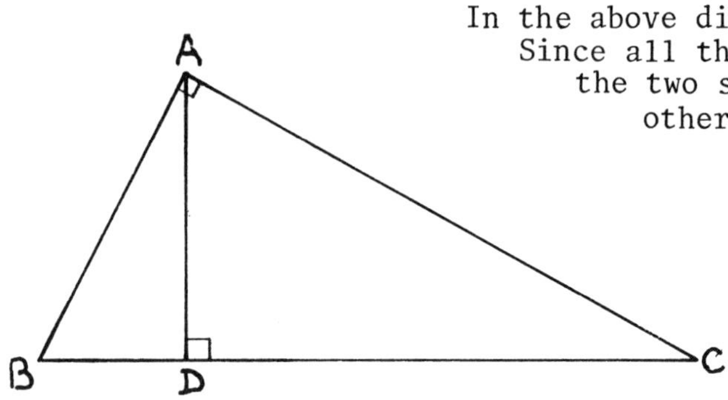

fig. 5

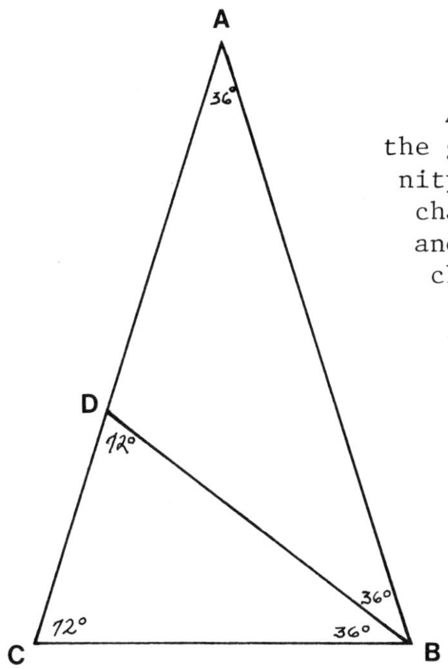

Any triangle can be subdivided so that one part is the gnomen of the other. (You will have the opportunity to do this in the problem set at the end of the chapter.) However, the Golden Triangle is a unique and elegant case. You will recall from the previous chapter that the Golden Triangle is the isosceles triangle whose angles measure 36°, 72°, 72° respectively.

Since the vertex angle measures exactly one-half a base angle, we can bisect one of the base angles to form another Golden Triangle within the original.

So $\triangle ACB \sim \triangle BCD$ and $\triangle ADB$ is the gnomon of $\triangle BDC$.

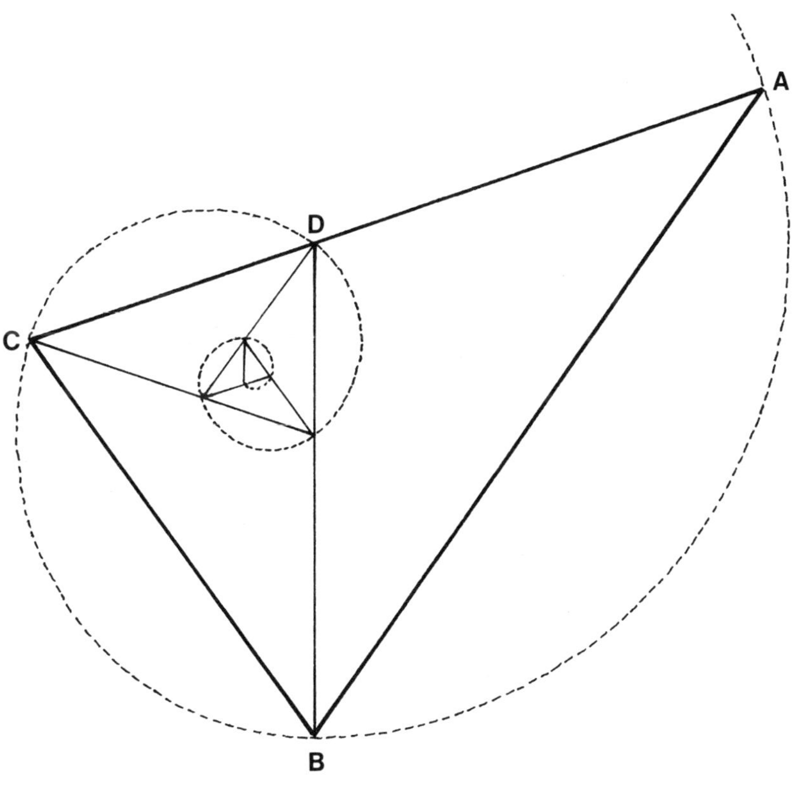

This process can be repeated to produce a series of smaller and smaller Golden Triangles and the resulting figure forms another structure for a logarithmic spiral which we shall examine in Chapter 6.

The process of adding a gnomen to a form can be carried on indefinitely giving us Nature's law of continued similarity. This is especially exemplified in growth patterns of structures such as shells and horns which, again, we shall study further in Chapter 6.

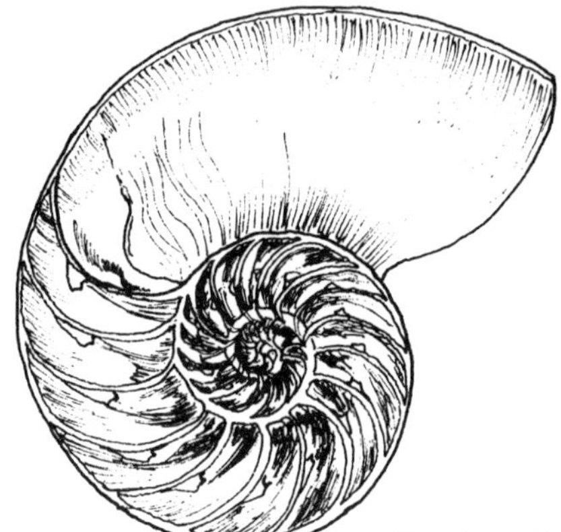

Chambered Nautilus

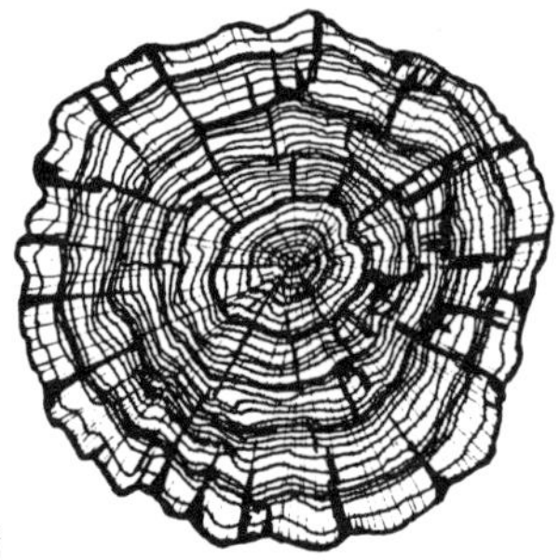

Stump of Baldwin Apple Tree

Northern White Cedar

We recognize many trees by their overall shapes, each species having characteristic growth patterns. It seems then that, allowing for the deviations caused by natural forces, the new growth each season is the gnomen for the tree's shape the previous year.

The concept of gnomon can also be applied to numbers. Although we tend to think of numbers as abstractions to name quantities, it is possible to examine them within a geometric framework. In fact, the Pythagoreans were intrigued by classifying numbers according to the shape that would most economically hold the given number of units. The Greeks could more easily identify with the idea of the number of units in a quantity since they indicated a number by use of pebbles or marks in the sand.

For example, the following illustrates the first four triangular numbers.

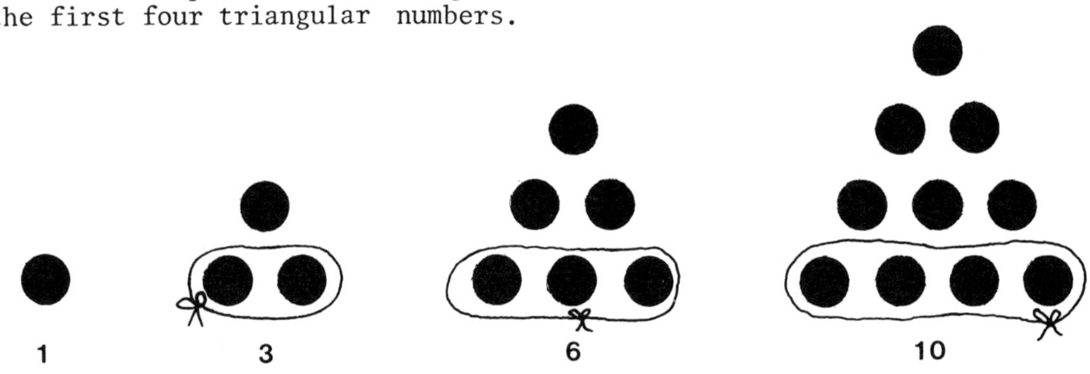

In the diagram, the row circled becomes the gnomon for the previous triangular number. If we look at the numbers named by these rows, we see that they are 2, 3, and 4. If the process is carried on we have gnomons named by 5, 6, 7, 8, etc. Thus the set of gnomons for the triangular numbers is the set of <u>natural numbers</u> with the exception of 1.

The numbers we know as perfect squares are so named because the Pythagoreans saw them in the shape of squares.

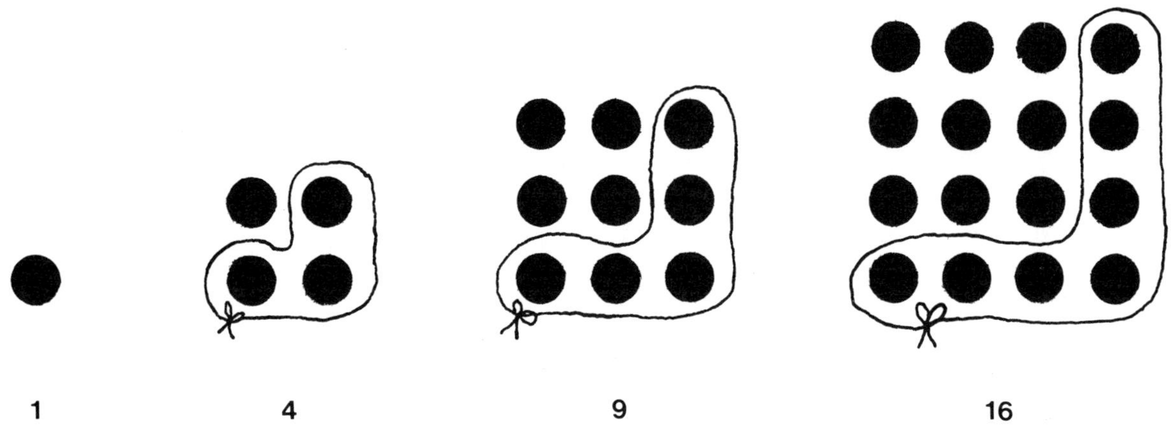

Again, the gnomon for the preceding square number is seen in the enclosed area in the shape of the carpenter's square. This time we have the sequence 3, 5, 7, etc. or the set of odd numbers with the exception of 1.

Let us return to the concept of gnomons in rectangles and look at the now familiar Golden Rectangle which is a special case. In Chapter 2 we learned that we could construct any number of Golden Rectangles simply by adding a square to the length. This means that the square is the gnomon of the Golden Rectangle and it is the <u>only</u> rectangle with that unique property. In fact, the Golden Rectangle is sometimes referred to as the <u>"Whirling Square" Rectangle</u>, since upon repeated division of the rectangle into reciprocal and its square gnomon, the squares appear to whirl around the point of intersection of the main diagonal and the diagonal of the reciprocal.

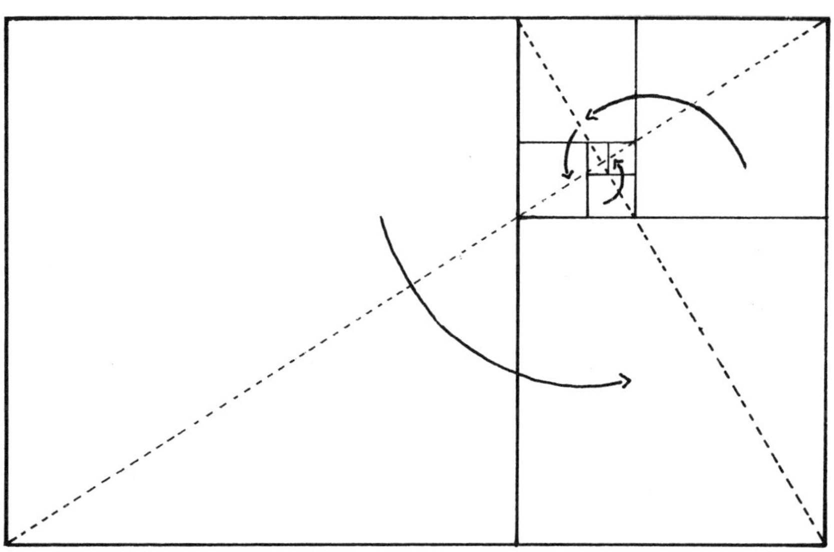

THE ARMATURE OF THE RECTANGLE

To this point we have dealt primarily with non-objective design. However, throughout the history of art, many painters were, and are, concerned with representing three-dimensionality on a two-dimensional surface. For this problem the artist needs a framework to provide the scaffolding on which to erect his composition, be it abstract, representational, planar or three-dimensional. When working with complex content and form more than random placement is necessary for strength of design.

Twentieth century painter Charles Bouleau, in his book The Painter's Secret Geometry, has examined paintings through the ages for the underlying structures of the various works. He analyzes the works of such masters as Vermeer, Botticelli, Cezanne, and Durer. In each case he discovers composition related to specific rectangular divisions which are a part of what he has chosen to call "the armature of the rectangle."

We shall briefly touch on two important ideas concerning division of the rectangle for placement of key forms within a representational work. Again, the diagonal plays the most important role in the evolution of the armature. By using various diagonals the artist-designer can subdivide the entire rectangle into a series of smaller sections that relate to the entire composition. The viewer may not be consciously aware of these diagonals but they help to create the underlying order of a work.

In the following diagrams the evolution of the armature is shown in sequence.

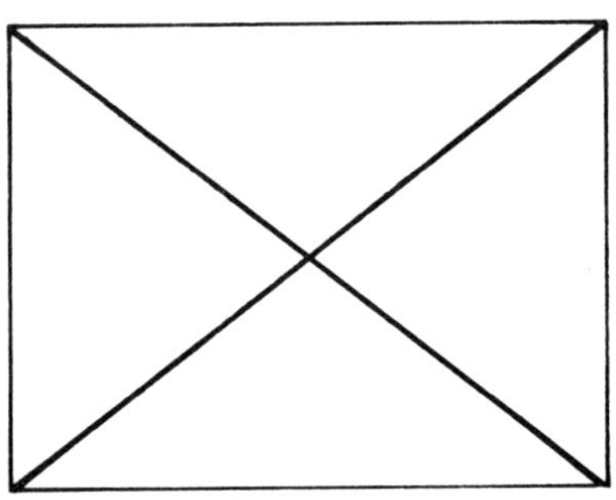

The two main diagonals are drawn.

fig. 1

Line segments are drawn from the midpoints of the longer sides of the rectangle to the ver tices of the rectangle.

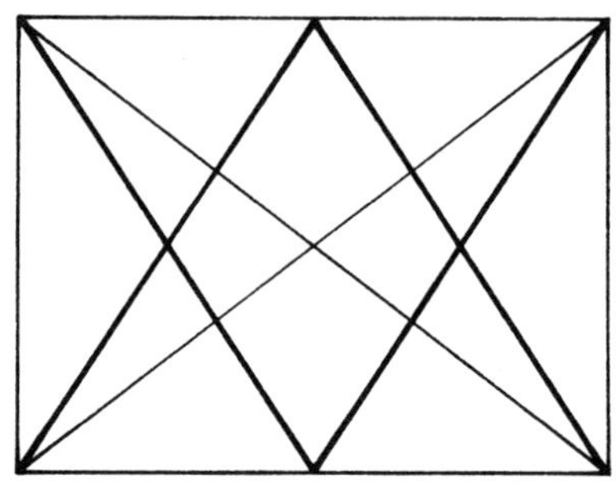

fig. 2

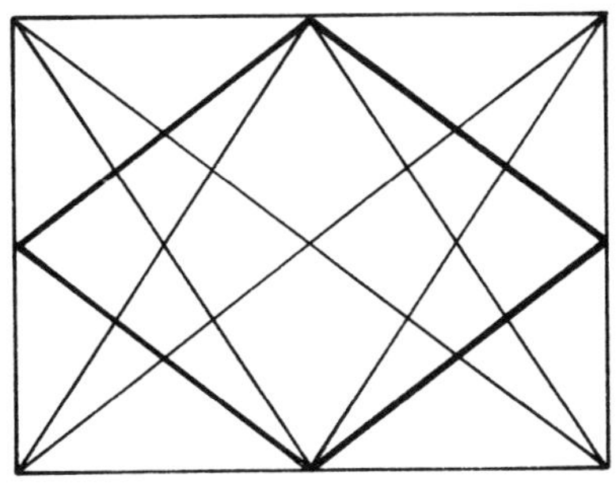

Line segments are drawn connecting consecutive midpoints of all the sides.

fig. 3

Line segments are drawn from the midpoints of the shorter sides of the rectangle to the vertices of the rectangle.

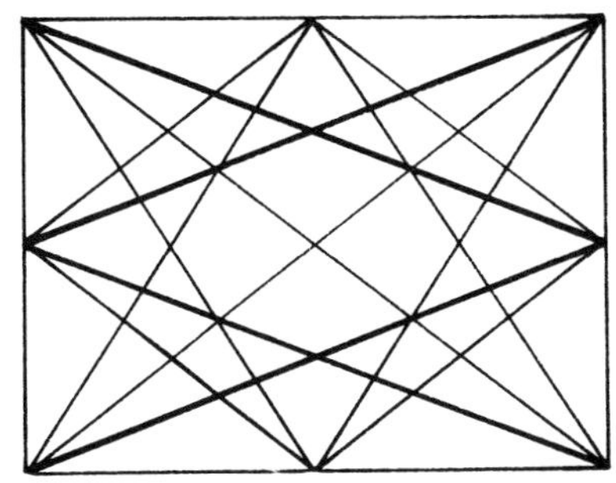

fig. 4

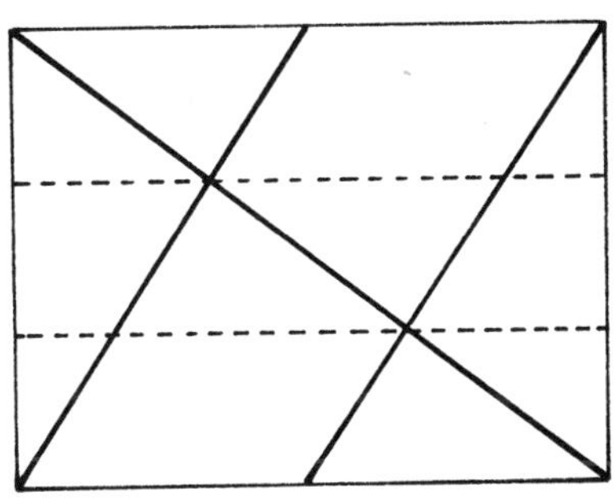

Fig. 5 shows the horizontal subdivision when parallels are drawn to the longer side.

Some of the points of intersection of these diagonals give us reference points for interesting rectangular subdivisions of the space which may be particularly useful when working with still life or landscape.

fig. 5

The rectangle may be divided into <u>thirds</u> by drawing parallels to the sides through the points of intersection of a main diagonal and the segments joining a vertex to the midpoints of the longer sides.

Fig. 6 shows the vertical subdivision when parallels are drawn to the shorter side.

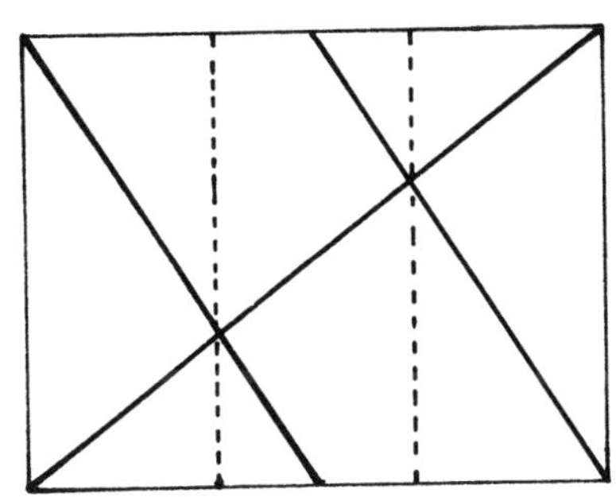

fig. 6

The rectangle may also be divided into fourths by using points of intersection of the main diagonals and line segments joining consecutive midpoints of the sides.

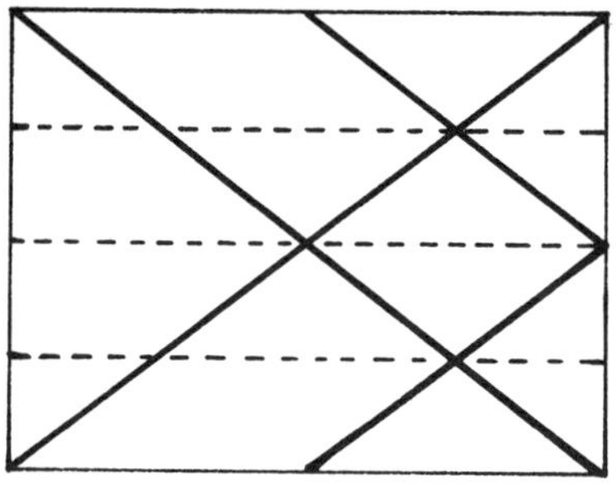

Fig. 7 illustrates the horizontal subdivision when parallels are drawn to the longer sides.

fig. 7

Fig. 8 shows the vertical subdivision when parallels are drawn to the shorter side.

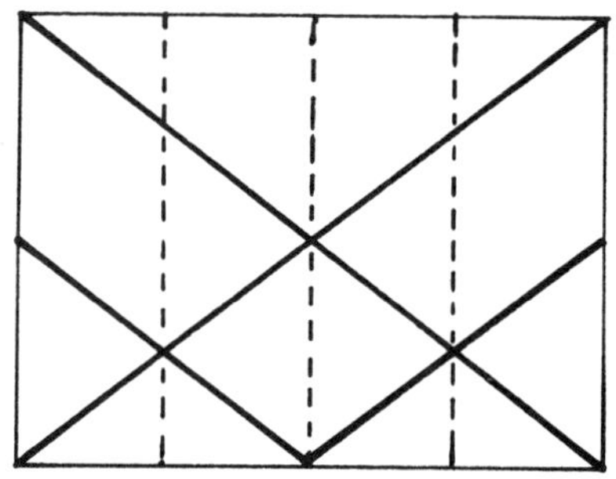

fig. 8

Although the sides of the rectangle may be divided into thirds or fourths by the method described in Construction 8, the diagonals themselves play an important role in providing the framework for a composition. The points of intersection of the various diagonals become focal points for images in the composition.

Here is a sampling of some of the ways the Dynamic Rectangles may be subdivided to provide a workable armature.

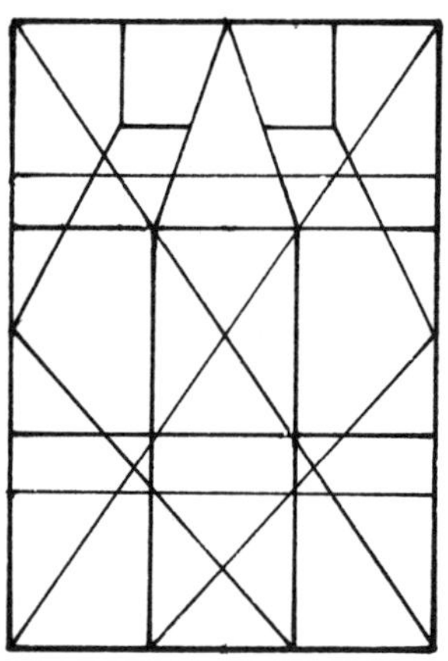

$\sqrt{2}$ Rectangle

$\sqrt{3}$ Rectangle

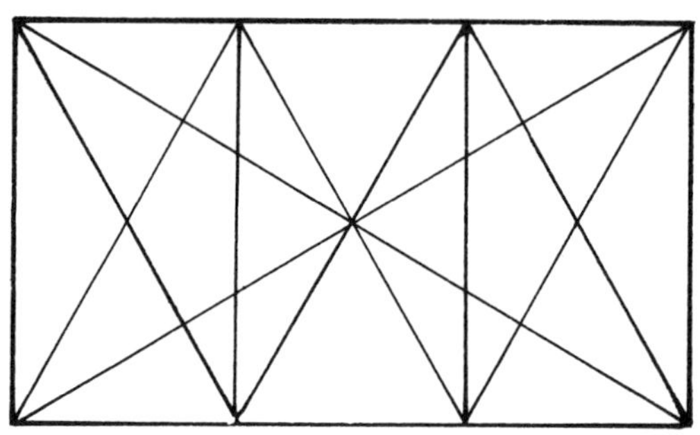

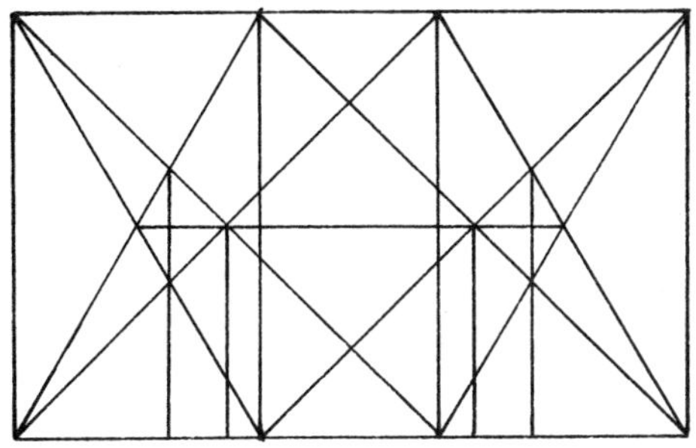

Golden Rcctangle

$\sqrt{5}$ Rectangle

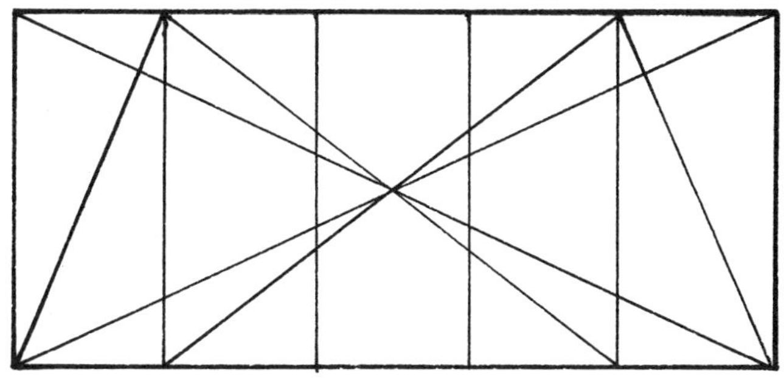

A way to subdivide a Dynamic Rectangle into smaller rectangles similar to itself is to divide each side into n equal units (see Construction 8) where n is any number. Connect opposite sides by line segments joining the points of division.

In the following diagram n equals 5.

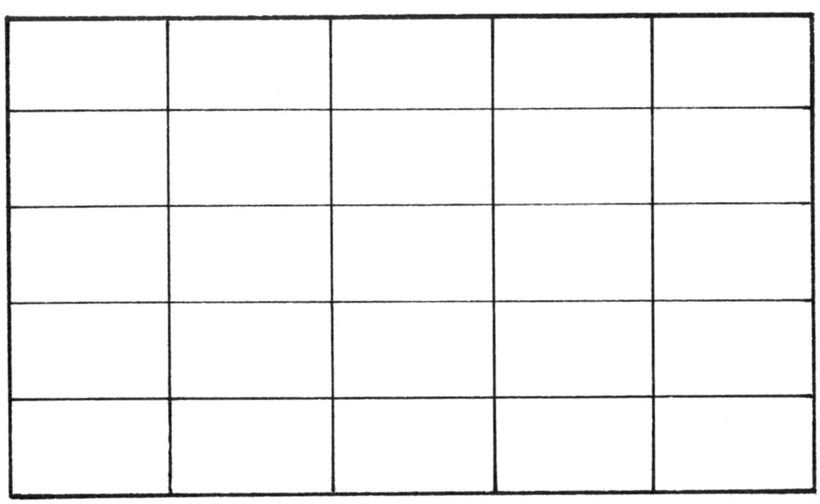

$\sqrt{3}$ Rectangle

Within these divisions diagonals can again be used to focus interest within a composition.

THE DYNAMIC PARALLELOGRAMS

With a little mathematical sleight-of-hand we can transform the Dynamic Rectangles into geometric figures that have even more dynamic qualities of implied motion. The authors have named these figures the <u>Dynamic Parallelograms</u>.

A parallelogram is formed whenever opposite sides of a quadrilateral are parallel. There are no restrictions on the angles as with a rectangle. However, it turns out that consecutive angles are always <u>supplementary</u>. This means that a rectangle can be "tipped" such that the sides remain the same length but one angle of a consecutive pair will measure between 0^0 and 90^0 and the other will measure between 90^0 and 180^0.

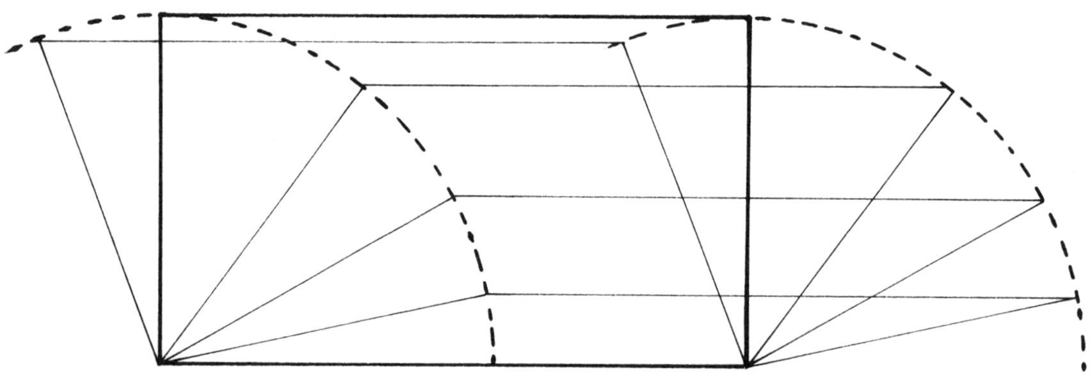

The above diagram illustrates a few of the infinite number of $\sqrt{3}$ Parallelograms that can be constructed from a $\sqrt{3}$ Rectangle. The same procedure could be followed for other Dynamic Rectangles. Note that the Dynamic Rectangles are themselves members of the family of Dynamic Parallelograms.

On the following page you will find a chart that makes comparisons between each of the Dynamic Rectangles and other plane figures, particularly the Special Triangles which we have already discussed.

DYNAMIC RECTANGLES

IN RELATION TO SPECIAL TRIANGLES AND OTHER PLANE FIGURES

$\sqrt{2}$ Rectangle

Harmonic Triangle

(Harmonic Rectangle)

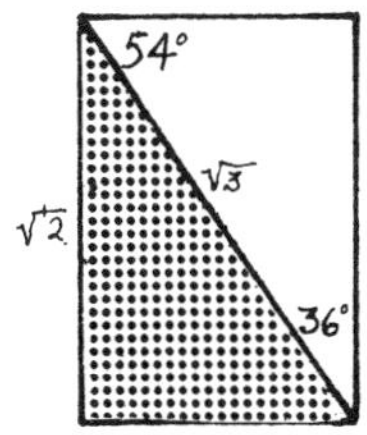

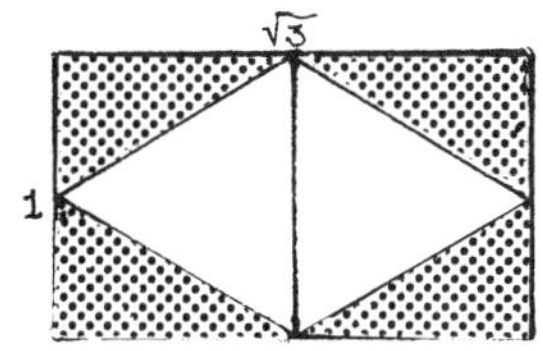

$\sqrt{3}$ Rectangle

Double

Equilateral Triangles

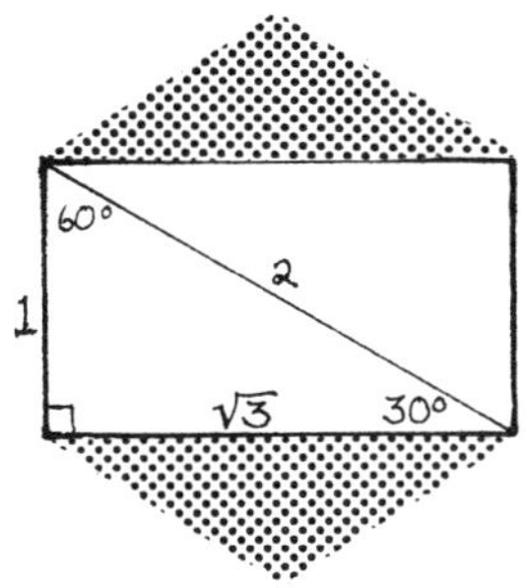

$\sqrt{3}$ Rectangle

30:60:90 Triangle

Regular Hexagon

$\sqrt{4}$ Rectangle

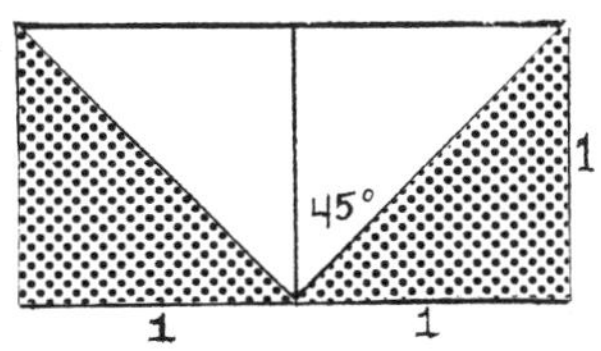

Four Isosceles

Right Triangles

Double Square

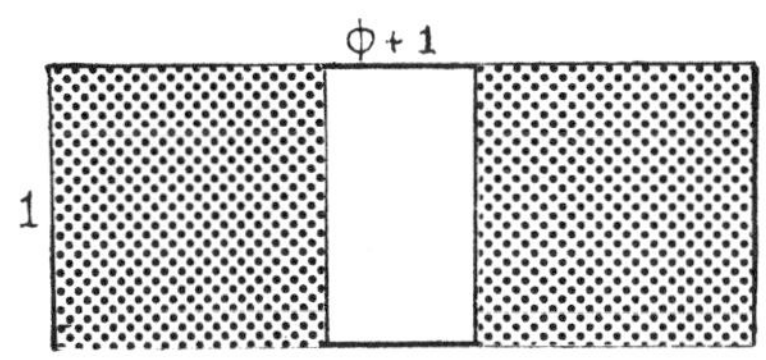

$\sqrt{5}$ Rectangle

One Square and Two Golden Rectangles

or Two Overlapping Golden Rectangles

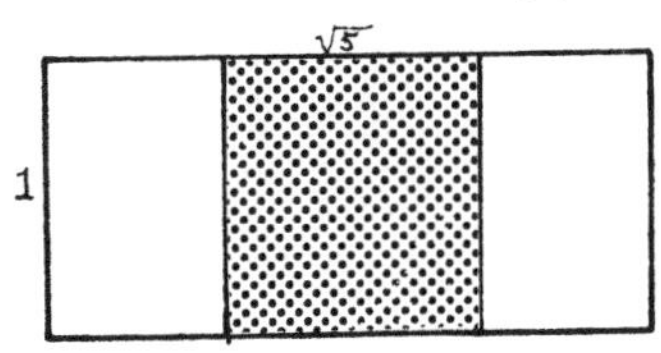

$\emptyset+1$ Rectangle

Two Squares and One Golden Rectangle

or Two Overlapping Golden Rectangles

Golden Rectangle

Golden Triangle

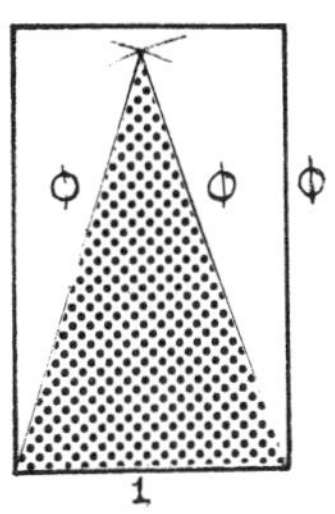

$\sqrt{\emptyset}$ Rectangle

Triangle of Price

(Rectangle of Price)

expands to

Golden Rectangle

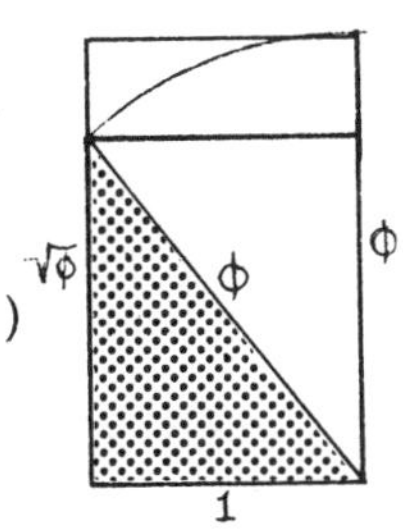

PROBLEM **1**

Draw a scalene triangle that is not a right triangle. Subdivide it into two triangles in such a way that one of the triangles is the gnomon of the other.
Hint: two triangles are similar if two angles of one are congruent to two angles of the other. Remember that one of the small triangles must be similar to the larger one.

PROBLEM **2**

Find reproductions of two paintings from different periods of Art History. Photocopy them. Using tracing paper, construct an armature for each. Compare and contrast them in a written statement.

PROBLEM **3**

Do an analysis of actual leaf proportions by escribing a rectangle around a leaf so that its sides are tangent to the rectangle. Do a minimum of five. Measure the sides of each rectangle and find the ratio of the length to the width. Do any of the rectangles approximate Dynamic Rectangles? If so, name them.
Hint: It might be easier to do rubbings of the leaves using lightweight bond paper and a soft pencil.

PROBLEM **4**

Construct at least six different $\sqrt{2}$ Parallelograms and six different Golden Parallelograms.

PROBLEM **5**

Construct a square whose sides measure 8 inches. Using your knowledge of the Golden Section, create an harmonious subdivision of the square.

PROBLEM **6**

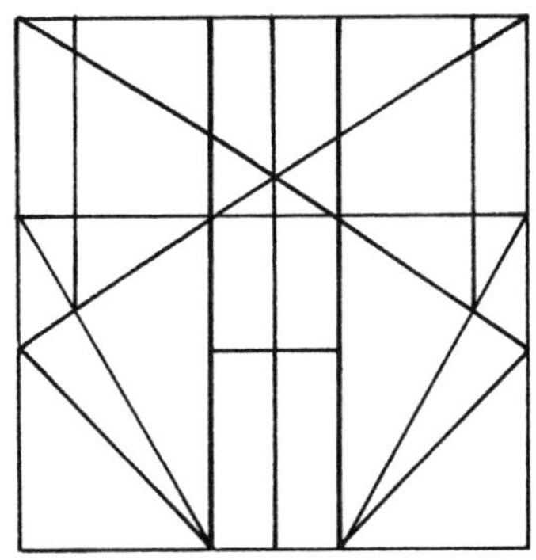

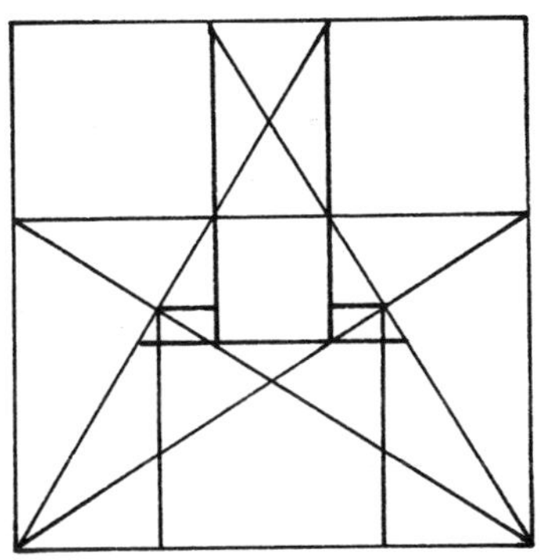

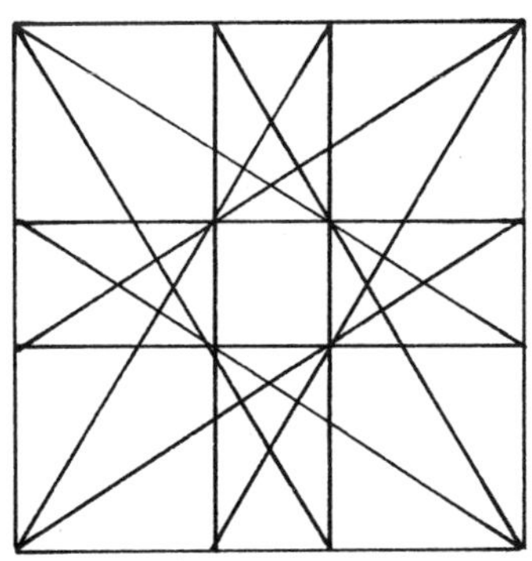

Using compass and straightedge and the techniques described in this chapter, reconstruct one of the above subdivisions using a square that measures 6" on a side.

PROBLEM **7**

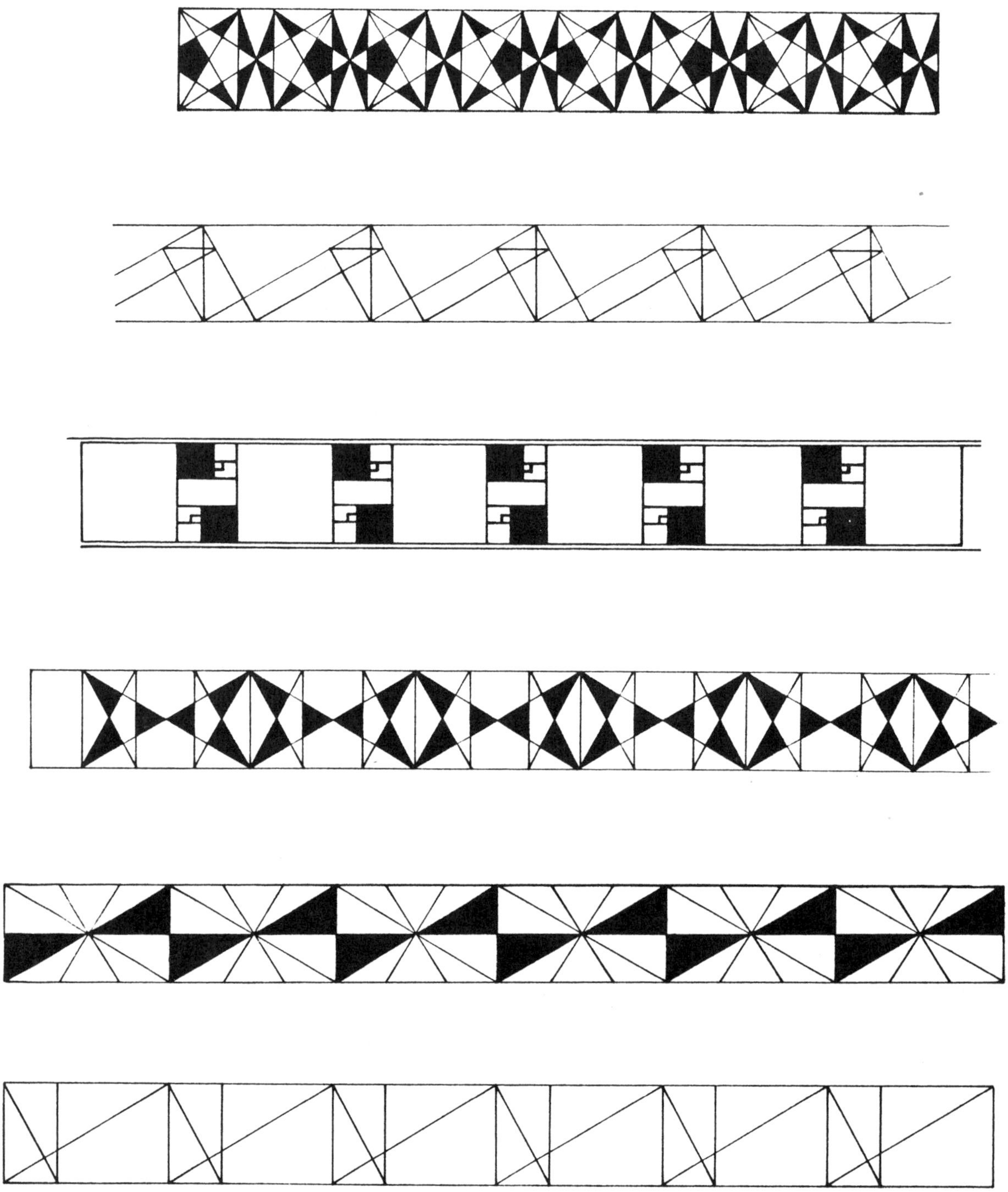

Analyze the method of construction of one of the above border designs and duplicate it on a piece of paper.

PROJECT 1

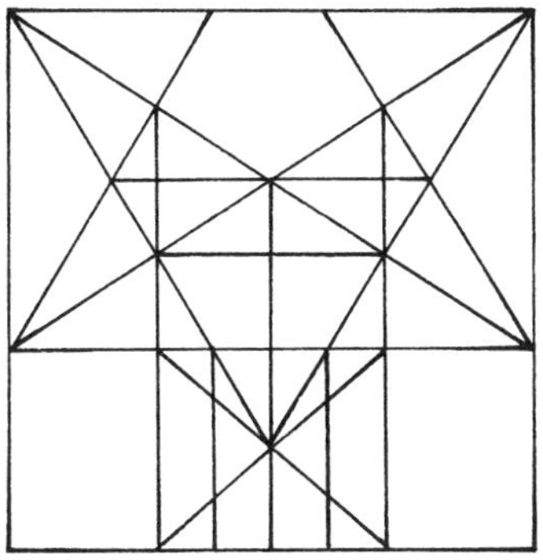

A

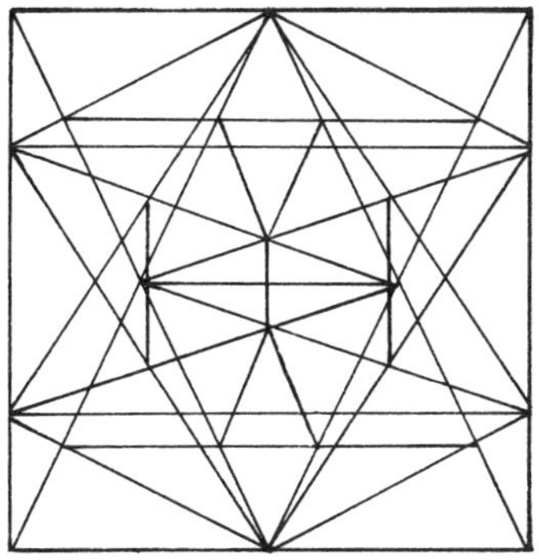

B

Using either A or B as the basis for your composition and a minimum size of 12" do:

a) a black and white variation in two-dimensions
or
b) a black and white variation in three-dimensions
or
c) a monochromatic variation in either two or three dimensions
or
d) only the primary colors plus black and white in either two or three-dimensions
or
e) a pair of complementary colors in either two or three-dimensions.

PROJECT 2

Take the square you constructed in Problem 5 and embellish it with your choice of material, color and technique.

PROJECT 3

Using the subdivision you constructed in Problem 5 as an armature, create a collage or drawing (landscape, still life, etc.)

PROJECT **4**

Create a border design using one of the Dynamic Rectangles and the construction techniques you have learned. Take the same rectangle and turn it into a Dynamic Parallelogram. Subdivide it by the same method used for the rectangle and repeat the border design. Create a third variation of this design by skewing the parallelogram even further and using the same subdivision.

PROJECT **5**

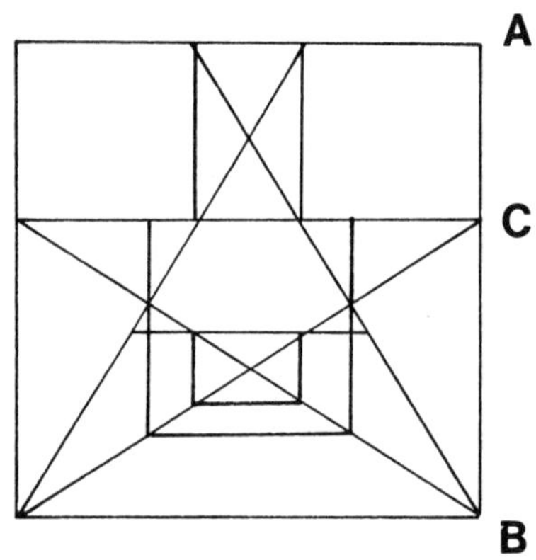

a) Using a 3" or 6" square, reconstruct the above diagram given that C is the Golden Section of AB.

b) Photocopy a group of them and create an overall pattern from the group.

c) Mount on a sturdy board.

d) Color them with black and white only or use a personal color scheme.

PROJECT **6**

Based on your understanding of Northwest Coast Indian design and the use of Dynamic Rectangles therein, create your own blanket or box design study.

PROJECT **7**

If the results of Problem 3 yield leaves that can be escribed by at least two different Dynamic Rectangles that can be used harmoniously in composition, create a plane design with those leaves which incorporates the following materials and techniques: line drawing; crayon rubbing; or tissue collage; or use your choice of materials and technique.

PROJECT **8**

Create a fantasy fish, leaf or bug that has internal proportions that fit within one or more of the rectangles discussed in this chapter. Incorporate your creation into a design in either two or three-dimensions.

PROJECT **9**

Using the results of Problem 7 as a resource:
a) do a color variation of the same border,
b) using multiples of the same design, do an allover pattern,
c) do a border that encloses a space on all four sides.

PROJECT **10**

Choose any Dynamic Rectangle or Dynamic Parallelogram and construct a minimum of ten congruent figures that are large enough to manipulate comfortably. (The photocopier is a wonderful tool for constructing congruent figures!) Using one as a module, fold, bend or score, but do not cut, to produce a three dimensional unit. Treat the others the same way and join them to form a modular sculpture.

PROJECT **11** SCULPTURAL FORM

This net shows how to lay out twelve rhombuses so that they may be folded into a three-dimensional figure called a Rhombic Dodecahedron. The rhombus that works for this net is the one whose angles measure 70°32' and 109°28'. (For your net, use a protractor and approximate the angles by measuring halfway between 70° and 71° and halfway between 109° and 110°.) In this rhombus the ratio of the lengths of the diagonals is $\sqrt{2}$ to 1, the same as the ratio of the lengths of the diagonal to a side in the unit square. The size of the rhombus to be used is the choice of the designer, but all faces must be congruent. Therefore, it is helpful to construct one rhombus as accurately as possible and use it for a template, or photocopy it for the net.

While the surface is still in the plane, enhance the faces using one of the following:

1. variations on harmonic subdivisions of the square.
2. divide a diagonal or side in Ø proportions and connect to points on a corresponding subdivision.

All twelve faces may be in black and white or line only, or may utilize a monochromatic, analogous or triadic color scheme.

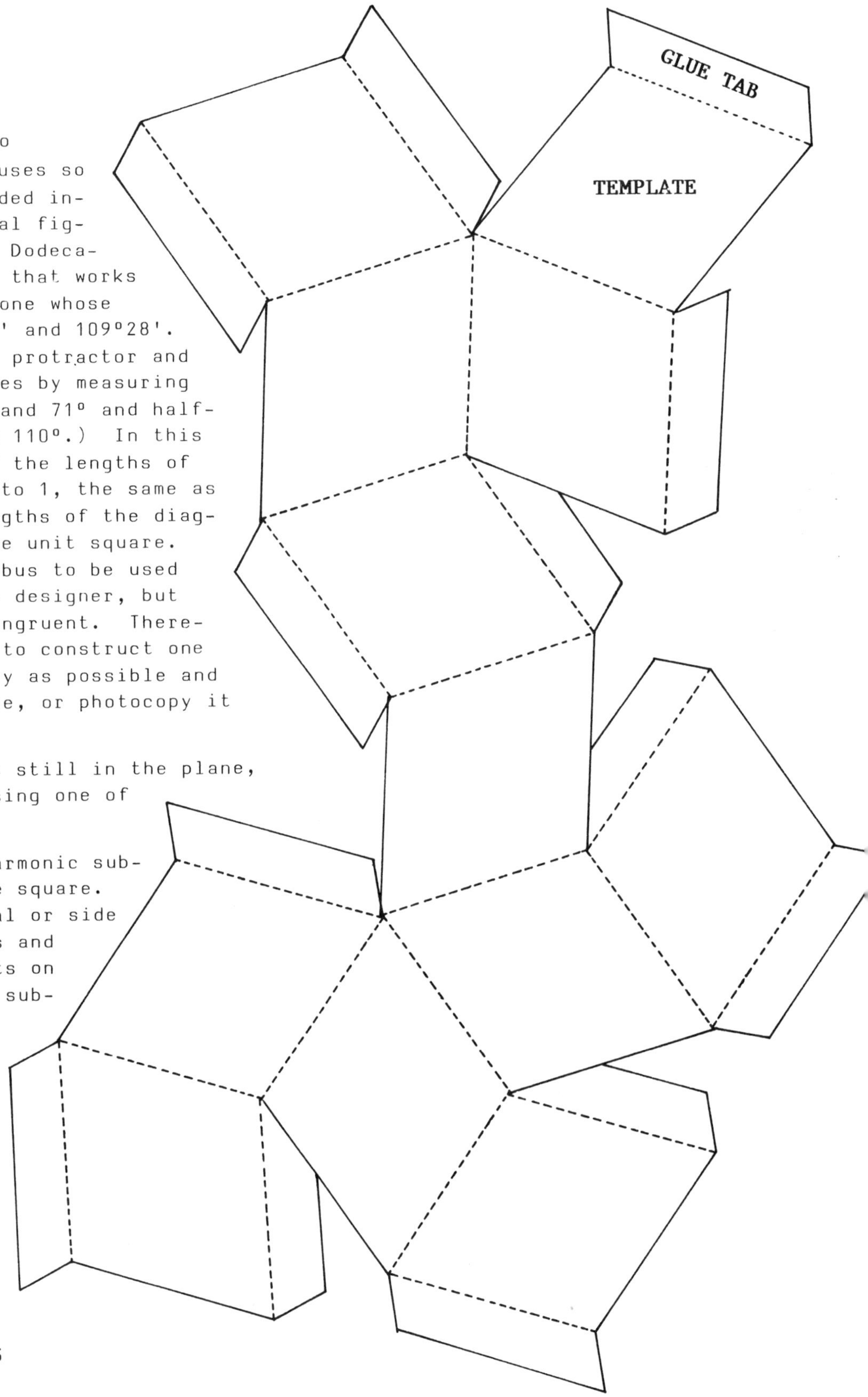

CUT ON SOLID LINES

FOLD ON DOTTED LINES

Further Reading

Bouleau, Charles. The Painter's Secret Geometry. New York: Harcourt, Brace and World, 1963.

Doczi, Gyorgy, The Power of Limits. Boulder: Shambala Press, 1981.

Edwards, Edward B. Pattern and Design with Dynamic Symmetry. New York: Dover Publications, Inc., 1967.

Hambidge, Jay. The Elements of Dynamic Symmetry. New York: Dover Publications, Inc., 1967.

Hambidge, Jay. The Parthenon and Other Greek Temples. New Haven: Yale University Press, 1924.

Hambidge, Jay. Practical Applications of Dynamic Symmetry. New Haven: Yale University Press, 1932.

Waddington, C.H. Behind Appearance. Cambridge: MIT Press, 1969.

Fibonacci Numbers

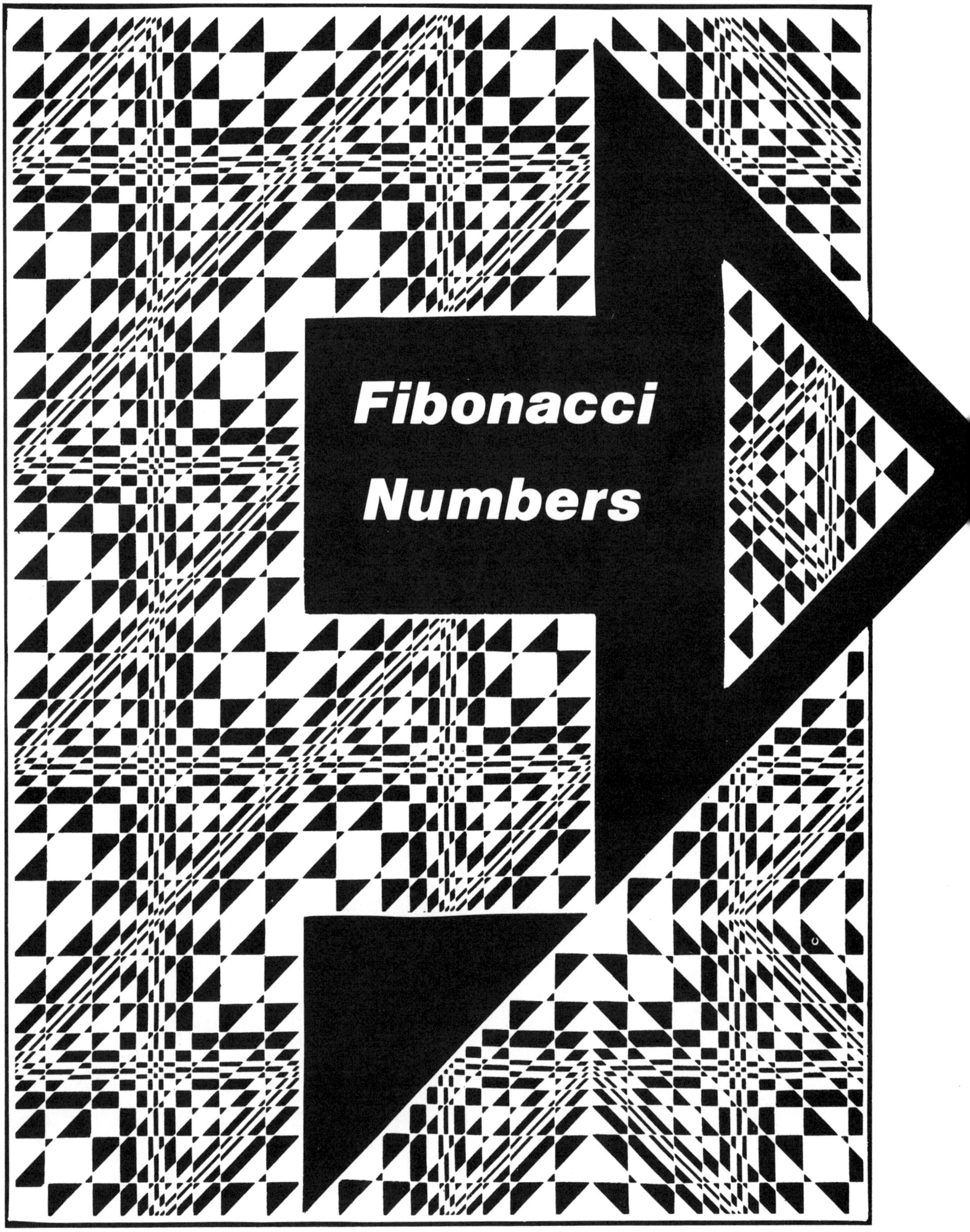

5

Once upon a time, way back in the 13th century, there lived an Italian boy named Leonardo, son of Bonaccio, later to become known as Fibonacci (1170-1250). He grew up in the city of Bugia in North Africa where his father was a customs official. As a youth he studied with Muslims who taught him the Arabic system of numerals which he realized was much less cumbersome than Roman notation. He wrote several books on geometry and trigonometry but he is most well known for the book he wrote at age 27 entitled *Liber Abaci*, The Book of Abacus, which became the prime vehicle for introducing Arabic-Hindu numbers to the European world. He is considered the greatest mathematician of the Middle Ages.

It was, however, a small section of *Liber Abaci* with a particular theoretical problem that has kept his name alive these many centuries. His solution to the problem offered a special series of numbers that was to have implications in the fields of art, botany, biology, and music. Today, still, no one can fully explain this special group of numbers.

First, let us look at the problem he set for himself. Then, let us examine the conclusions. Keep in mind that this problem is more idealistic than realistic in terms of animal reproduction.

THE RABBIT PROBLEM

How many rabbits would you have at the beginning of each month if you start with a single pair and:

a. each adult pair gives birth to a new pair every month,
b. each pair reproduces from the second month on, and
c. no rabbit dies.

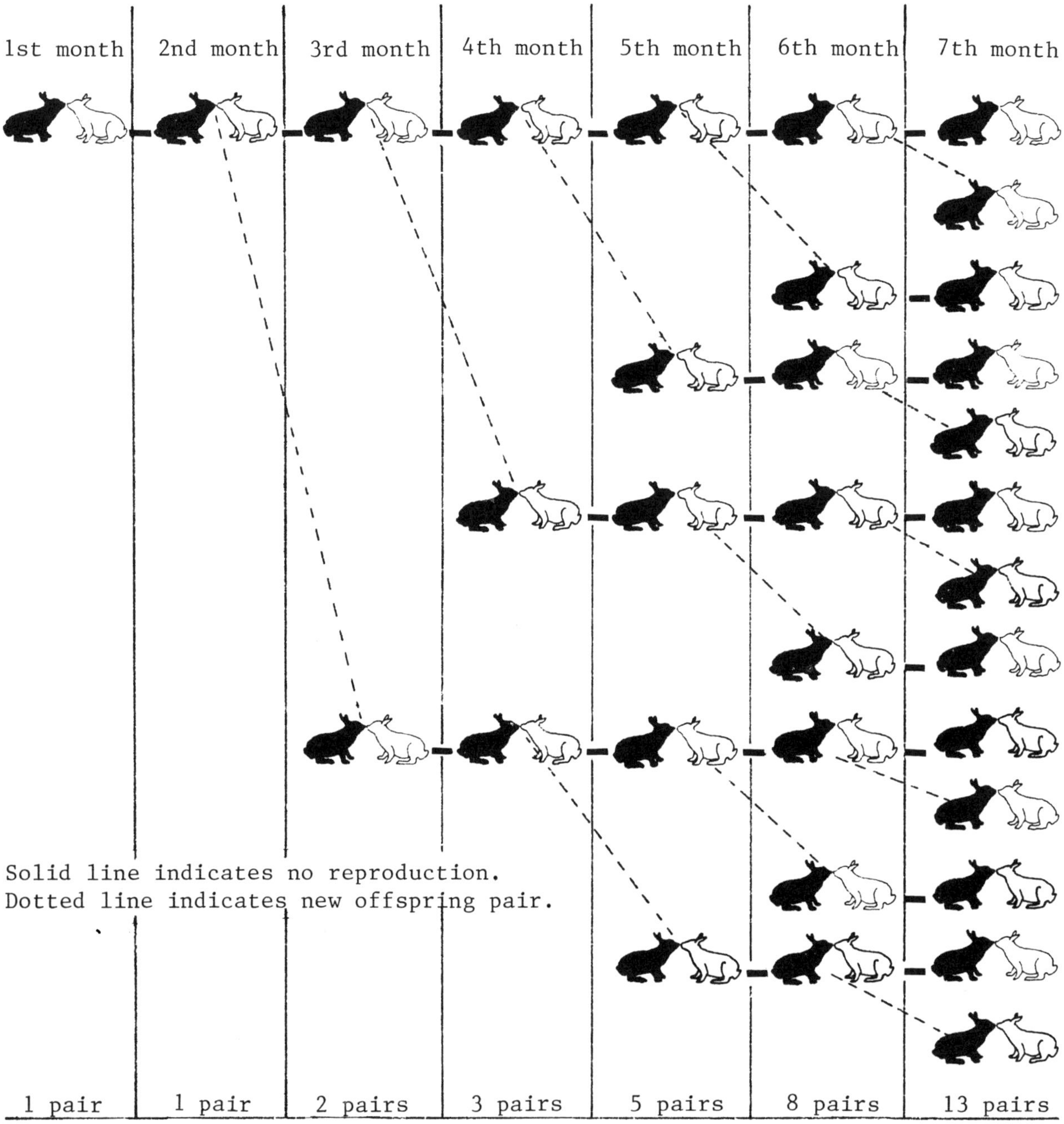

You will notice that a pattern emerges giving rise to a numerical <u>sequence</u> whose terms are 1, 1, 2, 3, 5, 8, 13. If we continue watching these rabbits, (and none gets stewed!) the pattern would continue to give 1, 1, 2, 3, 5, 8, 13, 21, 34, 55, ... Each term in this sequence is the sum of the two preceding terms. In honor of the man who set the problem, this sequence has come to be known as the Fibonacci Sequence.

Looking again at the natural world we find a situation in which this sequence occurs in the genealogy of the male bee. A male bee has a single parent, only a mother, whereas a female bee has both a mother and a father. We trace the ancestry of the male in the following diagram

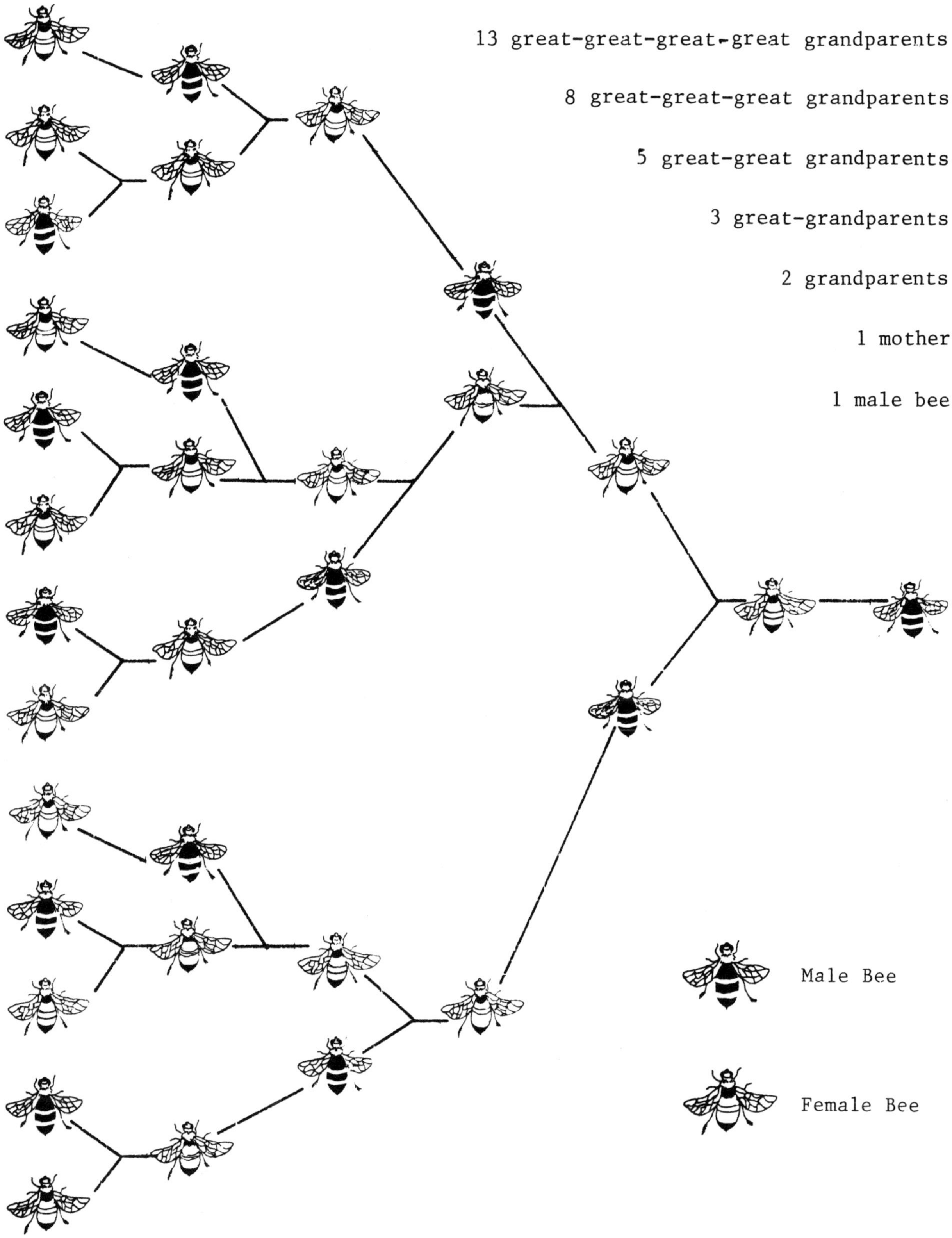

Another example of the appearance of the Fibonacci Sequence in nature is in branching patterns. If we observe the branches, leaves, and flowers of the Achillea ptarmica (sneezewort), we find several occurrences of Fibonacci numbers. The simplified diagram below illustrates the manner in which the branch and leaf patterns yield the Fibonacci sequence. In each case consider the number of leaves or branches cut by various horizontal planes.

In the field, Nature does not follow this pattern absolutely, but the tendency for these numbers to appear is too great to discount.

MATHEMATICAL PROPERTIES OF THE FIBONACCI NUMBERS

This sequence of numbers has some special properties that have intrigued mathematicians throughout the years and continue to do so. We will examine a few of these properties here.

The following chart shows the first 50 terms in the Fibonacci Sequence.

n indicates the position of the term in the Fibonacci Sequence

F_n indicates the Fibonacci number in the nth position

for example: the 32nd Fibonacci number is 2,178,309

THE FIRST FIFTY FIBONACCI NUMBERS

n	F_n	n	F_n	n	F_n
1	1	18	2,584	35	9,227,465
2	1	19	4,181	36	14,930,352
3	2	20	6,765	37	24,157,817
4	3	21	10,946	38	39,088,169
5	5	22	17,711	39	63,245,986
6	8	23	28,657	40	102,334,155
7	13	24	46,368	41	165,580,141
8	21	25	75,025	42	267,914,296
9	34	26	121,393	43	433,494,437
10	55	27	196,418	44	701,408,733
11	89	28	317,811	45	1,134,903,170
12	144	29	514,229	46	1,836,311,903
13	233	30	832,040	47	2,971,215,073
14	377	31	1,346,269	48	4,807,526,976
15	610	32	2,178,309	49	7,778,742,049
16	987	33	3,524,578	50	12,586,269,025
17	1,597	34	5,702,887		

By looking at the chart, you will notice how quickly the numbers get large. F_{50} has 11 digits. Obviously, generating the terms by hand beyond this point is cumbersome and time consuming. However, the process is simplified radically by the use of the computer, and mathematicians have been able to examine the larger terms with considerable ease. For example, F_{19137} has been computed and was found to be 4000 digits long.

We have already noted that each term in the sequence is the sum of the two preceding terms. There are several other sums of interest as well.

For example:

$1 + 1 + 2 = 4 \qquad 4 = (5) - 1$

$1 + 1 + 2 + 3 = 7 \qquad 7 = (8) - 1$

$1 + 1 + 2 + 3 + 5 = 12 \qquad 12 = (13) - 1$

$1 + 1 + 2 + 3 + 5 + 8 = 20 \qquad 20 = (21) - 1$

$1 + 1 + 2 + 3 + 5 + 8 + 13 = 33 \qquad 33 = (34) - 1$

Notice that in each case the sum of the terms is a number that is one less than another Fibonacci number. In fact, if the first n terms (where n is any number) are added, i.e. $F_1 + F_2 + F_3 + \ldots + F_n$, then the sum is one less than F_{n+2} (the second term after F_n).

$$\text{In general } F_1 + F_2 + F_3 + \ldots + F_n = F_{n+2} - 1$$

Now let us see what happens when we add only the <u>even numbered</u> terms or only the <u>odd</u> <u>numbered</u> terms.

The even numbered terms are:

$F_2, F_4, F_6, F_8, F_{10}, \ldots$ or, 1, 3, 8, 21, 55, ...

$1 + 3 = 4 \qquad 4 = (5) - 1$

$1 + 3 + 8 = 12 \qquad 12 = (13) - 1$

$1 + 3 + 8 + 21 = 33 \qquad 33 = (34) - 1$

$1 + 3 + 8 + 21 + 55 = 88 \qquad 88 = (89) - 1$

In general, $F_2 + F_4 + F_6 + \ldots + F_{2n} = F_{2n+1} - 1$ or the sum is one less than the next term in the sequence.

The <u>odd numbered</u> terms are:

$F_1, F_3, F_5, F_7, F_9, \ldots$ or 1, 2, 5, 13, 34,...

$1 + 2 = 3$

$1 + 2 + 5 = 8$

$1 + 2 + 5 + 13 = 21$

$1 + 2 + 5 + 13 + 34 = 55$

In general $F_1 + F_3 + F_5 + \ldots + F_{2n-1} = F_{2n}$ or the sum is the next term in the sequence.

Another pattern that is evident in this sequence is that the square of any term differs from the product of the terms on either side of it by ± 1.

For example, consider three consecutive terms such as 8, 13, and 21.

$13^2 = 169$

$(8)(21) = 168$

and $13^2 - (8 \times 21) = 1$

If we had chosen 5, 8, and 13 we would have:

$8^2 = 64$

$(5)(13) = 65$

$8^2 - (5 \times 13) = -1$

In general $(F_n)^2 - (F_{n-1})(F_{n+1}) = \pm 1$

The aforementioned properties are but a few of those that set the Fibonacci Sequence apart from other sequences. For the mathematically curious reader, these properties are proved in Appendix B.

FIBONACCI AND PHI

The authors have found the most interesting property of the Fibonacci Sequence to be the one that connects it to $\emptyset$. Notice in the table of Fibonacci numbers that as n gets larger, the ratio of any pair of consecutive terms in the sequence gets closer to $\emptyset$.

$$\frac{F_4}{F_3} = \frac{3}{2} = 1.50000 \qquad \frac{F_{13}}{F_{12}} = \frac{233}{144} = 1.61806$$

$$\frac{F_8}{F_7} = \frac{21}{13} = 1.61538 \qquad \frac{F_{15}}{F_{14}} = \frac{610}{377} = 1.61804$$

$$\frac{F_{11}}{F_{10}} = \frac{89}{55} = 1.61818 \qquad \frac{F_{16}}{F_{15}} = \frac{987}{610} = 1.61803$$

BEYOND F_{15} THE RATIO OF ANY PAIR OF CONSECUTIVE FIBONACCI NUMBERS IS THE SAME AS $\emptyset$ TO FIVE DECIMAL PLACES.

In fact, the computer has enabled us to examine $\emptyset$ to many more decimal places. In an article for the Fibonacci Quarterly, Murray Berg describes the calculation of $\emptyset$ to 4600 decimal places by dividing F_{11004} by F_{11003}, each of which contains 2300 digits. Obviously, this goes beyond hand calculation as it would be more than slightly inconvenient to divide a 2300 digit number by another 2300 digit number and carry out the division to 4600 places. This particular exercise was completed in 20 minutes on an IBM 1401 computer. The calculation was checked by interchanging the divisor and dividend to compute the reciprocal of $\emptyset$. With the exception of the number one before the decimal, the two decimal representations were identical up to the 4598th place.

In addition to this direct connection to $\emptyset$, we can generate another sequence that relates both to the Fibonacci numbers and to $\emptyset$.

The <u>Golden Sequence</u> is the summation sequence whose initial terms are 1 and $\emptyset$.

$$1,\ \emptyset,\ 1+\emptyset,\ 1+2\emptyset,\ 2+3\emptyset,\ 3+5\emptyset,\ 5+8\emptyset,\ldots$$

Notice the appearance of a double Fibonacci Sequence. Beginning with the third term, the arrows pointing upward indicate numbers in the Fibonacci Sequence.
Beginning with the second term (since $\emptyset = 1\,\emptyset$), the arrows pointing downward show that the <u>coefficients</u> of $\emptyset$ are in Fibonacci Sequence.

If we examine this sequence further we see that this is one we have seen before.

Remembering that $\emptyset$ is the solution to the equation $x^2 - x - 1 = 0$, we have $\emptyset^2 - \emptyset - 1 = 0$.

$$\therefore \emptyset^2 = 1 + \emptyset$$

We can rewrite the first three terms of the sequence $1,\ \emptyset,\ \emptyset^2$

Since this is a summation sequence, and we know from Chapter Two that $\emptyset^n + \emptyset^{n+1} = \emptyset^{n+2}$, we can rewrite this sequence:

$$1,\ \emptyset,\ \emptyset^2,\ \emptyset^3,\ \emptyset^4,\ldots$$

That is, the Golden Sequence is the sequence of the powers of $\emptyset$ that we examined in Chapter Two.

Fibonacci numbers translate in a practical way into dimensions of length and width important for the visual artist. Since the ratio of two consecutive Fibonacci numbers rapidly approaches Ø as n increases, using a pair toward the beginning of the sequence offers a tool for constructing a rectangle that approximates the Golden Rectangle but with sides whose lengths are commensurable.

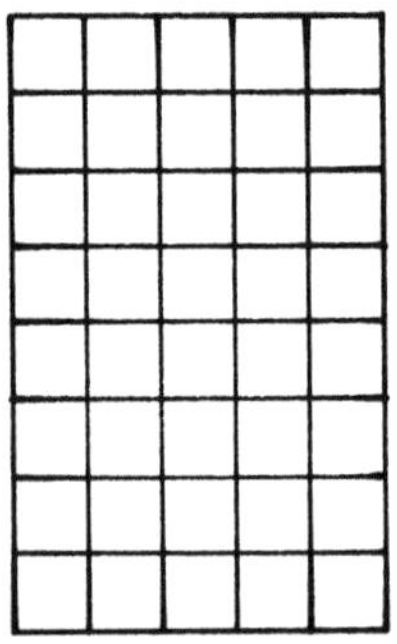

5:8

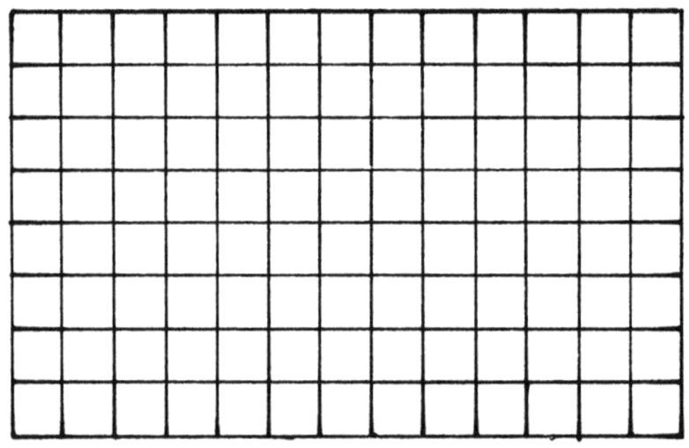

8:13

$$\frac{8}{5} = \frac{1.60000}{1} \qquad \frac{13}{8} = \frac{1.62500}{1}$$

Let us examine another way in which the Fibonacci Sequence is beneficial to the designer. It provides a way to harmonically divide line segments and bounded portions of the plane in a manner such as illustrated in the following figure.

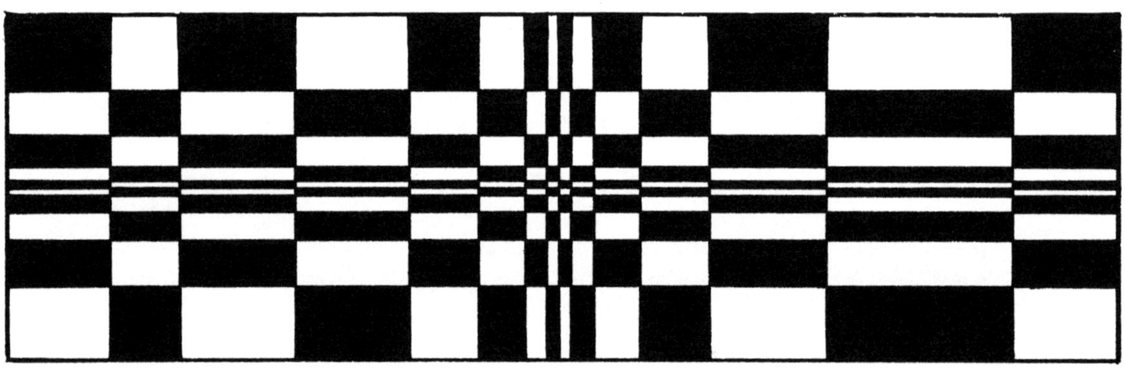

In this figure the sides of the rectangle have been divided into lengths that are in Fibonacci sequence, beginning at the midpoints and working one's way toward the endpoints. The points of division on one side are then connected to the corresponding points on the opposite side by line segments. The design is completed by darkening alternate areas.

The construction for this type of division follows:

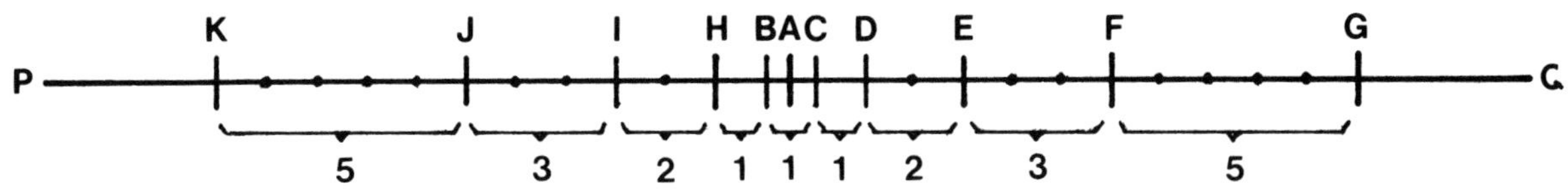

DIVIDE A LINE SEGMENT INTO LENGTHS CORRESPONDING TO PALINDROMIC SEQUENCES OF FIBONACCI NUMBERS.

Given $\overline{PQ}$, divide $\overline{PQ}$ into lengths corresponding to palindromic sequences of Fibonacci numbers.

1. Choose a unit length.
2. Locate the midpoint of $\overline{PQ}$ (point A).
3. Mark off $\frac{1}{2}$ unit on either side of A (points B and C).
4. Mark off units in sequence in both directions from B and C.
5. With BC=1, move to the right marking the points that correspond to lengths measuring 1, 2, 3, and 5 units respectively (points D, E, F, G)
6. Repeat to the left to locate points H, I, J, and K.

The line segments may be extended as far as desired in either direction. If the process were continued the segments to the right of G and to the left of K would each measure eight units.

Fibonacci divisions are not limited to rectangles. In fact, as the number of sides of the polygon increases, so do the number of design possibilities.

Note that the design possibilities become endless with the introduction of color, curves, and sequential divisions of line segments beginning at other points than the midpoint. If multiple units of a design are required for a project, a photocopier is a very useful tool.

Student Work
Chapter 5, project 1

Lana Yonkers

FIBONACCI AND PHYLLOTAXIS

In the natural world the arrangement of leaves, buds, and seeds is termed phyllotaxis. Its connection with Ø and Fibonacci is evident when we examine the phenomenon of gnomonic growth, in which the shape remains the same as the size is altered, exactly as we discovered in Chapter Two when we created successively larger Golden Rectangles.

In organic forms that grow from the center outward, a spiral formation occurs because the spiral fills space efficiently in a perfectly regular manner and is capable of infinite expansion. The most striking example of spiral phyllotaxis occurs in the head of the common sunflower. The rhombic seed pockets form a series of intersecting spirals in a plane. Under normal circumstances of growth these provide a practically constant ratio of 55 long curves travelling in one direction intersecting 89 shorter ones travelling in the other direction. Some heads may be found to have 89 longer curves intersecting 144 shorter ones.

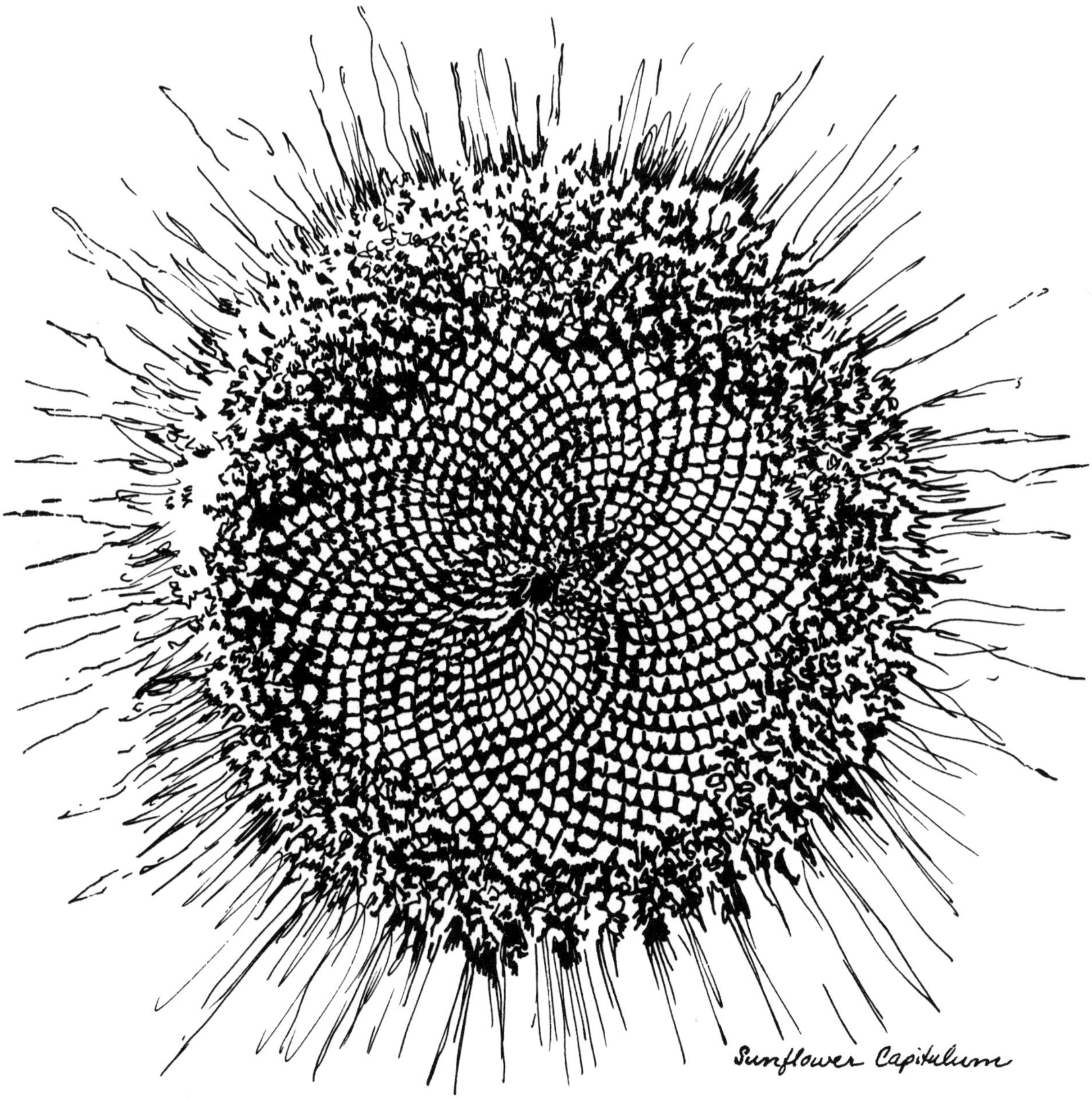

Sunflower Capitulum

If we examine these ratios closely we see that both $\frac{89}{55}$ and $\frac{144}{89}$ give close approximations to Ø, and that, in fact, 55, 89, and 144 are all terms in the Fibonacci Sequence.

Like the sunflower, the head of a daisy has two sets of intersecting spirals, utilizing another pair of Fibonacci numbers with a ratio approximating Ø. Usually there are 21 seed spirals in one direction intersecting 34 seed spirals in the other.

We also find that the number of petals on particular flowers is frequently a Fibonacci number.

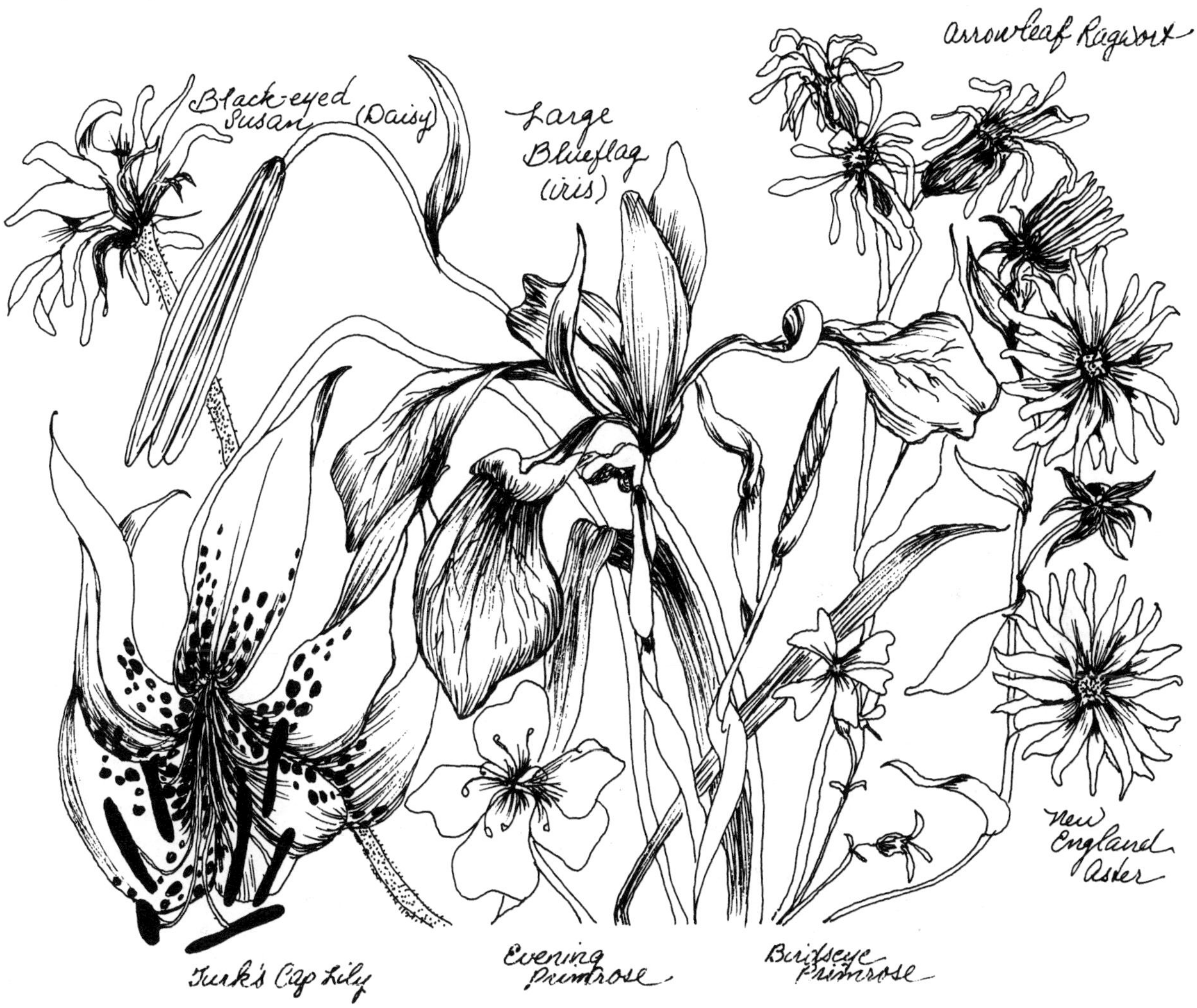

Number of petals	type of flower
3	irises, lilies
5	buttercups, primroses
8	some delphiniums
13	ragwort, some marigolds
21	some asters, marigolds
34,55,89	different types of daisies

Pine needles tend to grow in clusters of 2, 3 or 5 depending on the species.

Fibonacci spirals into the third dimension in the phyllotaxis of pineapples and the cones of evergreen trees. The pineapple is more mathematically analogous to the cylinder while evergreen cones are more analogous to conical shapes. The hexagonal scaly plates of the pineapple are thus arranged in helical spirals whereas the seed pockets on the cone form a different kind of curve called concho-spiral which comes closer to approximating the logarithmic spiral. We shall examine these in detail in the next chapter.

Apart from the shape of the curves, the spirals still appear in Fibonacci numbers. If we consider the pineapple as a cylinder, cut it vertically and flatten it, we can see the spirals transformed into straight lines in the plane.

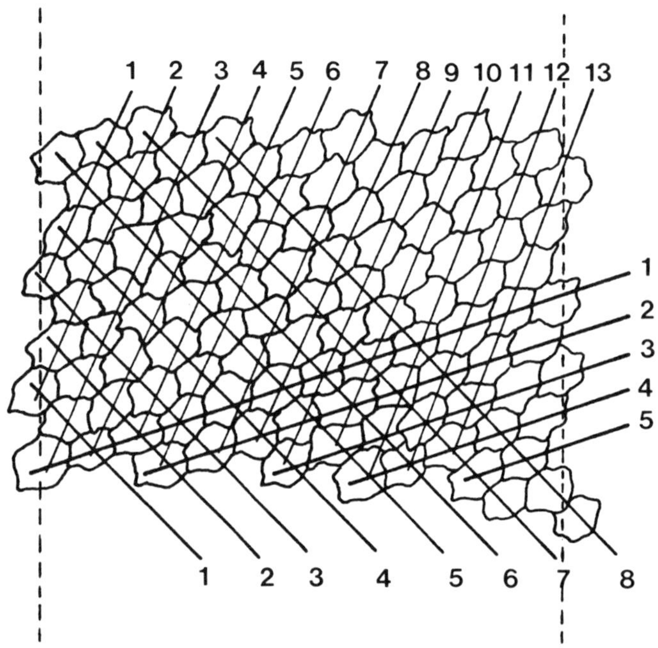

In the diagram the vertical lines represent the cut and the portion between them the flattened view of the scales on the skin. We see 5 sharp curves to the right, 13 steep curves to the right again, as well as 8 more gently sloping curves to the left.

Translated once again into three dimensions, we can see how the straight lines from the illustration of the flattened section wrap around the fruit.

In the cones of evergreen trees the seed protecting scales form two groups of intersecting spirals. These frequently appear in the ratios of $\frac{8}{5}$ or $\frac{13}{8}$. In addition to the fact that the number of spirals in either direction is a Fibonacci number, it is often the case that the number of seeds in any one spiral is also a Fibonacci number.

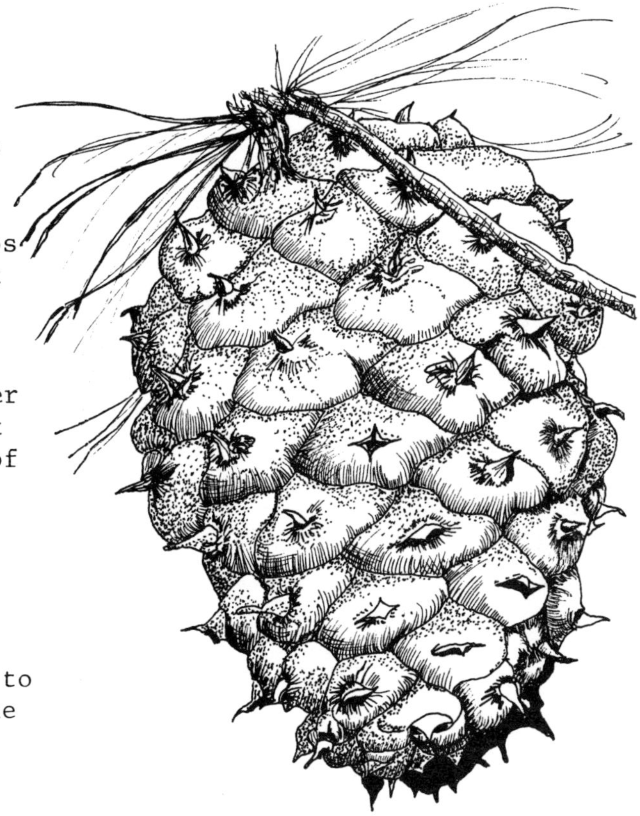

The ratios of the groups of curves in spiral phyllotaxis seem to frequently come from members of the following sequence:

$$\frac{1}{1}, \frac{2}{1}, \frac{3}{2}, \frac{5}{3}, \frac{8}{5}, \frac{13}{8}, \frac{21}{13}, \frac{34}{21}, \frac{55}{34}, \frac{89}{55}, \frac{144}{89},$$

Progressing from the arrangement of seeds to the arrangement of leaves around a stem, we again encounter a phenomenon that suggests a sequence of Fibonacci numbers. The leaves of a majority of higher plants are arranged in a spiral sequence growing out and up the cylindrical stem. If we choose a particular leaf on a stem and travel up the stem passing through consecutive leaf nodes, we will eventually arrive at a leaf situated directly above the first leaf. If **t** is the number of complete turns made around the stem, and **n** the number of leaves passed along the way, then the fraction $\frac{\mathbf{t}}{\mathbf{n}}$ is said to be the divergency constant in that particular species.

In this sequence, each term is composed of a pair of consecutive Fibonacci numbers such that, with the exception of the initial term, the sequence formed by the numerators and the sequence formed by the denominators are both Fibonacci Sequences, and as the terms progress the ratio comes closer and closer to Ø.

In the diagram on the following page, leaf number six is situated directly above leaf number one. Before arriving at leaf number six, five leaves are passed through in two complete turns around the stem. Thus, the divergency constant for this type of plant is $\frac{2}{5}$.

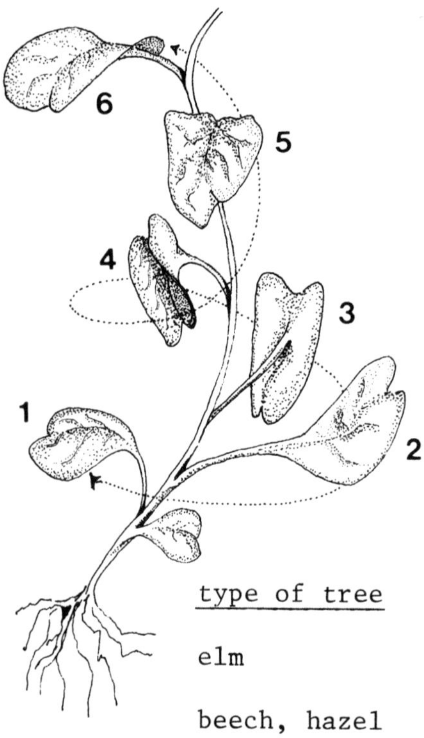

Divergencies are generally found to be terms from the following sequence:

$$\frac{1}{2}, \frac{1}{3}, \frac{2}{5}, \frac{3}{8}, \frac{5}{13}, \frac{8}{21}, \frac{13}{34}, \frac{21}{55}, \ldots$$

Again, the numerators and denominators are in Fibonacci sequence. This time, however, the individual terms are composed of fractions of the form $\frac{F_n}{F_{n+2}}$

The following table gives divergencies for some specific types of trees.

type of tree	divergency constant
elm	1/2
beech, hazel	1/3
apple, oak, apricot, poplar	2/5
pear, weeping willow	3/8
willow, almond, pussy willow	5/13

Nature uses this arrangement to provide the proper amount of sunlight for the plant. This way of distributing leaves seems to give the necessary amount of exposure to sunlight as well as protection when needed.

Under perfect conditions, which the mathematician can create, but Nature never does, a plant will obtain maximum exposure to sunlight and minimum overlapping of leaves if each leaf is separated from the succeeding one by an angle of about 137.5°. At this angle, provided the shoot be kept perfectly straight, no two leaves would ever be exactly one over the other. This angle is called the Ideal Angle or sometimes the Fibonacci Angle and bears a close relationship to ∅.

If we multiply the total number of degrees in a circle, 360°, by the fraction 1/∅ (the reciprocal of ∅), we get an angle of about 222.5°. The angle needed to complete the circle measures 137.5°, the Ideal Angle.

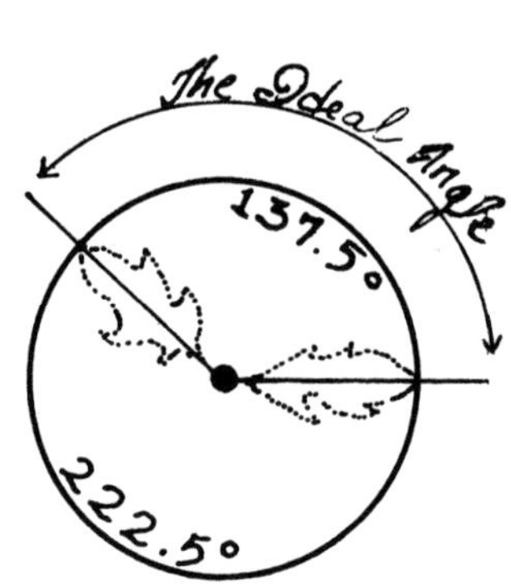

Maximum exposure to sunlight is not always a healthy condition for a plant. Since the needs of plants vary, the Ideal Angle of divergence would not provide optimal growing conditions for every species, hence the variety of divergency constants.

FIBONACCI AND MUSICAL COMPOSITION

We are greatly indebted to J. Michael Pope, one of our students, who, at the time of this writing, is in the process of obtaining a Master's Degree in Music Composition. Exposure to the Fibonacci Sequence in our course sparked a desire to utilize these ideas in his own work with musical form. He sees great potential in the Fibonacci Sequence as a musical structure. In the following section he offers an example of his work with an explanation.

A composer of music may use the Fibonacci sequence to proportionally segment the dimension of his or her art form: time. For example, in determining a skeletal design of the duration of the movements of a symphony, the composer may utilize the numbers of the sequence to create time frames that are in Ø proportion to each other. The duration of individual sections of each movement may also be divided proportionally using the Fibonacci sequence. Or, a composer may use the sequence to produce rhythms within the individual sections.

The following excerpt from the chamber work, Yes Yes Men Live, for two violins and bassoon, utilizes the Fibonacci sequence to generate rhythmic change.

At this point in the piece, the composer creates a rhythmic pattern that gradually slows (ritardando) the melodic alternation. At its longest duration the melodic material reverses and the rhythmic pattern gradually increases the speed of alternation (accelerando) in proportion to the slowing. A reduction of the passage to a single line, displays the use of the Fibonacci sequence.

J. Michael Pope

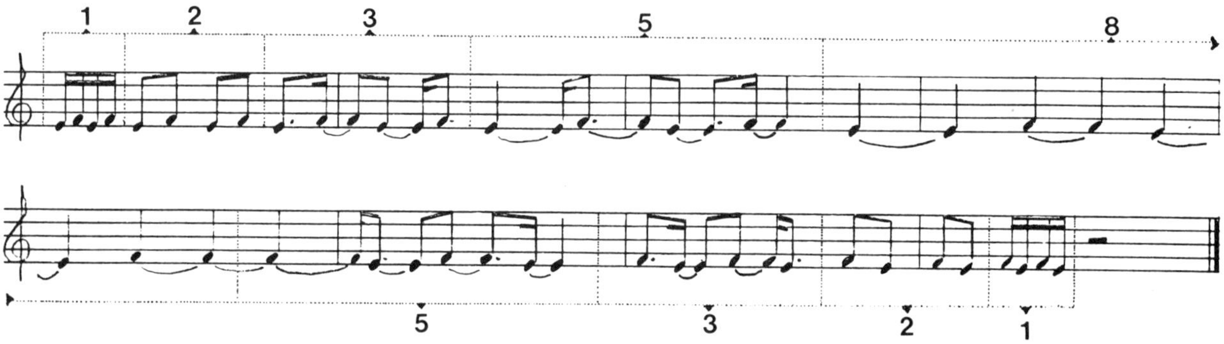

The key to understanding this structure is to know that the sixteenth note is considered the unit measure. The following shows how the Fibonacci numbers are related to musical time. Note that the numerators are in Fibonacci progression.

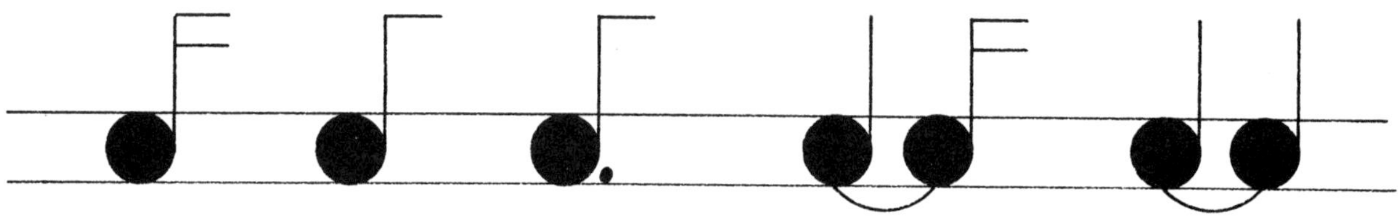

$\frac{1}{16}$ $\qquad$ $\frac{1}{8} = \frac{2}{16}$ $\qquad$ $\frac{3}{16}$ $\qquad$ $\frac{1}{4} + \frac{1}{15} = \frac{5}{16}$ $\qquad$ $\frac{1}{4} + \frac{1}{4} = \frac{8}{16}$

VISUAL AND AURAL MUSICAL PERCEPTION

Michael has translated his musical composition into the following visual variations, given his feeling that we can "hear with our eyes and see with our ears." In each case the dimensions of the rectangle are in a ratio of 3 to 2, and the subdivision of the interior space is also in the Fibonacci progression. Beginning on the far left and moving from top to bottom, the three horizontal bands represent the instrumental parts of the two violins and the bassoon respectively. The central portion in each depicts, for him, the fusion of all three parts. Vamp #1 is his minimal interpretation, while Vamp #2 he considers a "mindscape."

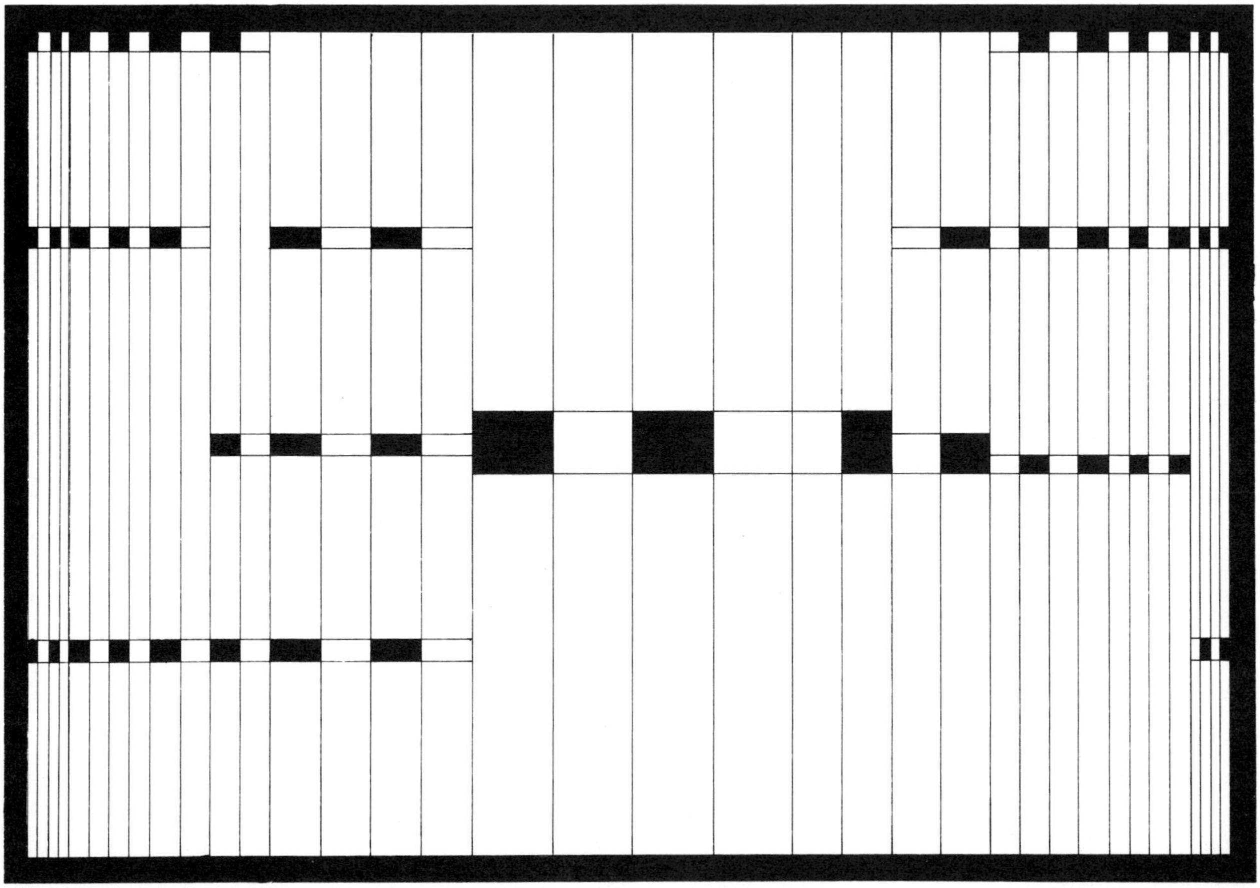

Vamp #1 Minimal — J. Michael Pope

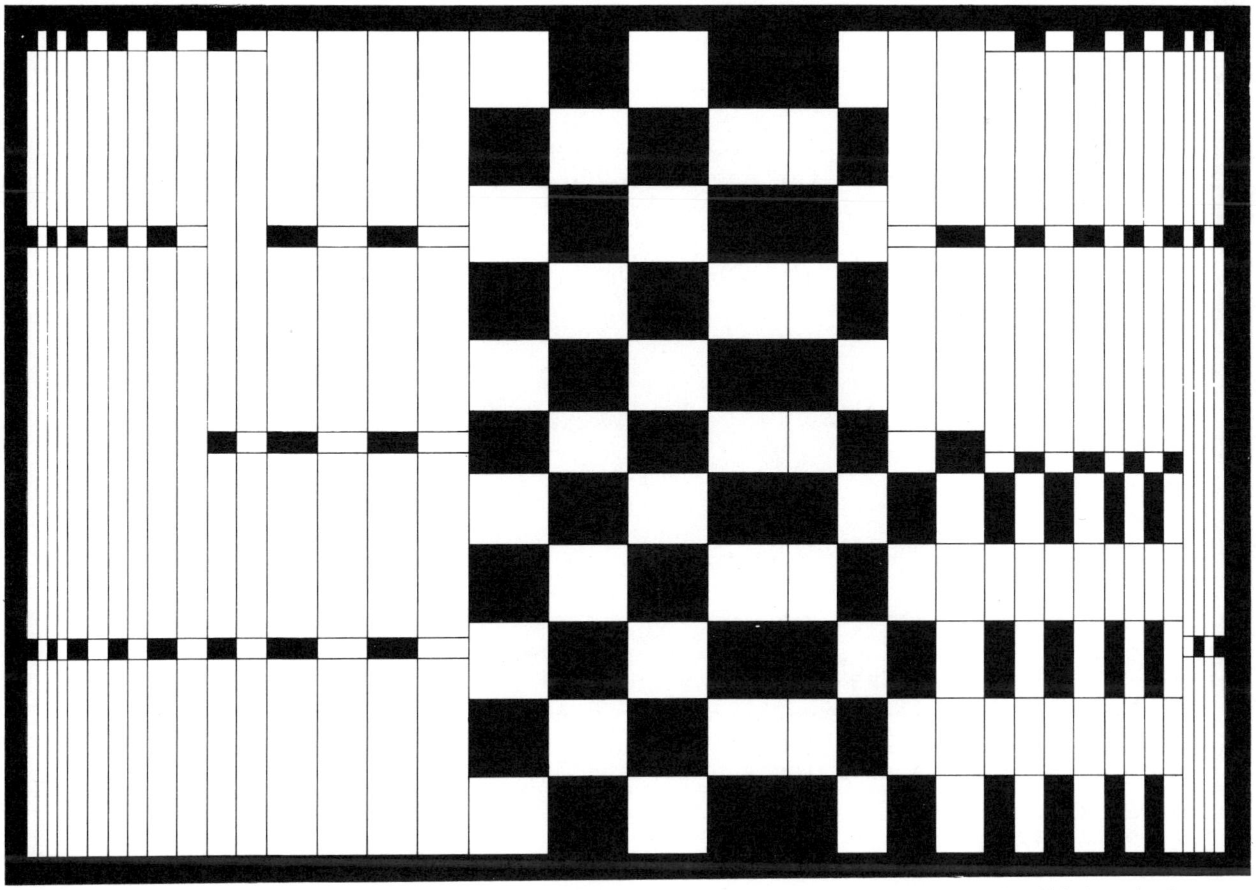

Vamp #2 Mindscape — J. MIchael Pope

FIBONACCI, PHI AND PHOOD

In our research, we have found Fibonacci numbers springing up in strange places. Harriet Lincoln, one of our students, provided us with this recipe for Eastern pastry. We share it with you for your gustatory delight.

Fibonacci Baklava

Pastry:

1½ lbs. Filo dough (found in refrigerator cases at supermarkets)
½ cup sugar
1 lb. butter, melted
3 cups walnuts, finely chopped (may be done in a blender)
1 tsp. cinnamon
½ lb. almonds, finely chopped
2 zweiback crackers, ground
1 tbs. water

Mix together the cinnamon, sugar, zweiback and chopped nuts.

Brush bottom of 9" x 13" pan with melted butter.
Place (2) filo sheets on the bottom of pan. Brush with melted butter. Add (3) more filo sheets. Layer with some of the nut mixture. Follow with (3) buttered filo sheets alternating with nut mixture (5) more times or until all but (8) filo sheets are used. Finally add the last (8) buttered sheets. Sprinkle with water.

Cut into diamond shapes. Bake at 400 degrees for ¼ hour -- 350 degrees for ½ hour and 300 degrees for one hour, until golden. Place pan on wire rack to cool.

Make syrup from directions given below while the pastry bakes.

Honey Syrup

2 whole cloves
1 cup honey
1 tbs. brandy
1 small lemon
1 cup sugar
1 cup water
1 piece of cinnamon stick (2")

Remove lemon rind. Squeeze out 1½ tsp. of juice for reserve. Place lemon rind, sugar, water, cinnamon stick, cloves in a saucepan. Bring to boil. Reduce heat and cook 25 minutes without stirring. It will by syrupy. Stir in honey. Stir in reserve lemon juice and brandy.

Pour cooled syrup over cooled pastry. Let stand overnight and serve.

This recipe was invented by Sharon Maniates especially for this chapter.

Fibonacci Fudge

5 cups granulated sugar
13 ounces evaporated milk
1 stick of butter
13 ounces marshmallow (fluff type)
1 teaspoon salt
1 teaspoon vanilla
2 cups walnuts
2 large packages semi-sweet chocolate bits

Combine first 5 ingredients in large pan. Stir over medium heat until well blended. Boil for 5 minutes at medium-high heat. Over-cook rather than under-cook. Remove from heat and stir in last 3 ingredients until chocolate is melted and well blended. Pour into 2 buttered 8" x 8" pans. Allow to cool and cut into squares. Makes approximately 5 pounds.

Notice that there are 8 ingredients combined according to directions containing 8 sentences.

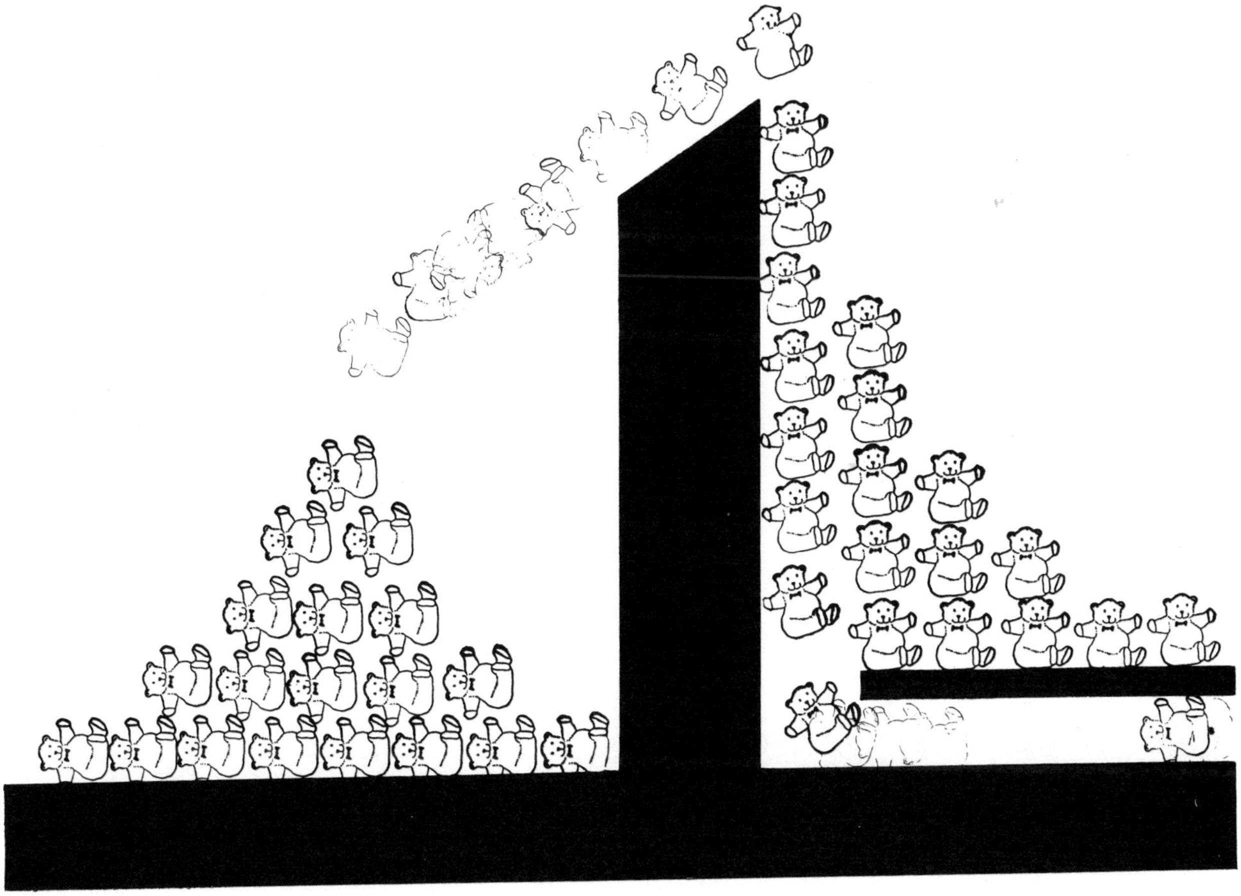

Peter Carey

Student Work
Chapter 5, project 3

PROBLEM 1

Figure the divergency constant for 5 different house plants. Do a table that gives the name of the plant together with its divergency constant. You may include drawings or photographs.

PROBLEM 2

Draw 3 line segments at least 8 inches long. Divide each one palindromically into Fibonacci lengths:

a. starting at the midpoint,

b. starting at the endpoints, and

c. starting at the two points that divide the segment into thirds. (See Construction 8). You will notice an irregularity at the midpoint in this case.

PROBLEM 3

Using cardboard, wood or plastic, devise a Fibonacci ruler to use for design purposes. Consider placement of initial term and size of unit for maximum effectiveness.

PROJECT **1**

Divide sides of one of the following shapes into lengths corresponding to palindromic sequences of Fibonacci numbers. Create a two or three dimensional object.

a. Pentagon

b. Any of the Dynamic Rectangles or Parallelograms

c. Any of the Special Triangles

PROJECT **2**

Using the concept of the leaf arrangement in a plant, create an abstract sculpture.

PROJECT **3**

Use a combination of Fibonacci numbers as the underlying structure to create a two or three dimensional work.

Walter Pasczko

Student Work
Chapter 5, project 1

Further Reading

Bicknell, Marjorie and Verner E. Hoggatt, Jr. Ed. A Primer for the Fibonacci Numbers. San Jose: The Fibonacci Association, 1973.

Cook, Theodore Andrea. The Curves of Life. New York: Dover Publications, Inc., 1979.

Huntley, H.E. The Divine Proportion. New York: Dover Publications, Inc., 1970.

Jacobs, Harold R. Mathematics A Human Endeavor. New York: W.H. Freeman and Company, 1982.

Stokes, William. Notable Numbers. Palo Alto, CA: William Stokes, 1974.

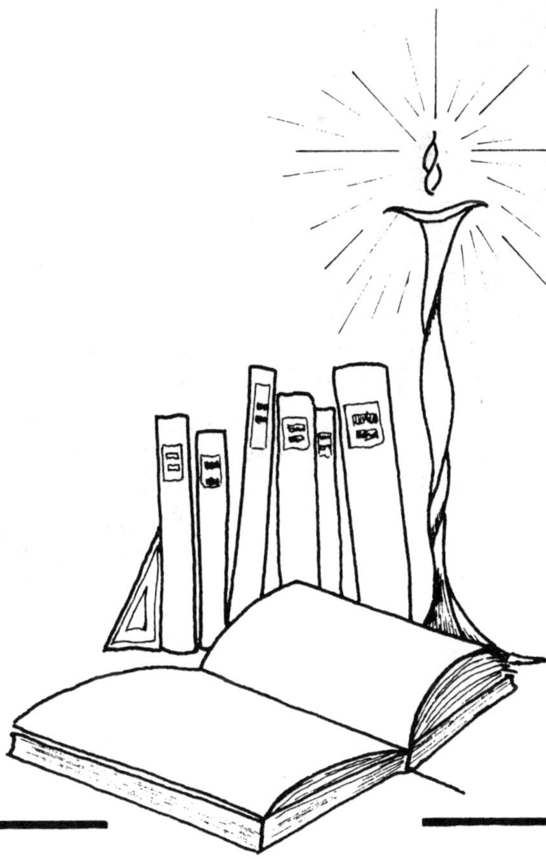

Spirals

6

Viruses and vines, webs and whelks, galaxies and gastropods share a versatile form that provides a spatial problem with a solution that is equally suitable for the microscopic as well as the macrocosmic. The problem is one of efficient growth in terms of energy, materials and expansion within space. The solution is the spiral which is the form that records the path of spinning material dispersing from a central point in either a right-hand or left-hand direction. As it moves outward, the spiral uniformly fills space in a perfectly regular manner as it curves around itself. It is capable of infinite expansion from small beginnings. According to Theodore Cook in his book, The Curves of Life, "There is little doubt that spirality is becoming more and more widely recognized as one of the great cosmic laws, and as being distantly hinted at even by the antipodal disturbances of the sun."

The concept spiral as well as the other patterns presented in earlier chapters are constructs of the human mind which act as explanations of phenomena after the fact of the natural occurrence. The spiral in Nature is never mathematically exact since there are so many conditions which would alter "mathematical perfection"; however, the concept of "spiral" allows us to find commonality of form in seemingly unrelated things. In Art, the spiral has long functioned as both a symbol and an ornament and has been considered to have magical properties. The Egyptians, the Greeks, the Africans, the Japanese and Chinese, and especially the Celts all have made use of this form.

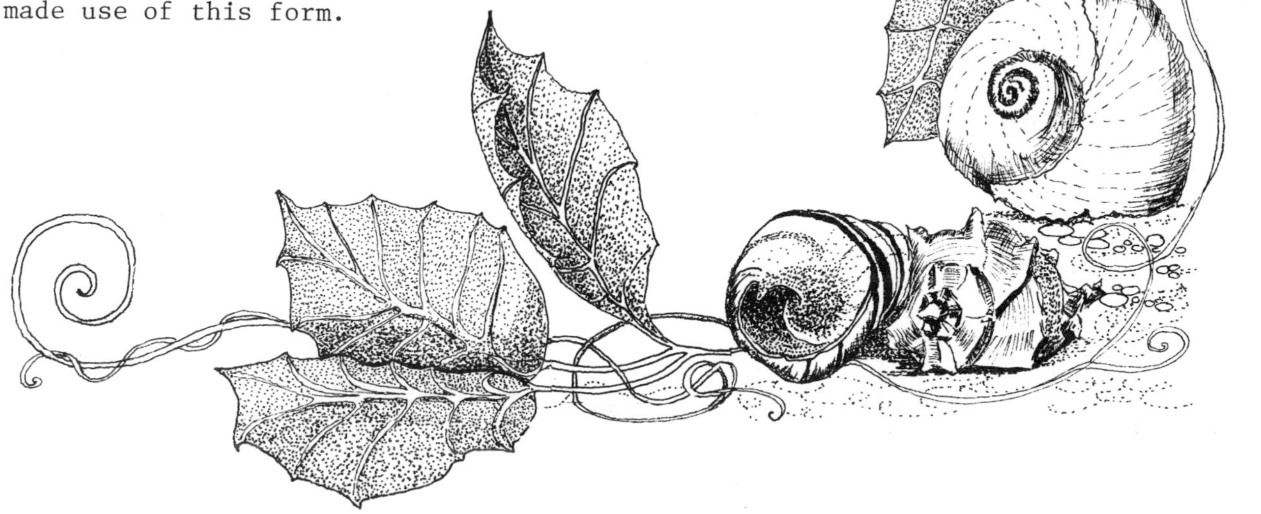

Drawing after George Bain
Pictish School of Celtic Art

As an abstract idea, the spiral has specific mathematical properties. It is a curve traced by a point continually moving around and away from a fixed point. This fixed point is called the pole and the manner in which the point moves away from it determines the particular type of spiral.

THE ARCHIMEDEAN SPIRAL

The Greek philosopher and mathematician Archimedes lived in the third century B.C. He attended the university at Alexandria where Euclid had taught, but returned home to Syracuse where he spent his time writing, solving problems of geometry, and inventing mechanical devices which were used in defense of the city when under attack by the Romans. Despite the cleverness of his designs, the Romans eventually took both the city and his life.

Although the political leaders sought him out for his solutions to practical problems, primarily military, he was not himself a political creature, but rather a thinker. Three works of Archimedes on plane geometry have survived the intervening centuries, one of which is on spirals.

The Archimedean spiral is the simplest spiral. It is the one exemplified by the grooves in a phonograph record (or for the modern generation, the coil of a spool of audio or video tape), or a coiled rope. It depends in the real world upon the uniform thickness of the material coiled as well as the uniform tightness of the coil. The diagram on the following page visually explains this curve.

From the Book of Kells

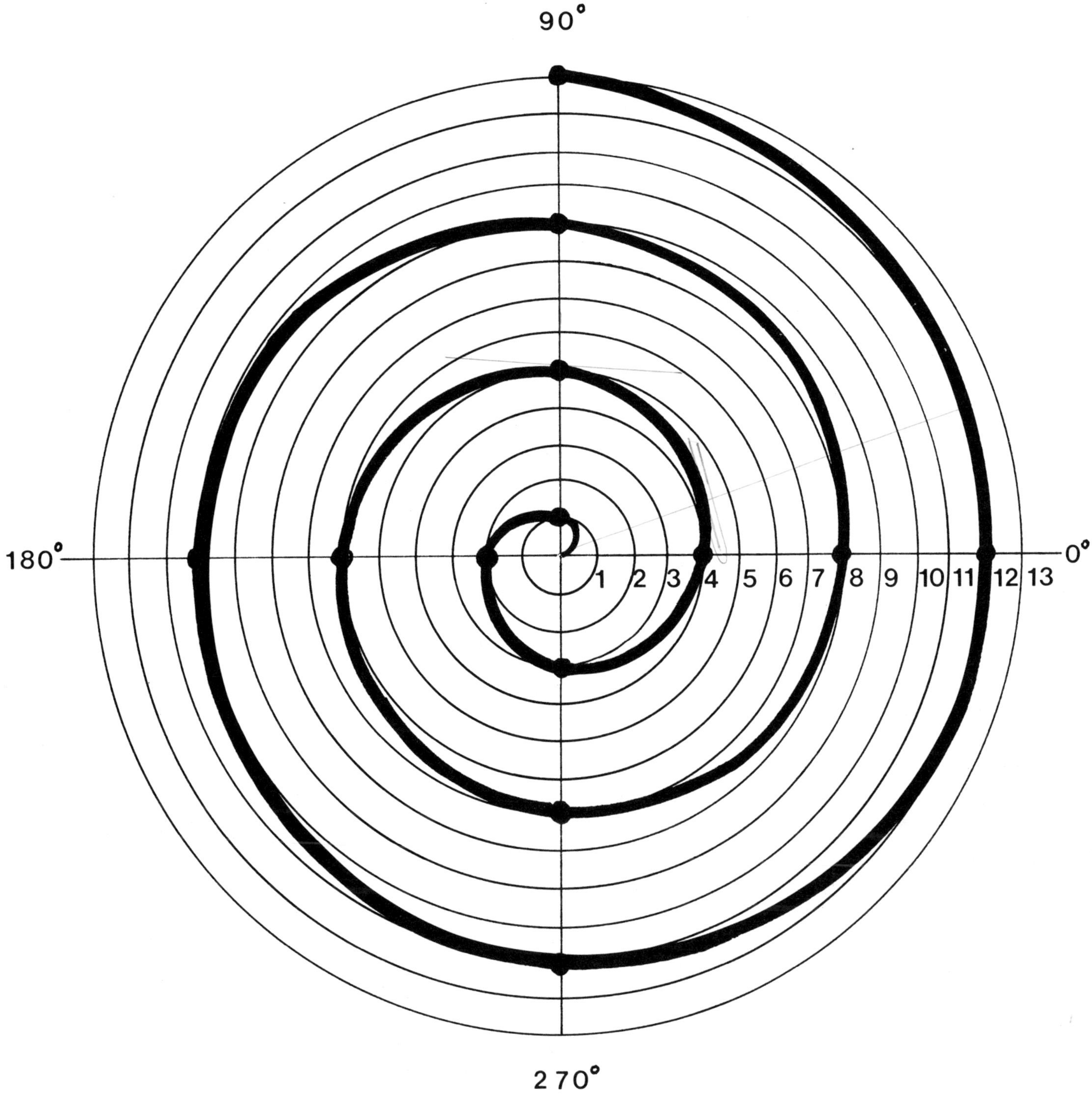

The concentric circles provide a framework for the spiral which moves at a uniform rate away from the pole. In this case, for each turn of 90 degrees the spiral moves one unit further away from the pole. Any spiral which lengthens at a constant rate per constant turn is an Archimedean spiral. When these two constants are determined, construction of the spiral can be accomplished fairly easily by the following method. Note that this "construction" differs from the previous ones in that the required tools are other than compass and straightedge.

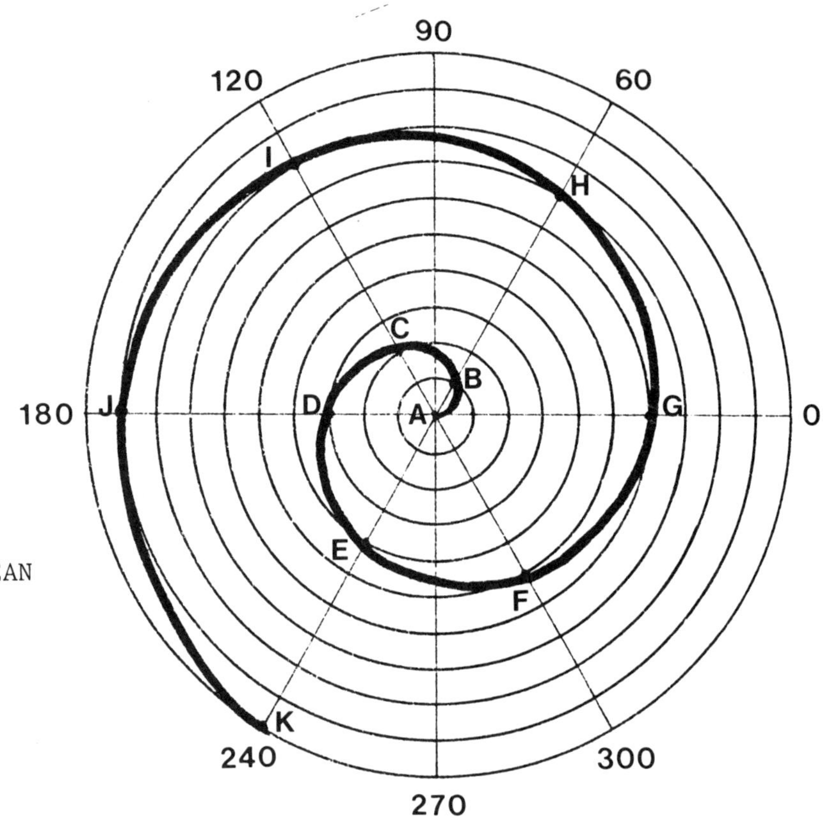

CONSTRUCT AN ARCHIMEDEAN SPIRAL

37

Given the rate at which the spiral moves away from the pole per constant turn, in this case 3/16" per turn of 60°, construct an Archimedean spiral.

1. Use a sheet of commercial polar graph paper or construct concentric circles such that the radii increase by the given constant rate.
2. Locate the pole and label it A.
3. Locate point B at the intersection of the first circle and the <u>radius vector</u> corresponding to 60°.
4. Locate point C at the intersection of the next larger circle and the radius vector corresponding to 120° (60° + 60°).
5. Continue the process, each time moving to the next larger circle for each subsequent turn of 60° to get points D,E,F, G,H,I,J,K.
6. Join the points in order with a smooth curve (a French curve is a helpful tool here).

Now the curve is an Archimedean spiral.

These diagrams exemplify three different Archimedean spirals, which can be constructed on a polar grid.

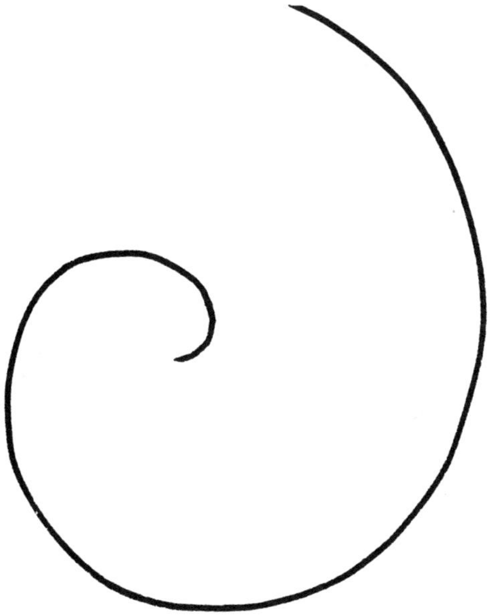

rate of $\frac{1}{4}$ unit per turn of 45°

rate of $\frac{1}{16}$ unit per turn of 30°

rate of $\frac{1}{6}$ unit per turn of 30°

A mathematical peculiarity can be observed if a radius vector is drawn in an Archimedean spiral. The spiral intersects the radius vector at equally spaced intervals, yet the angle formed by the spiral and the radius vector slowly changes, gradually approaching a right angle as the distance from the pole increases.

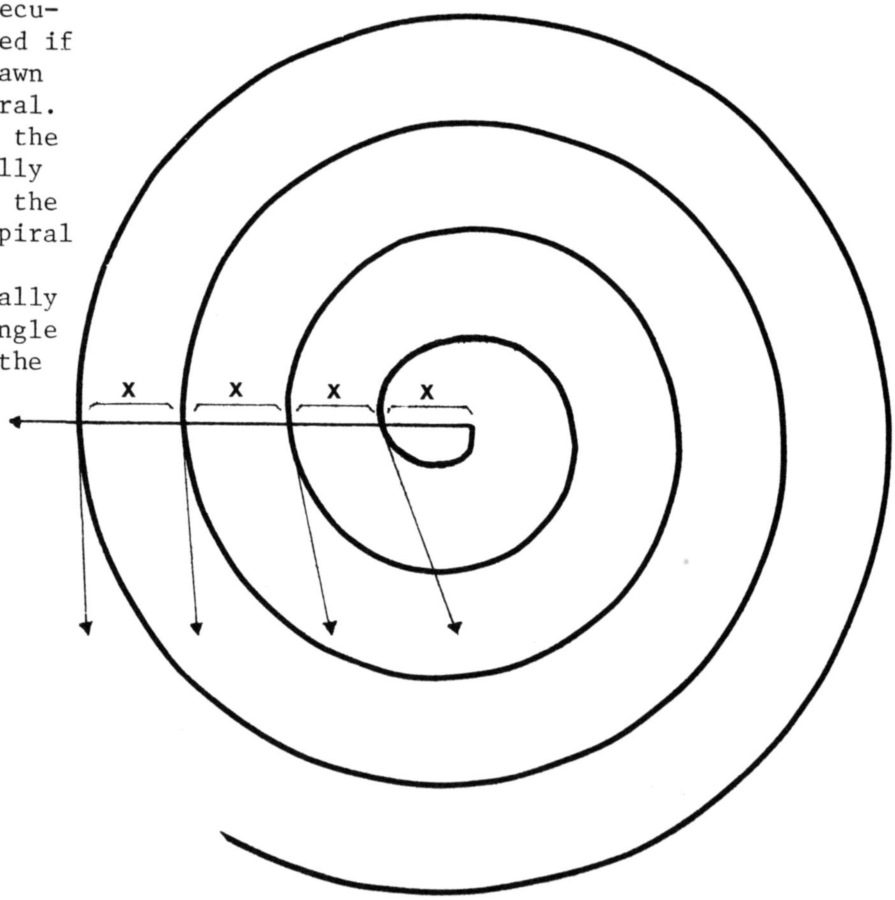

The distances from the pole of the successive whorls of the spiral along a radius vector are in arithmetic sequence: x, 2x, 3x, 4x,...

In Nature, some spiders use this spiral to spin their webs. They first lay out a framework of spokes which correspond to the radius vectors. They then move outward from the center to create a spiral by guiding themselves by the previous turn so that the distance between the turns is everywhere the same.

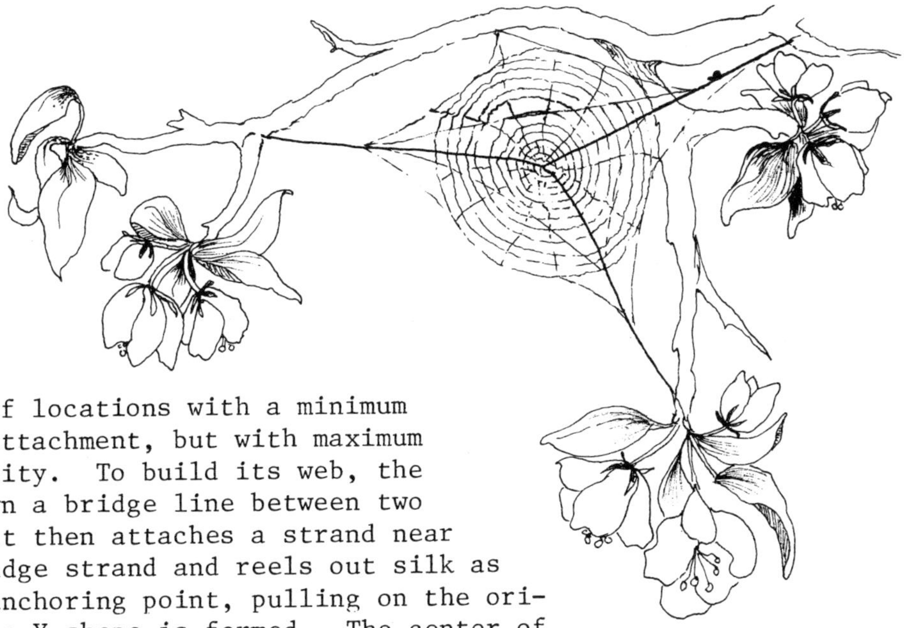

The Orb-Web Spider is particularly clever at conserving energy and material while covering the maximum area. Because it uses a triangular structure, its web can be placed in a variety of locations with a minimum number of points of attachment, but with maximum strength and flexibility. To build its web, the spider first lays down a bridge line between two vertical supports. It then attaches a strand near the middle of the bridge strand and reels out silk as it drops to another anchoring point, pulling on the original strand so that a Y shape is formed. The center of the Y becomes the center of the web so that there are three principal radii upon which the spider builds a spiral outwards.

Archimedean spirals also occur in the plant kingdom. They are seen in the whorl of young tomato plants, in the fiddlehead fern, and in the sago palm.

THE LOGARITHMIC SPIRAL

By far the most prevalent, and many think the most beautiful, spiral found in the natural world is the logarithmic spiral. It is the one found in the horns and teeth of animals; some spider webs; the seed cases and stems of plants; the beaks of birds; the shells of sea creatures; the umbilical cord, the cochlea of the ear, and the great bones in man; the galaxies of the Universe.

This is a spiral of growth that exemplifies Nature's laws of conservation. It fills space economically and regularly, affords strength with a minimum of materials, but has the exceptional property that while expanding it alters its size but never its shape. Because of the aforementioned properties, the spiral provides a perfect solution to the problems of growth from a single point, growth by accretion, and growth under resistance, all factors which divert growth from a direct path.

Sea shells present the clearest examples of logarithmic spirals, but it is the shell of the chambered Nautilus that comes closest to mathematical perfection. This 500 million year old creature can withstand pressure of one/half ton per square inch because of its chambered construction. It begins life with a single chamber half as large as a pea and continues to add successively larger chambers accumulating as many as thirty. The animal lives only in the outer chamber. The preceding ones are first filled with gas and then sealed off to provide buoyancy and allow for locomotion through its environment of depths of several thousand feet. The animal's shell maintains a spiral shape, each new chamber being the gnomon for the entire form, because it grows by accretion wherein secreted materials harden and can no longer change. Because these materials accumulate faster on the outside edge than on the inside, the shell curls in upon itself.

The shells of the adult Nautilus can be likened to a cone that has been rolled up wherein the earliest and oldest chambers are the smallest. A great many shells are essentially tubes with the differences between species resulting from the differing rates of expansion of the coiled tube.

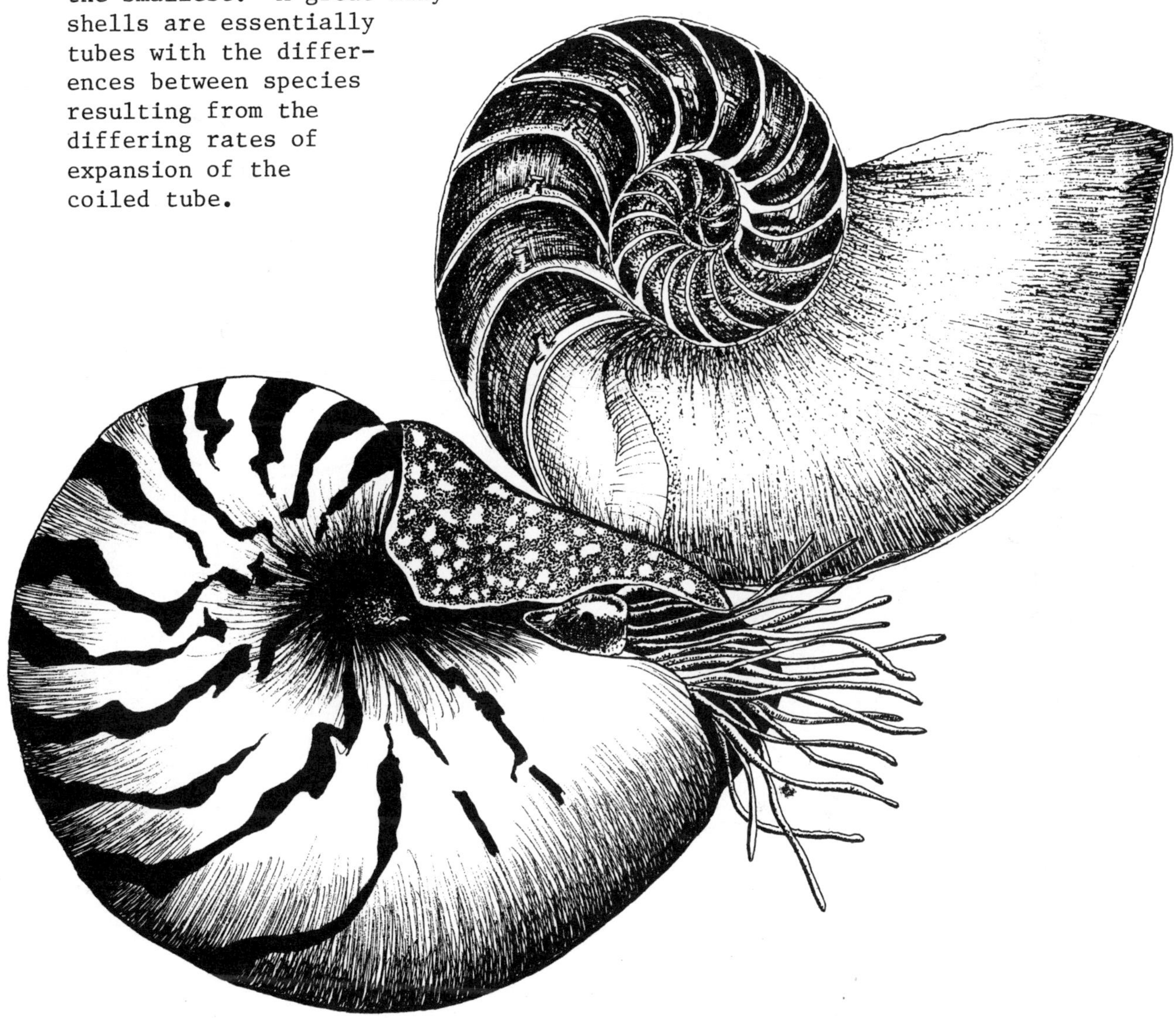

If we examine a cross section of this shell, we can see that it differs from the Archimedean Spiral in two ways. First, the spiral maintains a constant angle of intersection with the radius vectors, which prompted Descartes, in the 17th century, to name it the <u>equiangular spiral</u>.

In the 18th century, it was named the <u>proportional spiral</u> by Halley because of its other special property. That is, every logarithmic spiral is characterized by a particular geometric sequence, and the distances from the pole of the successive whorls along a radius vector correspond to the terms of that sequence. Thus the intervals increase proportionally further from the pole.

The 19th century mathematician Bernouilli called it both the Spira Mirabilis (miraculous spiral) and the logarithmic spiral; it is by the latter name that it is commonly known today.

$$\frac{AB}{BC} = \frac{BC}{CD} = \frac{CD}{DE}$$

$$\angle A \cong \angle B \cong \angle C \cong \angle D \cong \angle E$$

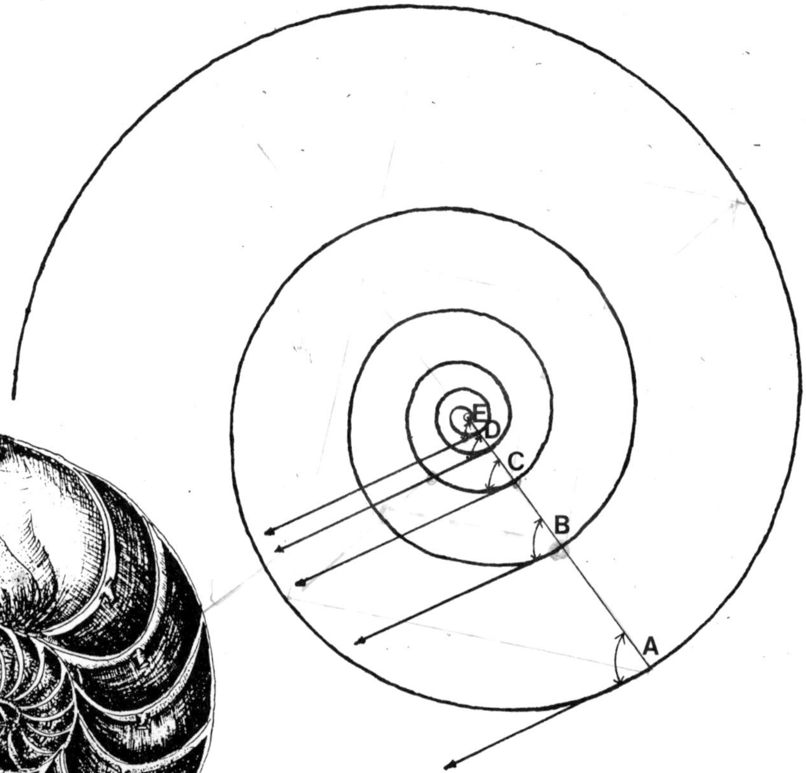

Every logarithmic spiral is related to a characteristic rectangle. Within each member of the family of Dynamic Rectangles can be drawn a straight line or angular spiral using the diagonals of the rectangle and its reciprocal as the armature. Segments parallel to the sides are drawn between the intersecting diagonals.

The following diagram illustrates this concept with some of the Dynamic Rectangles and extends the idea to encompass Dynamic Parallelograms and straight line spiral designs.

ANGULAR SPIRALS

IN DYNAMIC RECTANGLES

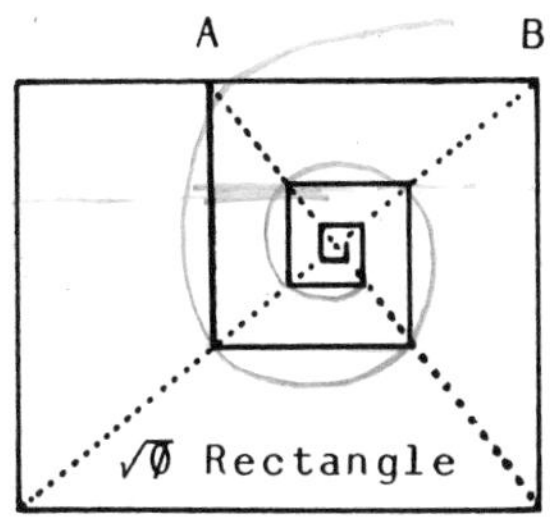

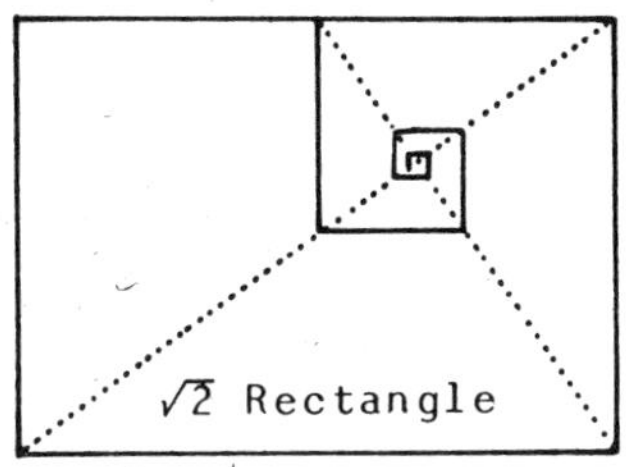

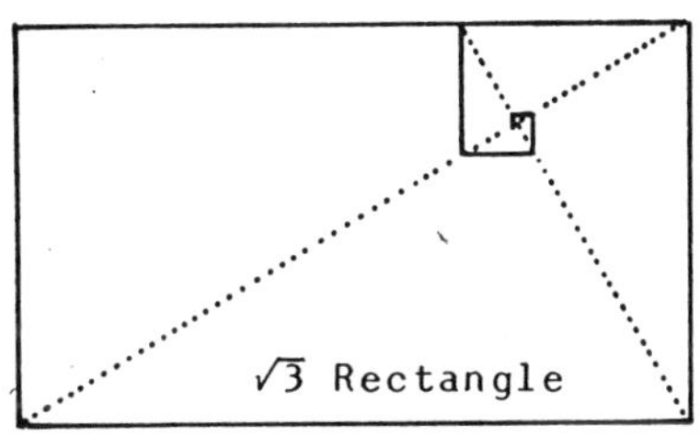

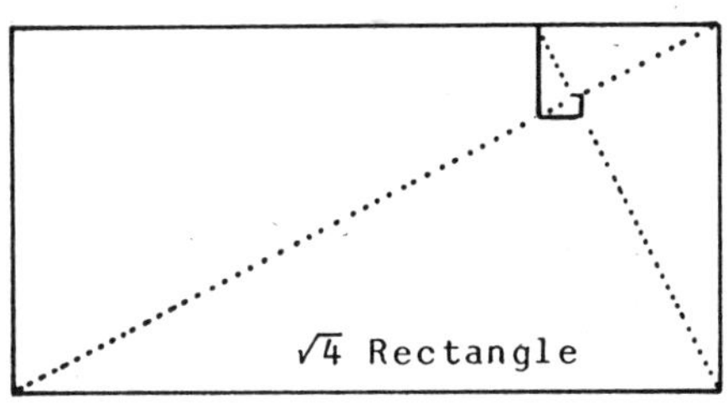

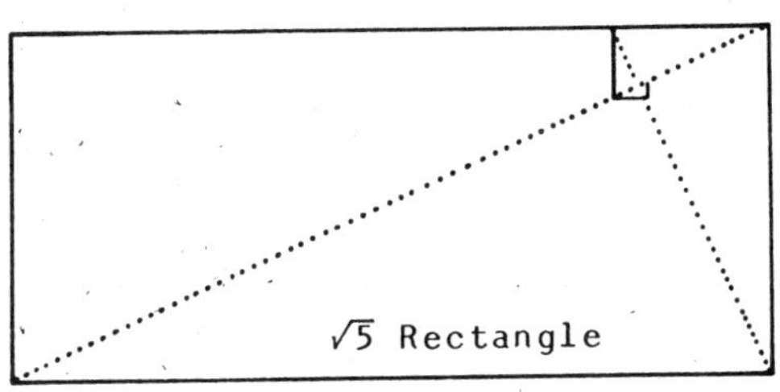

IN DYNAMIC PARALLELOGRAMS

$\sqrt{\phi}$ Parallelogram

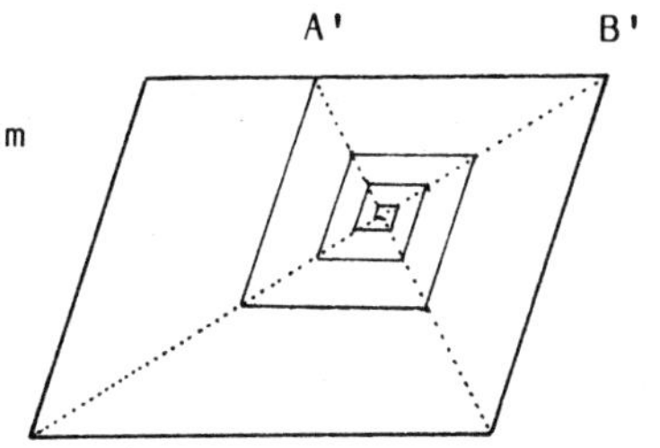

$\sqrt{2}$ Parallelogram

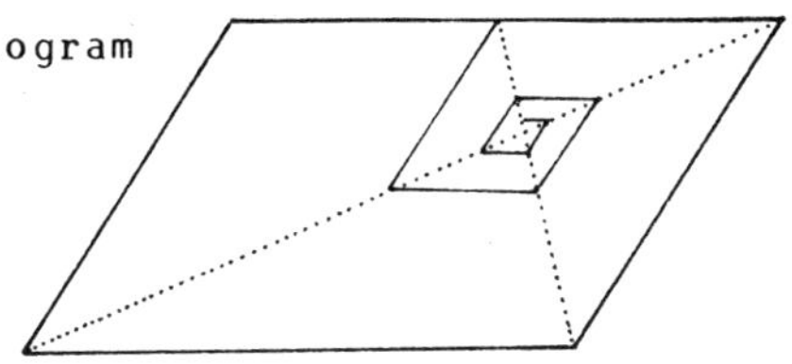

$\sqrt{3}$ Parallelogram

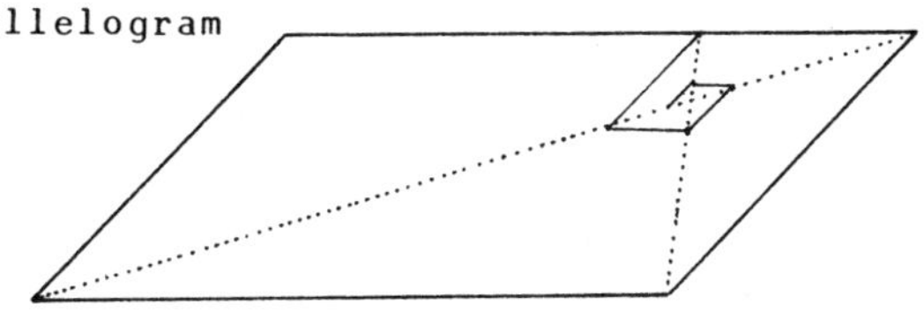

$\sqrt{4}$ Parallelogram

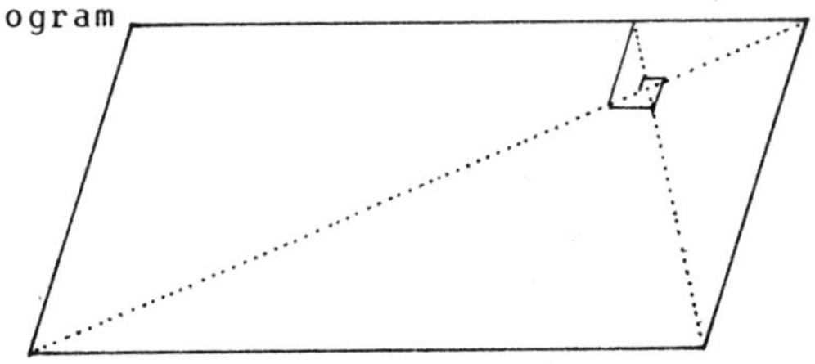

$\sqrt{5}$ Parallelogram

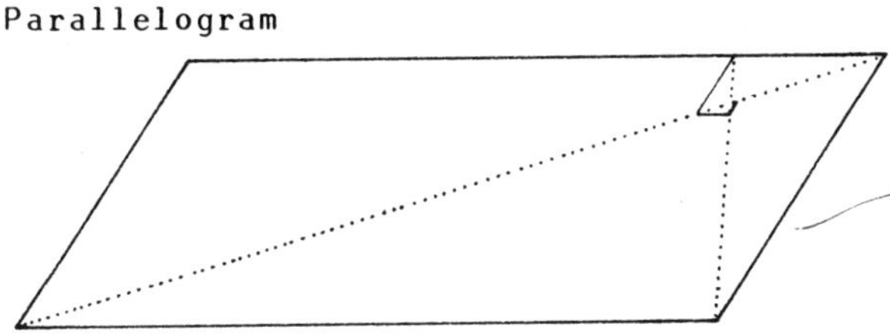

Note: The Dynamic Parallelograms may be "tipped" at <u>any</u> angle. Their reciprocals are gotten by using the same points of intersection as the parent rectangle. eg. AB = A'B'

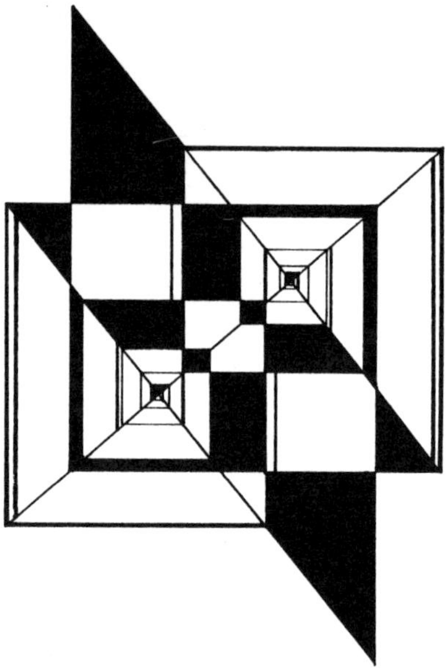

$\sqrt{\phi}$

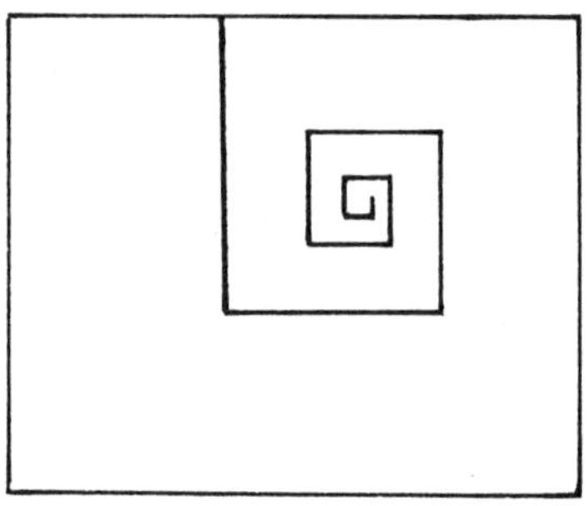

You can play with these. Enlarge them to a size that is comfortable for you to work on.

$\sqrt{2}$

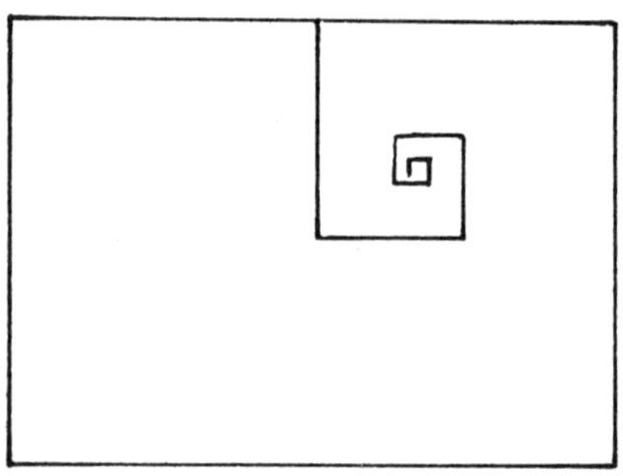

$\sqrt{4}$

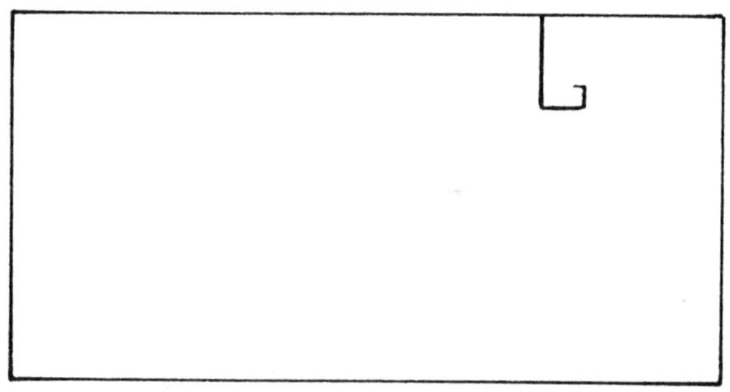

$\sqrt{3}$

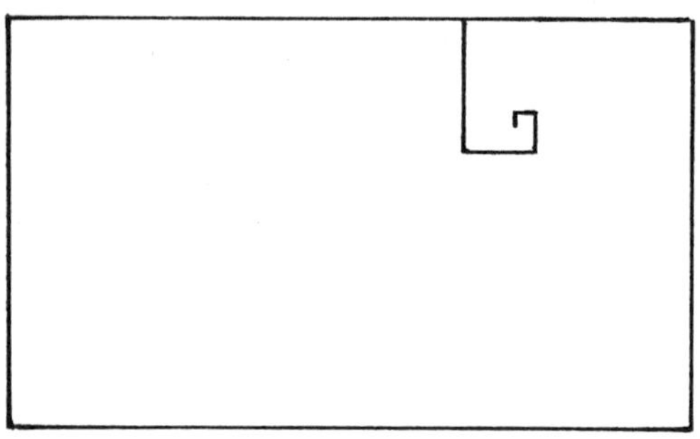

$\sqrt{5}$

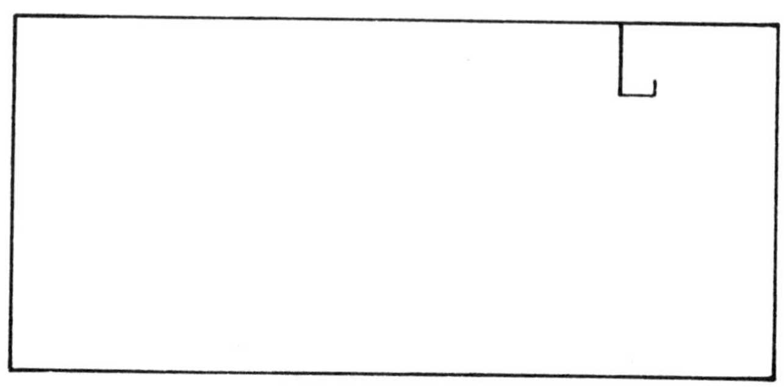

THE GREEK TECHNIQUE

Notice in the previous diagram that the greater the ratio of length to width, the fewer the whorls in the rectangle. That is, as the length increases in relation to a fixed width, the spiral moves away from the pole much more rapidly and therefore is less visible within the rectangle. The Classical Greeks were aware of this fact when creating spiral forms for the capitals of their temple columns. They experimented with different sized rectangles but found that the one that worked best for them was a 1.191 rectangle, which is one whose length lies between that of a square (1.000) and a $\sqrt{\emptyset}$ Rectangle (1.272). It seemed to allow for the right number of whorls in relation to the column it capped.

Constructing a logarithmic spiral given just the pole and radius vectors is not so simple as it is with the Archimedean one. However, in his book, Patterns and Design with Dynamic Symmetry, Edward B. Edwards describes in detail the method for creating logarithmic spirals in the Dynamic Rectangles using French curves and templates. We refer the reader to this book for more detailed information.

The method used by the Greeks is by far less taxing mathematically, and offers a product quite sufficient for most design purposes. We shall examine the steps involved here. This method may be used with any rectangle but we will demonstrate the procedure with the 1.191 rectangle which was the first choice of the Greeks.

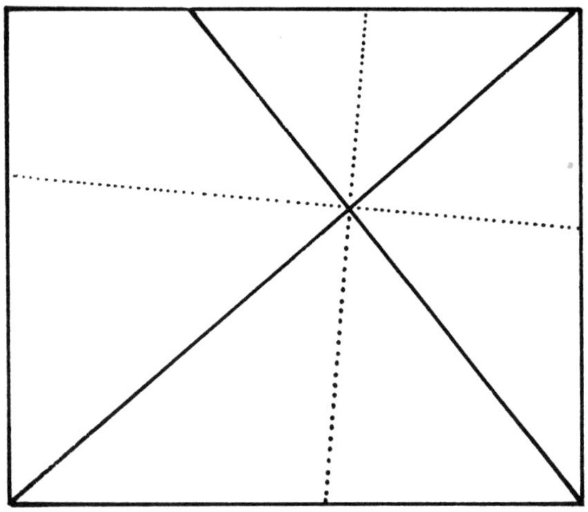

fig. 1

a. The main diagonal and the diagonal of the reciprocal are drawn (shown in solid lines).

The angles formed by their intersection are bisected (shown in dotted lines).

Note that the angle bisectors are <u>not</u> parallel to the sides.

b. Segments are drawn from the endpoints of the angle bisectors to the diagonals parallel to the sides of the rectangle.

Three squares are formed which provide the framework for the first part of the spiral. Arcs are drawn through each square using the innermost vertex of each as the point upon which the tip of the compass is placed.

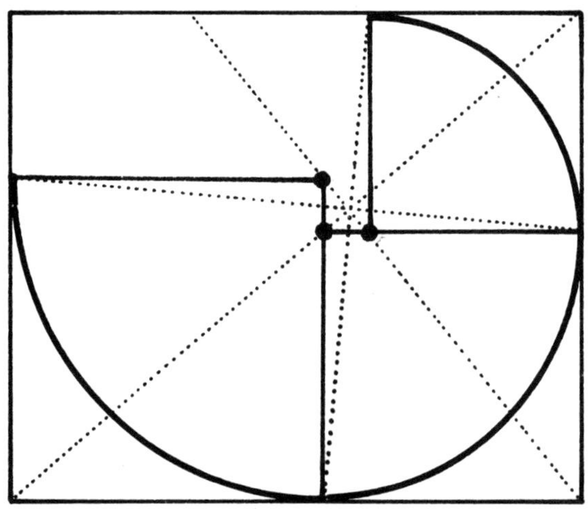

fig. 2

fig. 3

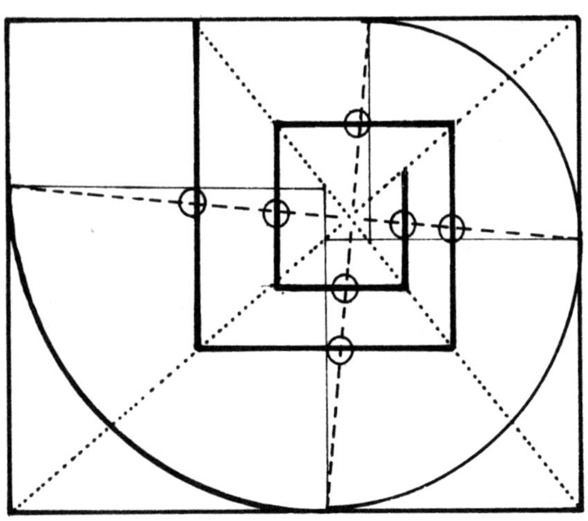

c. The straight-line or angular spiral is drawn as described previously. The spiral curve will be <u>tangent</u> to the sides of the angular spiral at the points where it intersects the angle bisectors drawn in part a. These points are circled in fig. 3.

d. To obtain the next series of squares, segments are drawn parallel to the sides from the circled points to the diagonals, i.e. $\overline{BO}$, $\overline{CP}$, $\overline{DQ}$, $\overline{ER}$.

Again, arcs are drawn through the squares by placing the compass tip on the innermost vertex of each square, i.e. place tip on O to cut $\widehat{AB}$, on P to cut $\widehat{BC}$, on Q to cut $\widehat{CD}$, and on R to cut $\widehat{ED}$.

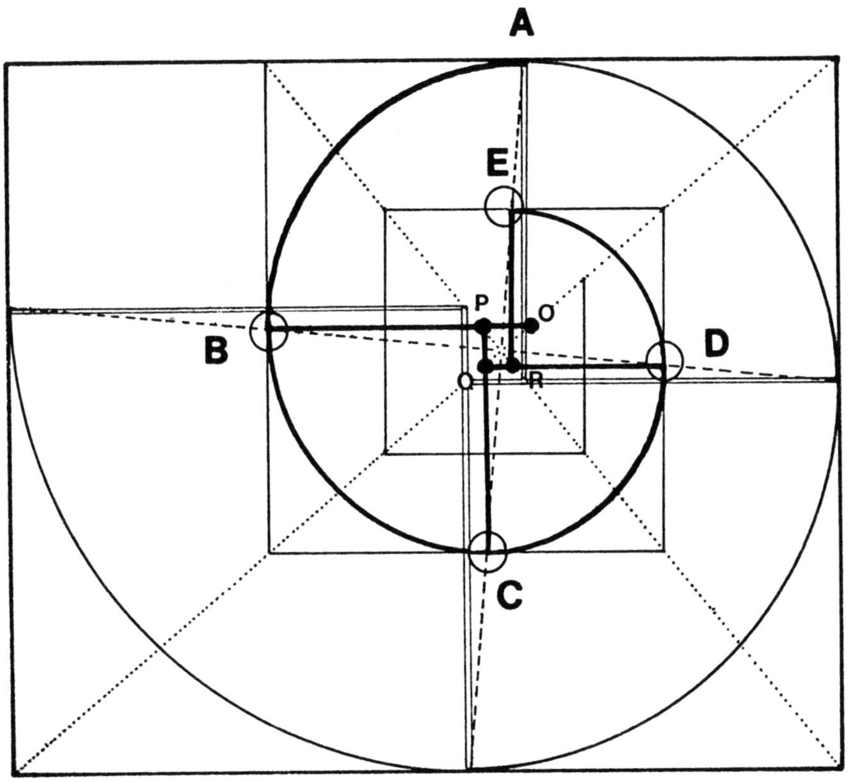

fig. 4

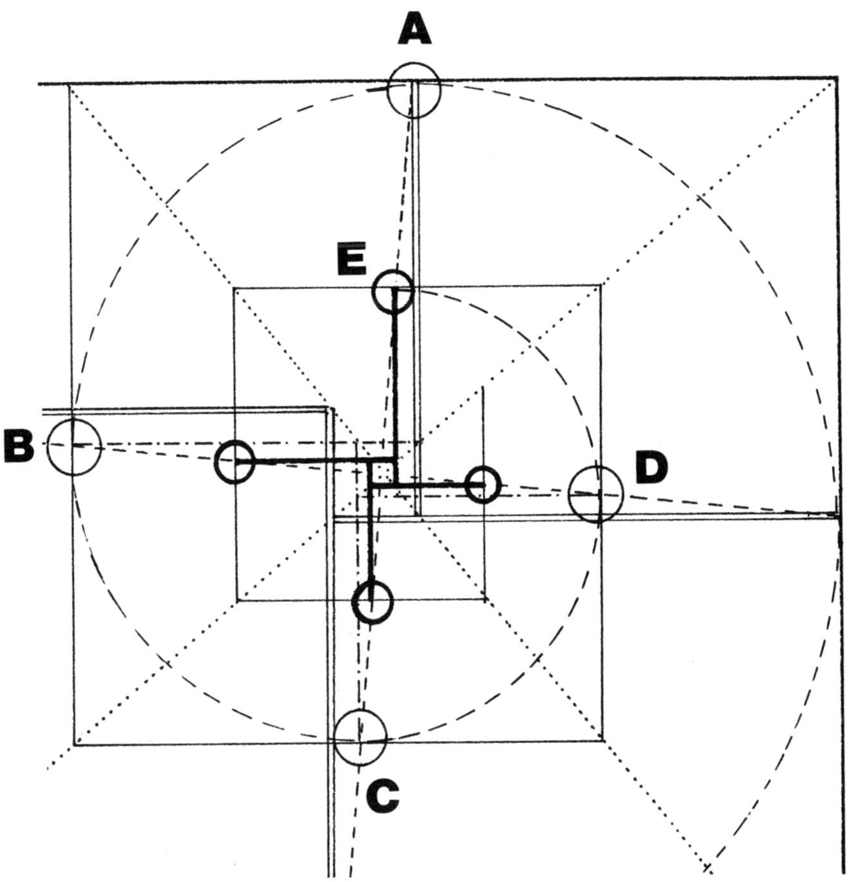

fig. 5

e. The next series of squares is formed precisely the same way as in part d. Again segments are drawn parallel to the sides from the circled points to the diagonals, and the compass tip is placed on those points on the diagonals to cut the quadrant arcs.

Fig. 5 shows the inner portion of the rectangular spiral, labeled as in fig. 4.

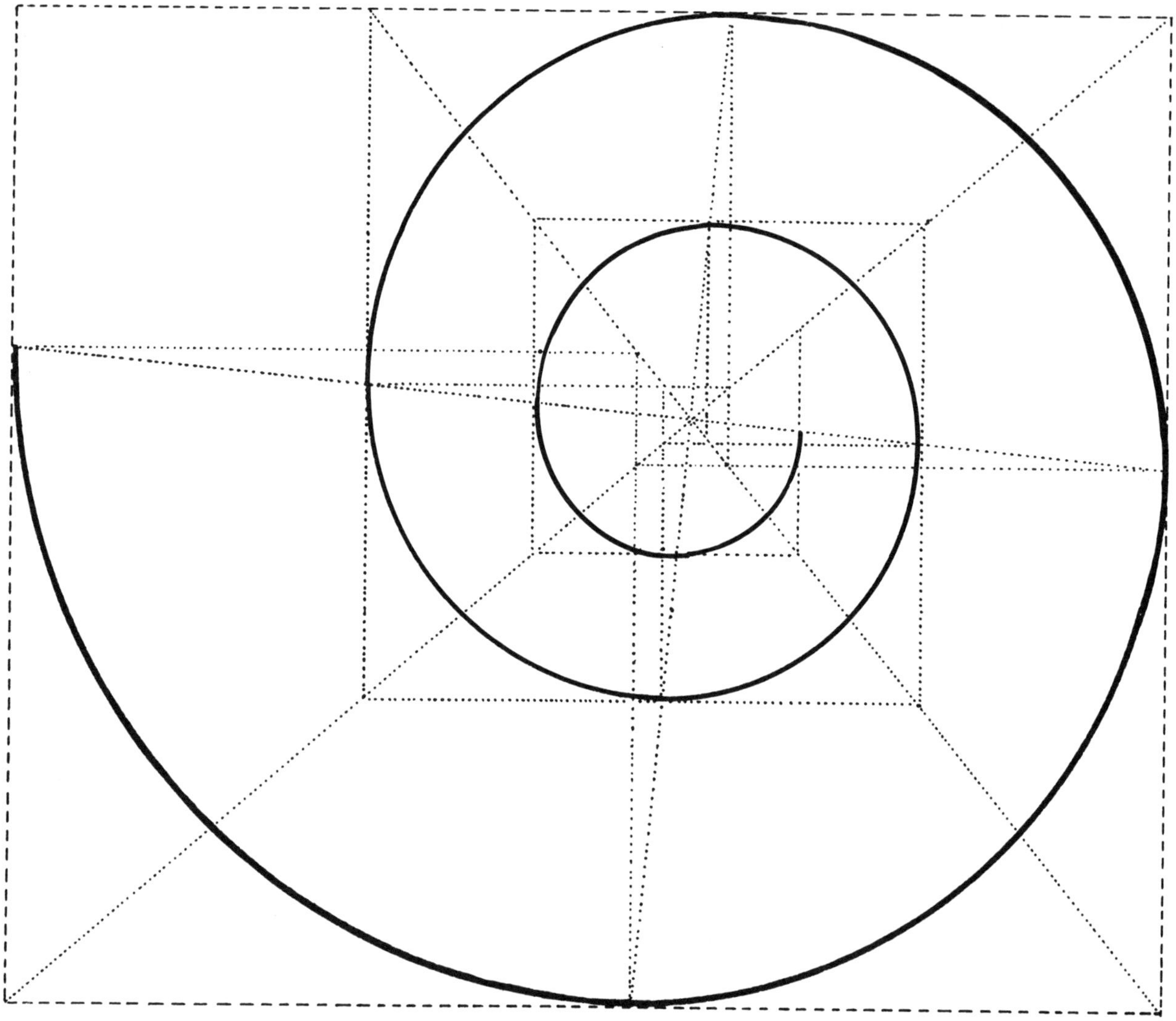

fig. 6

Fig. 6 shows the completed spiral.

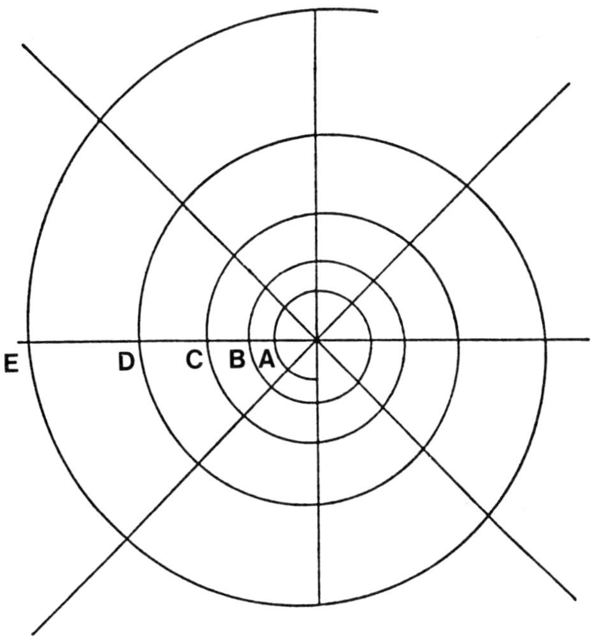

The spirals used for this purpose by the Greeks were but approximations of true logarithmic spirals due to their method of construction. The one that felt "right" for them may have done so because it very closely imitates a logarithmic spiral with harmonic proportions. This spiral is called the Phidian spiral, again named for the Greek sculptor, because the lengths of the segments cut by successive whorls of the curve along any radius vector are in Ø proportion.

$$\frac{ED}{DC} = \frac{DC}{CB} = \frac{CB}{BA} = \emptyset$$

Because of its inherent pattern of "whirling squares" the Golden Rectangle provides the framework for a logarithmic spiral which is quite easy to obtain.

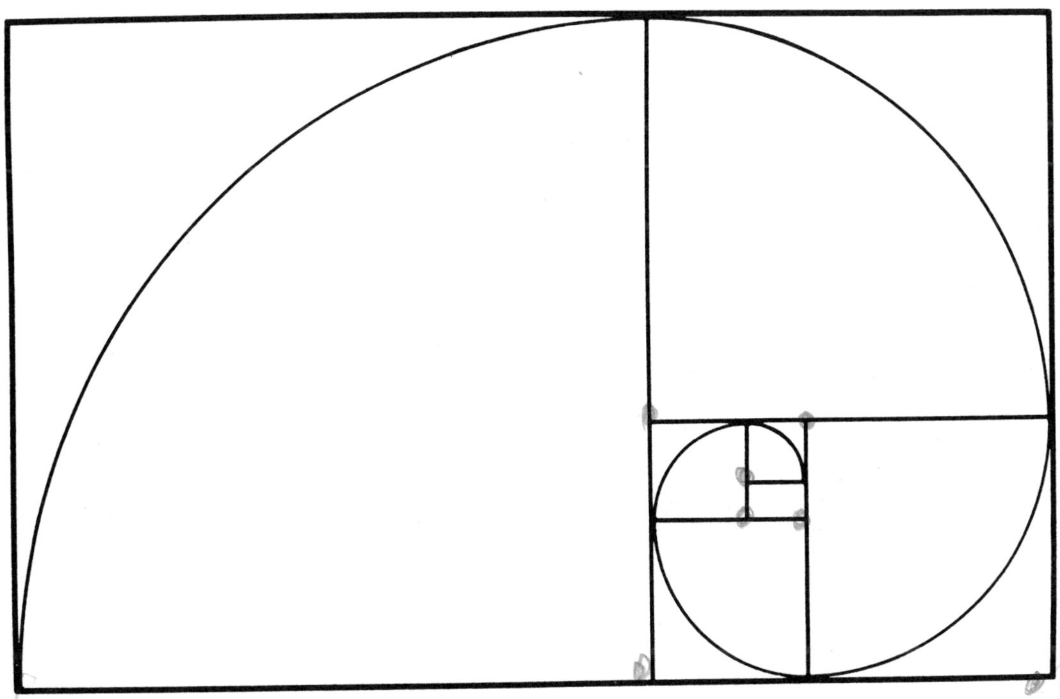

This particular spiral is called the Golden Spiral and the method for its construction follows.

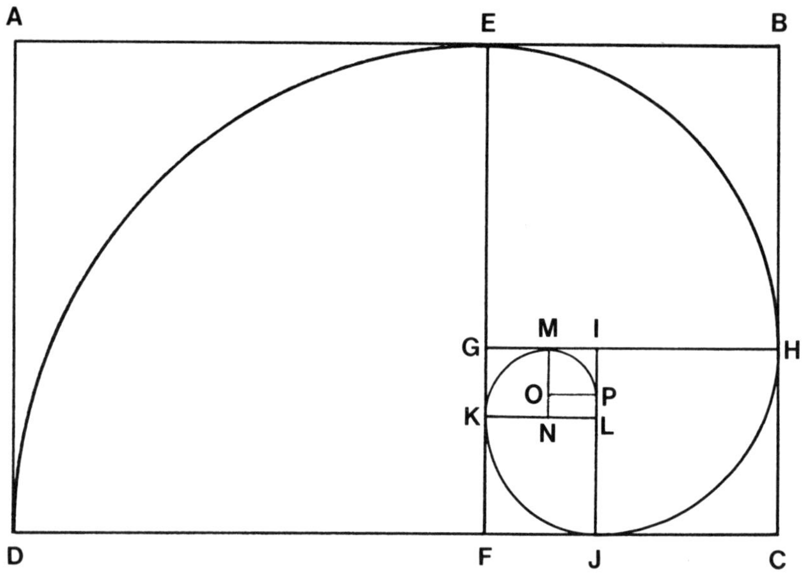

Given: ABCD a Golden Rectangle, construct a Golden Spiral.

CONSTRUCT A GOLDEN SPIRAL
GIVEN A GOLDEN RECTANGLE.

38

1. Open the compass to measure AD, place the metal tip on A and cut an arc on $\overline{AB}$.
2. Label the point of intersection E.
3. Without changing the setting, place the metal tip on D and cut an arc on $\overline{DC}$.
4. Label the point of intersection F.
5. Draw $\overline{EF}$.

Note: AEFD is a square.

Repeat steps 1-5 in succession to get:

6. Square BHGE from rectangle BCFE,
7. Square CJIH from rectangle CFGH,
8. Square FKLJ from rectangle FGIJ,
9. Square GMNK from rectangle GILK,
10. Square MIPO from rectangle MILN.

This is the pattern of "whirling squares".

11. Place the metal tip on O and draw $\overset{\frown}{PM}$.
12. Place the metal tip on N and draw $\overset{\frown}{MK}$.
13. Place the metal tip on L and draw $\overset{\frown}{KJ}$.
14. Place the metal tip on I and draw $\overset{\frown}{JH}$.
15. Place the metal tip on G and draw $\overset{\frown}{HE}$.
16. Place the metal tip on F and draw $\overset{\frown}{ED}$.

Now the spiral is a Golden Spiral.

Note: the spiral may be enlarged by increasing the size of the Golden Rectangle by the method described on pages 61 through 63 and continuing to cut the arc that joins the non-consecutive vertices of each new square.

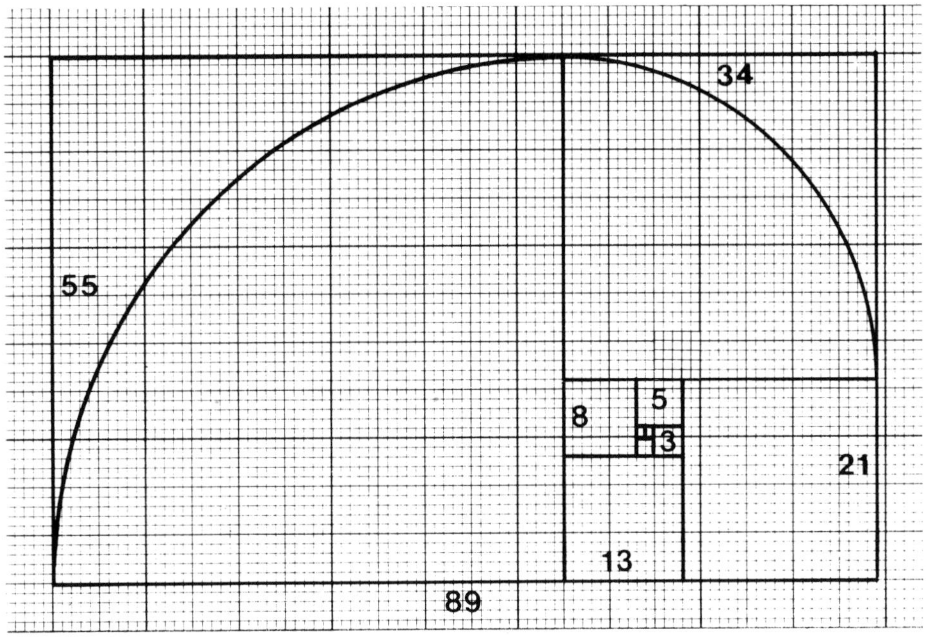

The Golden Rectangle and the Golden Spiral bear an intimate relationship to those Golden Numbers, the Fibonacci Sequence. We know from Chapter V that consecutive terms of the Fibonacci Sequence yield a ratio very close to Ø. If we begin with a rectangular grid and construct a rectangle whose length is 89 units and whose width is 55 units, we have a rectangle that closely approximates a Golden Rectangle. The length chosen for the sides are the eleventh and tenth terms of the Fibonacci Sequence (F_{11} and F_{10}). When the pattern of "Whirling Squares", as described in the previous construction, is established, the successively smaller squares have sides named by the Fibonacci number in reverse sequence.

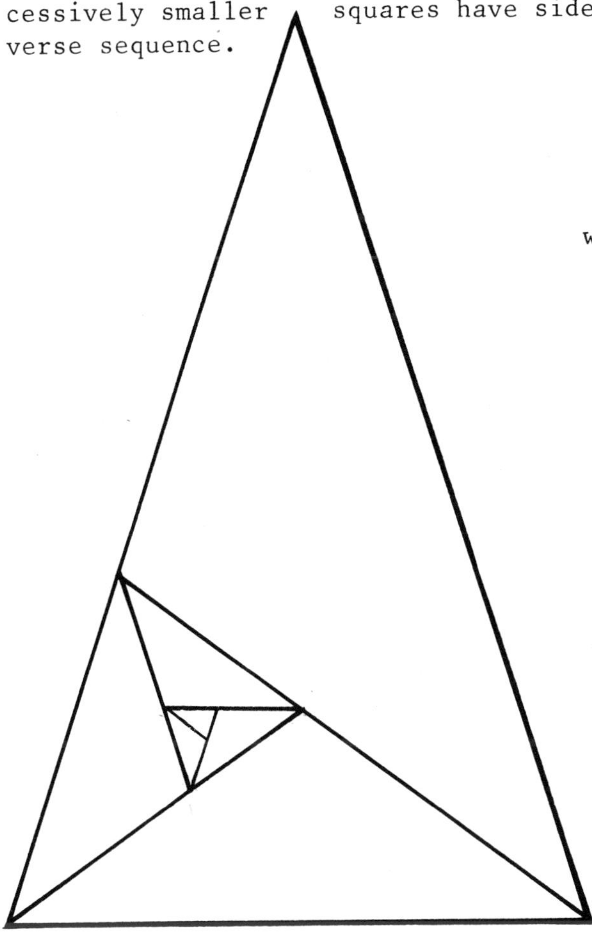

Rectangles are not the only figures which provide a framework for spirals. The Golden Triangle also affords the architecture for erecting a logarithmic spiral. You will recall from Chapter 4 that the Golden Triangle can be subdivided into a series of smaller and smaller Golden Triangles simply by bisecting successive base angles. Once this framework is established the related logarithmic spiral can be constructed as follows.

CONSTRUCT A LOGARITHMIC SPIRAL GIVEN A GOLDEN TRIANGLE.

39

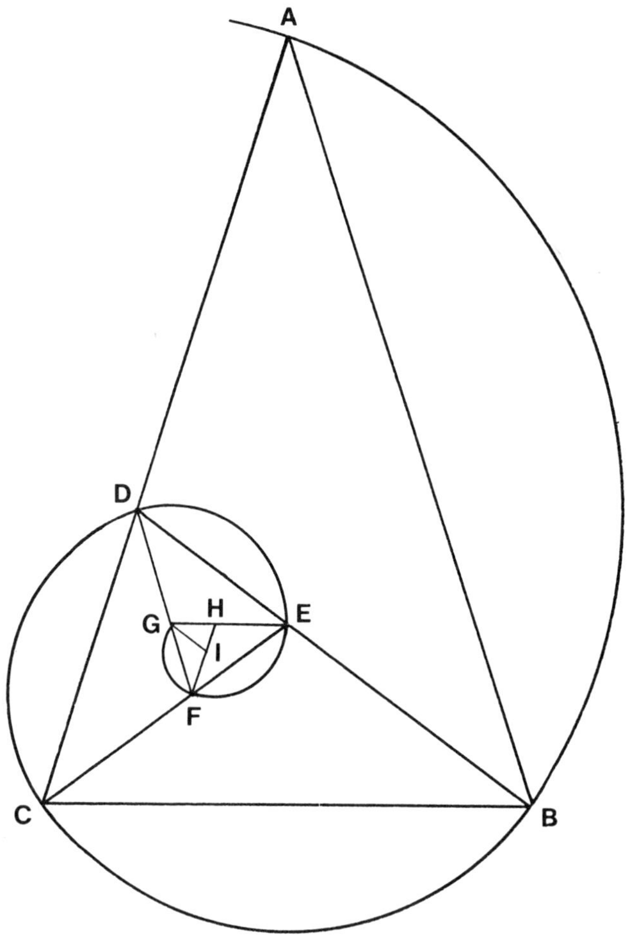

Given: △ABC, a Golden Triangle, construct a logarithmic spiral.

1. Subdivide the triangle by the method described on the previous page and label the points as shown.
2. Place the metal tip on D and draw $\overset{\frown}{AB}$.
3. Place the metal tip on E and draw $\overset{\frown}{CB}$.
4. Place the metal tip on F and draw $\overset{\frown}{CD}$.
5. Place the metal tip on G and draw $\overset{\frown}{DE}$.
6. Place the metal tip on H and draw $\overset{\frown}{EF}$.
7. Place the metal tip on I and draw $\overset{\frown}{FG}$.

Now the curve is a logarithmic spiral.

The Golden Triangle may be combined with other figures to form particularly striking designs. In the following example harmony abounds in Ø proportions with the use of two pentagons together with modified pentagrams formed by two rings of Golden Triangles. Note that the straight line spirals follow the lines of the same subdivision of the Golden Triangle which was used for the construction of the logarithmic spiral in Construction 39.

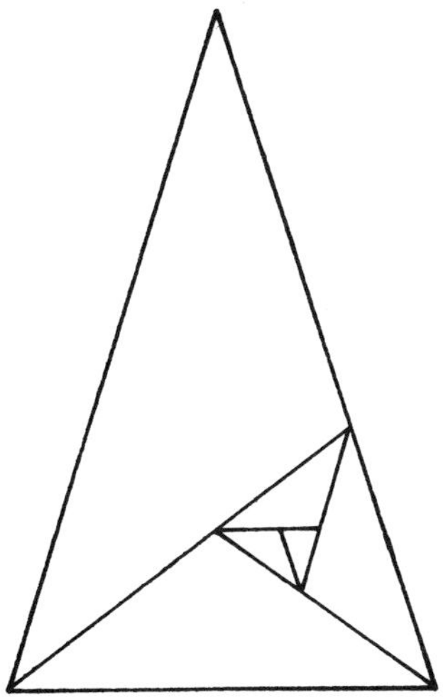

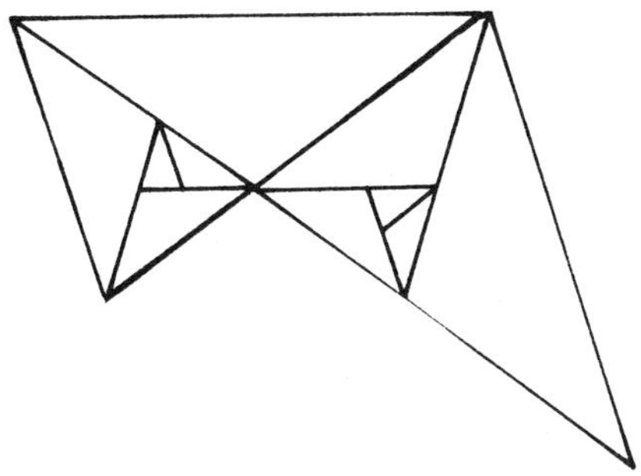

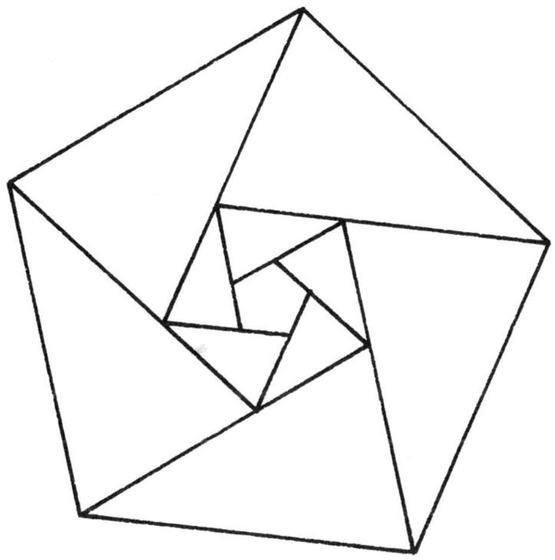

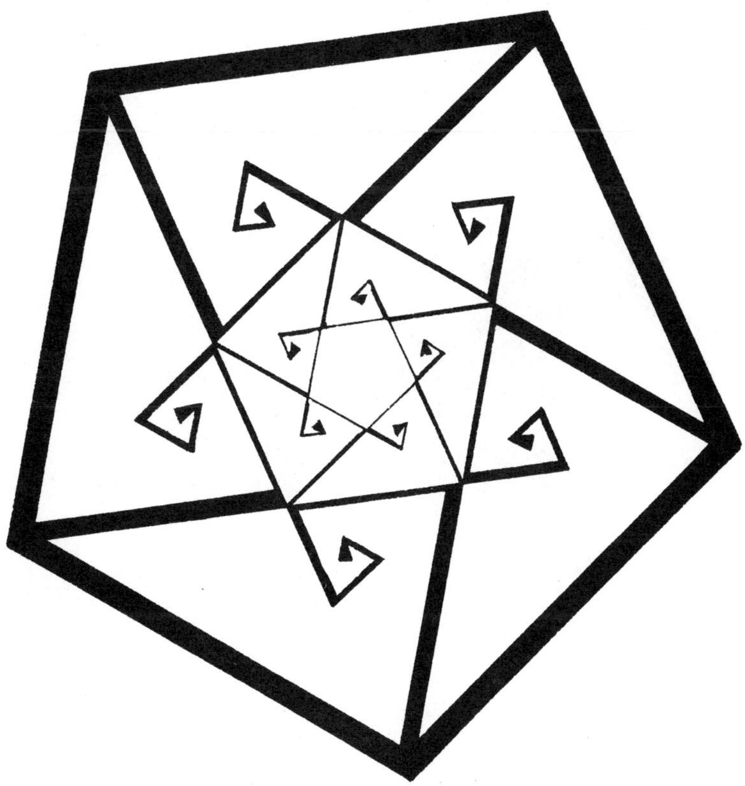

SPIRALS IN THE THIRD DIMENSION

Up to this point we have limited our discussion to spirals as they relate to the plane surface. As ours is a three-dimensional world, our analysis must be expanded to include the element of depth. Let us see what happens when the two spirals we have looked at are pulled up and out of the plane.

We shall first consider the case of the logarithmic spiral. In most natural examples the pole of the spiral is fixed in one plane and the curve expands not only away from the pole but through different planes as well. The resulting curve is called a concho-spiral and it is the one seen in many kinds of sea shells, in the scales of pine cones, and in some animal horns. When viewed head-on, the logarithmic spiral is clearly visable; when viewed from the side, the curve may pull sharply or gently away from the pole. The accompanying drawings illustrate these variations.

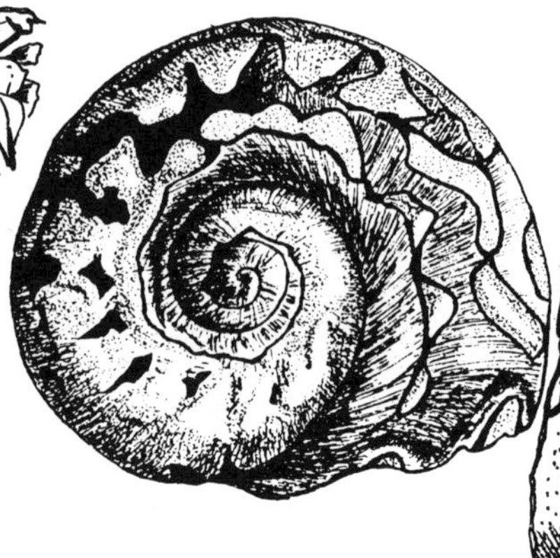

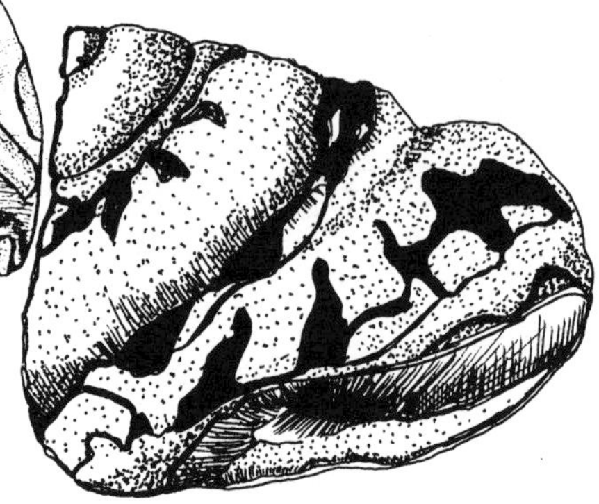

When the Archimedean spiral is pulled into the third dimension, a curve is formed for which the clearest example is the common carpenter's screw. This curve may be described as the uniform wrapping of a line around a <u>cone</u> from its vertex to its base. The curve is called a <u>conical helix</u> and it is found less frequently in Nature than the concho-spiral but is most useful in the man-made world.

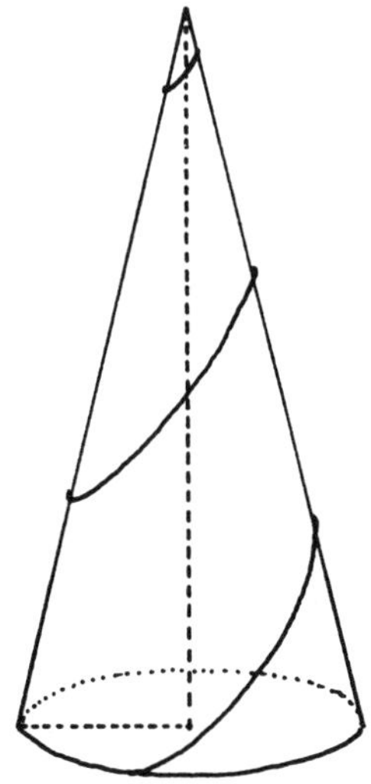

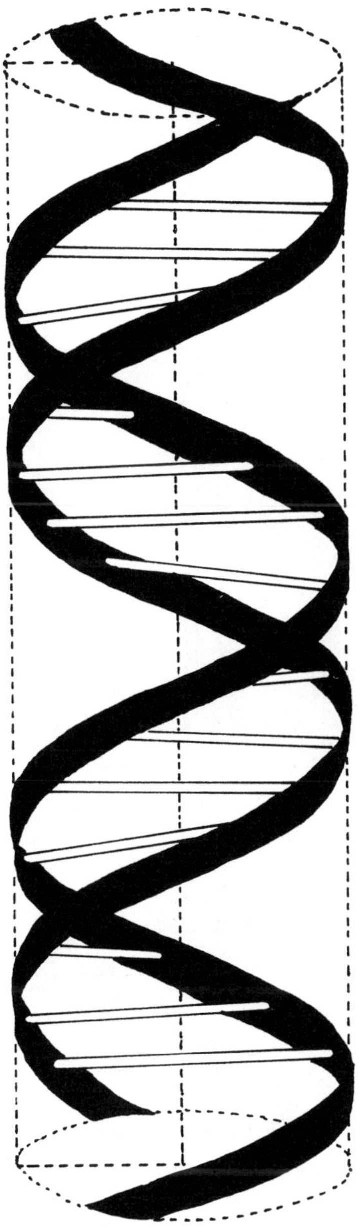

If the aforementioned line were to be wrapped around a <u>cylinder</u> rather than a cone, the result would be a cylindrical <u>helix</u>, or true helix. This curve, although spiral-like, is not truly a spiral since the diameter remains fixed as the curve lengthens. We include the helix in our discussion primarily because it is the curve fundamental to all life. The DNA molecule, which carries the genetic code for all living matter, is in the form of a double helix.

At a more visible level, we find the helix in the tendrils of hair and of vines. The helix provides strength when expansion is needed in a confined space using the least expenditure of organic materials.

In man's creations, the helix provides a practical as well as aesthetic solution to the problem of ascent in a space that is limited either horizontally or vertically. It gives us the spiral staircase as found in French chateaux and English cathedrals for it can assume a gentle gradient and affords an easy climb.

SPIRALS IN ART

The aesthetic sense recognizes and appreciates the fitness of certain forms and takes those as expressive of life, and, in the hands of the artist, gives it beauty. It appears that stylization of form in art comes a long time after the naturalistic treatment of the same object. The artist searches out the essentials and refines the form to its elemental principles. Throughout the centuries the artists of the great civilizations have sought to understand the inherent qualities of natural forms and to express that understanding through their craft.

The spiral is one of the oldest recurring symbols crossing time and culture as far back as pre-history. As a symbol it has the marvelous quality of holding many meanings simultaneously. It has been associated with the lotus blossom of Egypt where a great many of the ornamental patterns were based on this natural form.

An abstract variation of the spiral, the swastika, is one of the most ancient and widespread symbols. It connotes both positive and negative forces. As a sun symbol the arms follow the direction of the sun as it moves through the four seasons. In this form, it is taken to be a token of good luck and is found on ancient pottery and artifacts from Mycenae and Troy. It has also been found among the ruins in the Yucatan in Mexico and on the blankets of the Navajo Indians of North America. It is also known as the Saxon fylfot and the Greek gammadion (a cross form of four capital gammas.) In the twentieth century it was deliberately used in its negative form (with its arms twisting to the left) as a symbol of Nazi supremacy.

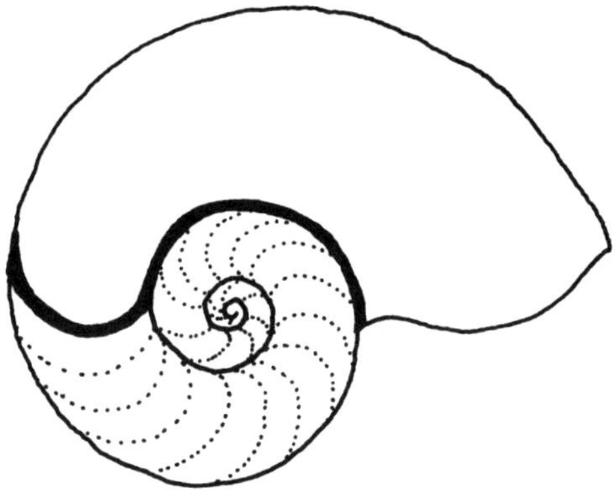

In his book, The Curves of Life, Theodore Cook suggests that the aforementioned swastika is a derivation of a curved line swastika which he feels evolved from the curves of the nautilus shell. The Oriental symbol of Yin-Yang, symbolizing the union of dark and light, male and female, positive and negative, may also be derived from this curve.

Simply drawing a "curly line" does not afford a believable facsimile of a spiral, and the artist-designer needs more concrete information at his disposal. The following pages demonstrate some of the ways that the spiral curve may be approximated without destroying its inherent characteristics.

Certain curves can be drawn which have the property of expanding in a regular fashion, but differ from a true spiral in that there is no single pole. Construction 40 demonstrates this type of curve generated by a square, hence having four "poles".

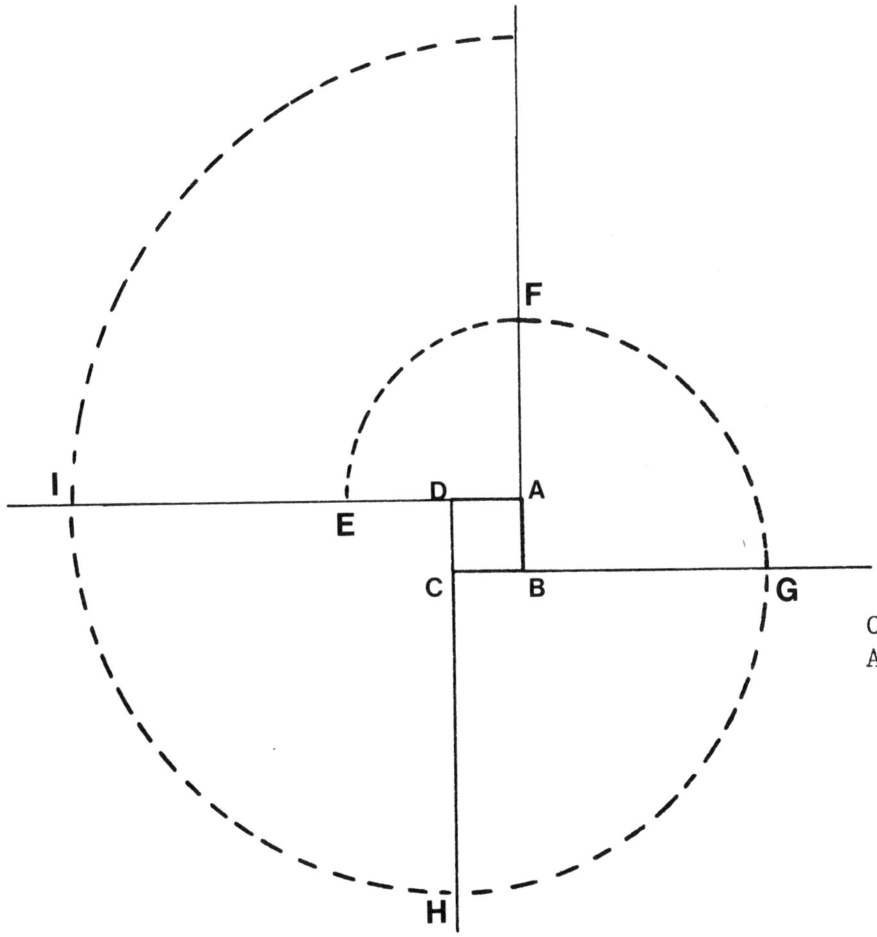

CONSTRUCT A "SPIRAL"
ABOUT A SQUARE

Given square ABCD, construct a "spiral" using the vertices as "poles".

1. Extend $\overrightarrow{AD}$, $\overrightarrow{BA}$, $\overrightarrow{CB}$ and $\overrightarrow{DC}$.
2. Open the compass to measure more than AD, place the metal tip on A, and cut an arc that intersects $\overrightarrow{AD}$ and $\overrightarrow{BA}$.
3. Label the points of intersection E and F respectively.
4. Place the metal tip on B, the pencil tip on F, and cut an arc that intersects $\overrightarrow{CB}$.
5. Label the point of intersection G.
6. Place the metal tip on C, the pencil tip on G, and cut an arc that intersects $\overrightarrow{DC}$.
7. Label the point of intersection H.
8. Repeat with the metal tip on D to cut $\overset{\frown}{HI}$.
9. Continue the process until the curve is the desired size.

Now the curve is spiral-like with four "poles".

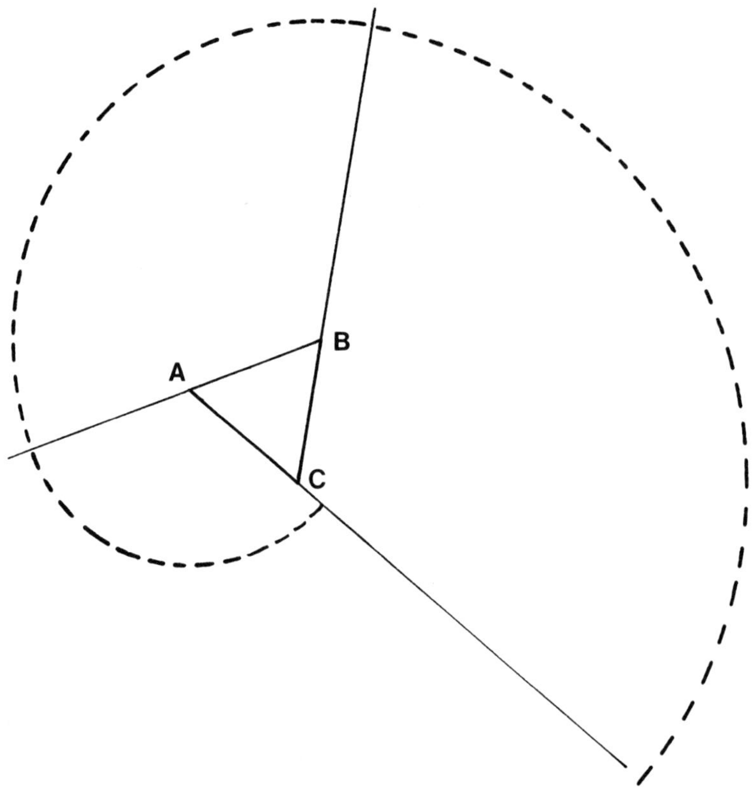

Fig. 1

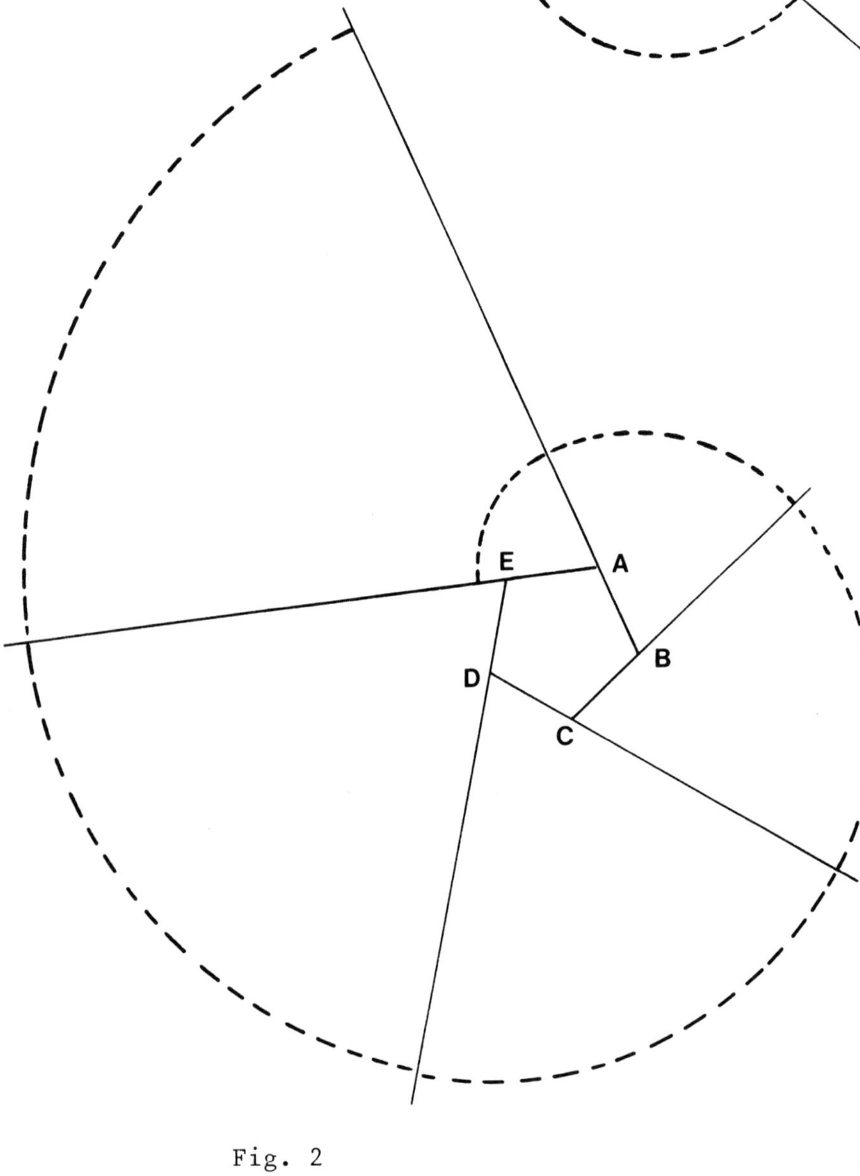

Fig. 2

The same technique as Construction 40 is used in Fig. 1 to create a "spiral" around an equilateral triangle (three "poles"), and, in Fig. 2, around a regular pentagon (five "poles"). Any regular polygon may be used.

Another way to obtain a curve similar to the ones produced by the method given in Construction 40 is to use two points on a line and alternate them as centers of semicircles as given in the following construction.

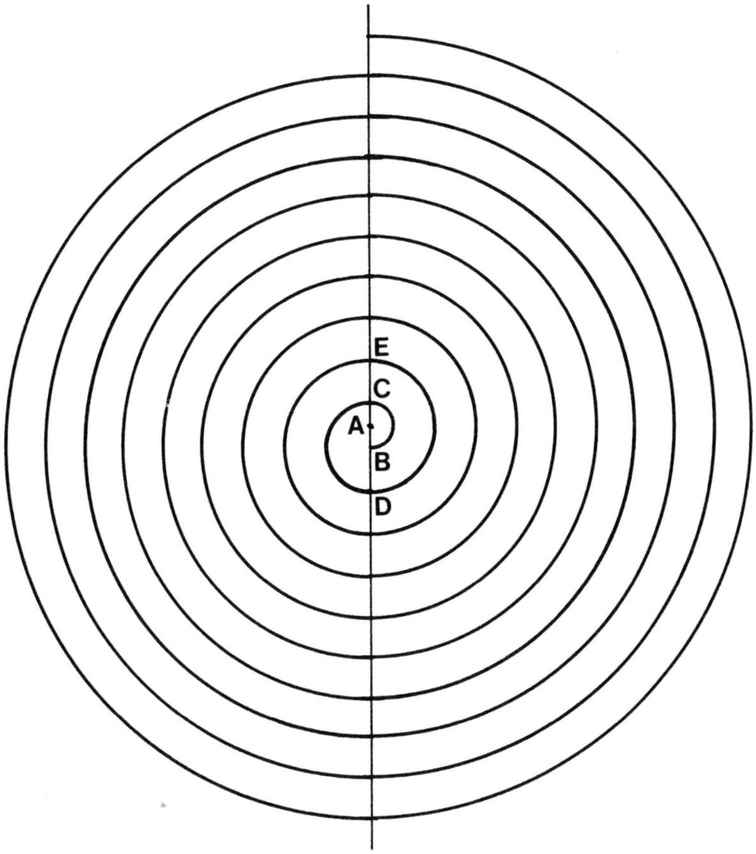

CONSTRUCT A "SPIRAL"
USING ALTERNATE "POLES"

Given points A and B, construct a "spiral" that uses these points as alternate "poles".

1. Draw $\overleftrightarrow{AB}$
2. Place the metal tip on A, the pencil tip on B, and cut semicircle BC.
3. Place the metal tip on B, the pencil tip on C and cut semicircle CD.
4. Place the metal tip on A, the pencil tip on D, and cut semicircle DE.
5. Continue to cut semicircles on either side of $\overleftrightarrow{AB}$ by alternately placing the metal tip on A and B until the curve is the desired size.

Now the curve is a "spiral" with two "poles".

One of the simplest of spiral forms to construct is the Baravelle Spiral. It is a straight line design based on connecting consecutive midpoints of sides of regular polygons. The following construction gives the methods for developing one within a square. The same concept may be applied to other regular polygons.

CONSTRUCT A BARAVELLE SPIRAL WITHIN A SQUARE.

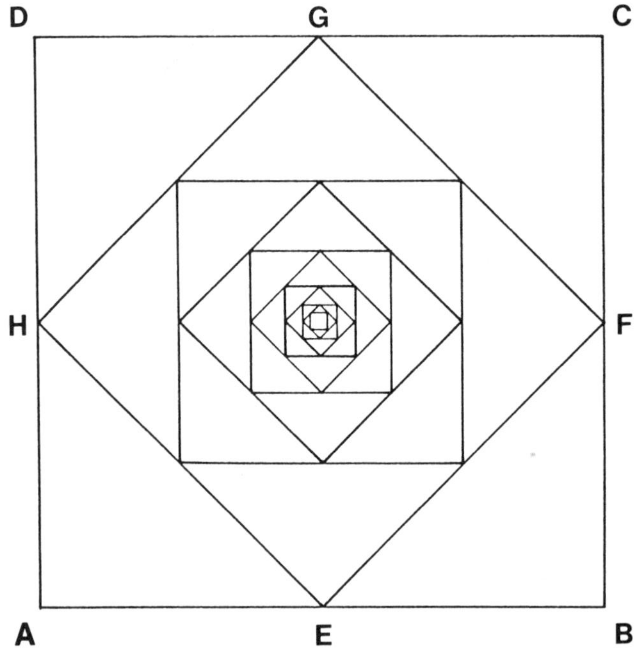

Given square ABCD, construct a Baravelle Spiral.

1. Locate the midpoints of $\overline{AB}$, $\overline{BC}$, $\overline{CD}$, and $\overline{DA}$.
2. Label them E,F,G, and H respectively.
3. Join them with line segments to form square EFGH.
4. In the same way, join the midpoints of the sides of EFGH to form another square.
5. Repeat the process until the final square is the desired size.
6. The "spiral" shape becomes more visible when the triangles are shaded as illustrated in the diagram.

Now the resulting figure is a Baravelle Spiral.

The accompanying demonstrate Baravelle Spirals in other polygonal shapes.

Baravelle Spirals have a very angular quality and without shading the "spiral" is not clearly evident. By choosing a point other than the midpoint of the sides, more subtle variations can be created which more nearly approximate a true curve. The closer the point is to an endpoint, the more curved the finished "spiral" will appear. In Construction 43, the point is located one-sixth of the distance from one vertex to the next. A greater or lesser distance may be chosen depending on whether a more angular look or a smoother appearance is desired.

Another advantage this type of "spiral" affords is that the designer is not restricted to regular polygons. The procedure is demonstrated in the square and other examples of straight-line "spirals" are given thereafter.

CONSTRUCT A STRAIGHT-
LINE SPIRAL IN A SQUARE.

43

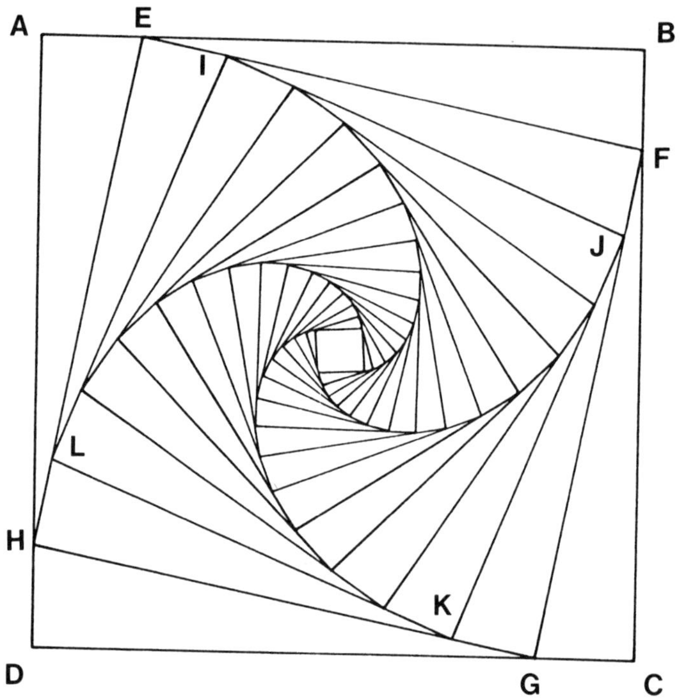

Given square ABCD, construct a straight-line spiral.

1. Locate points E, F, G, and H one-sixth of the distance from A to B, B to C, C to D, and D to A respectively. (Use Construction 8 to divide the line segment into sixths.)
2. Connect the points to form square EFGH.
3. Find one-sixth the distance from E to F and use the measurement to mark off points I, J, K, and L.
4. Connect the points to form square IJKL.
5. Repeat the process until the last square constructed is the desired size.

Now the figure is a straight-line spiral.

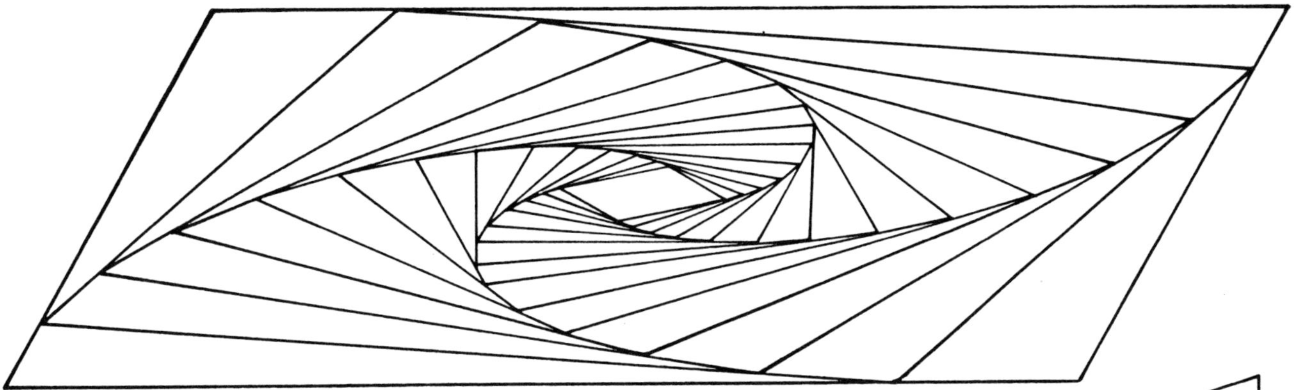

Ø+1 Parallelogram

Golden Triangle

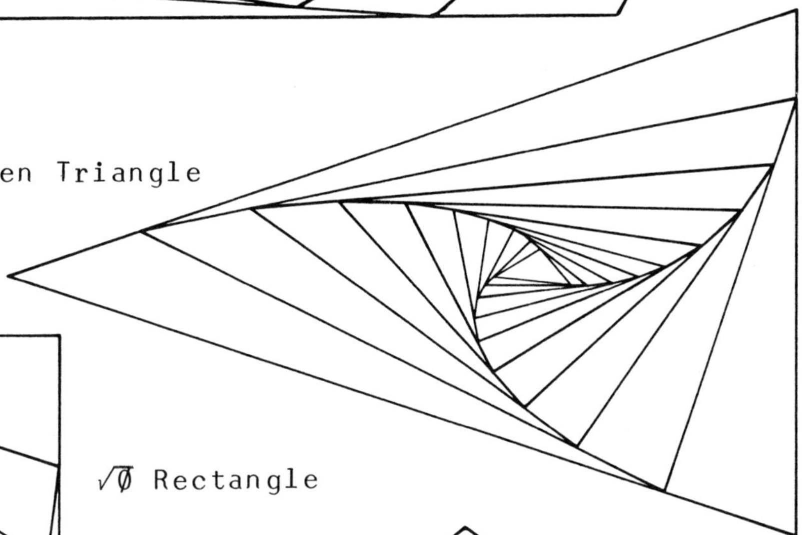

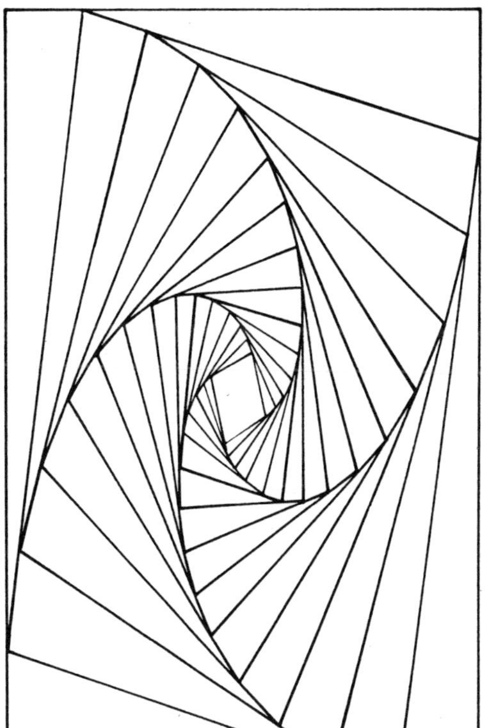

√Ø Rectangle

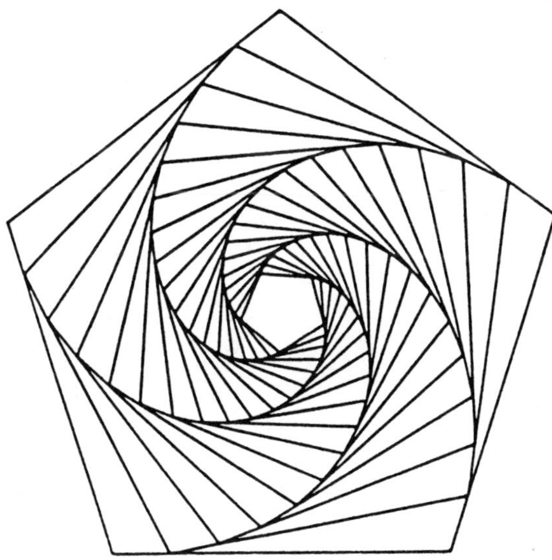

Pentagon

The angular spiral forms are but constructs for the artist-designer. However, Nature's harmony can be preserved in design by using these techniques within the Golden Triangle, the Dynamic Rectangles and Parallelograms, especially from the Ø-Family, and, of course, the omnipresent pentagon. Note that when an irregular figure is used, such as a rectangle, the same fractional portion of the width as well as the length must be used.

A last interesting straight-line spiral can be constructed using a square and the resulting progression of Dynamic Rectangles. You will recall from Construction 27 that the length of each of the Dynamic Rectangles depends on the length of the diagonal of the previous rectangle. In this case, the vertex of the original unit square becomes the pole about which the Dynamic Rectangles rotate. Each new rectangle is constructed directly on the diagonal of the preceding one, with its width the same as the length of the side of the square.

This spiral is an example of a case where going beyond the $\sqrt{5}$ Rectangle is essential. If the process were carried on indefinitely, all of the Dynamic Rectangles would be generated and the spiral would eventually approach a smooth curve.

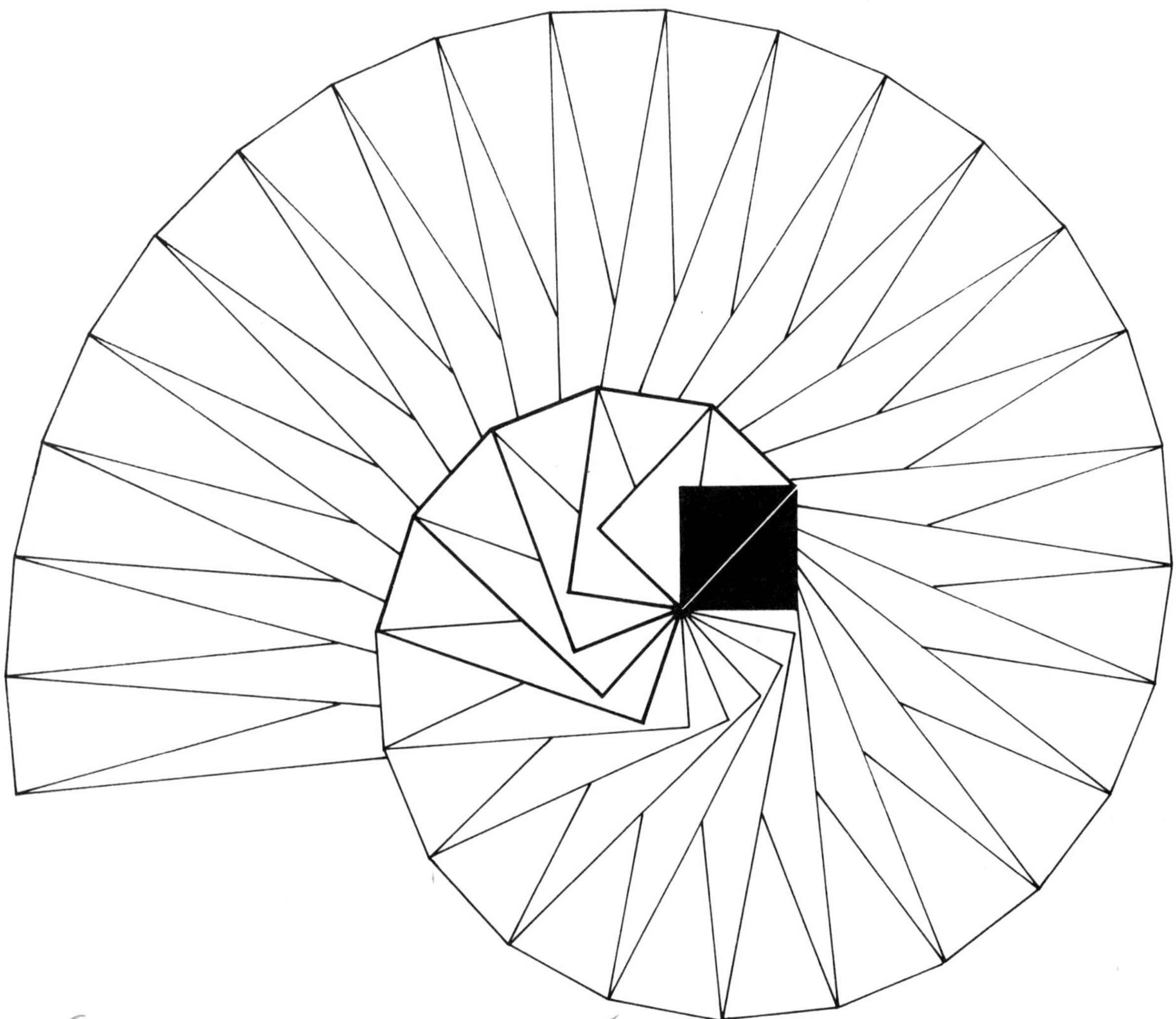

PROBLEM **1**

Using the polar graph, construct an Archimedean spiral that increases one unit for each turn of 45°.

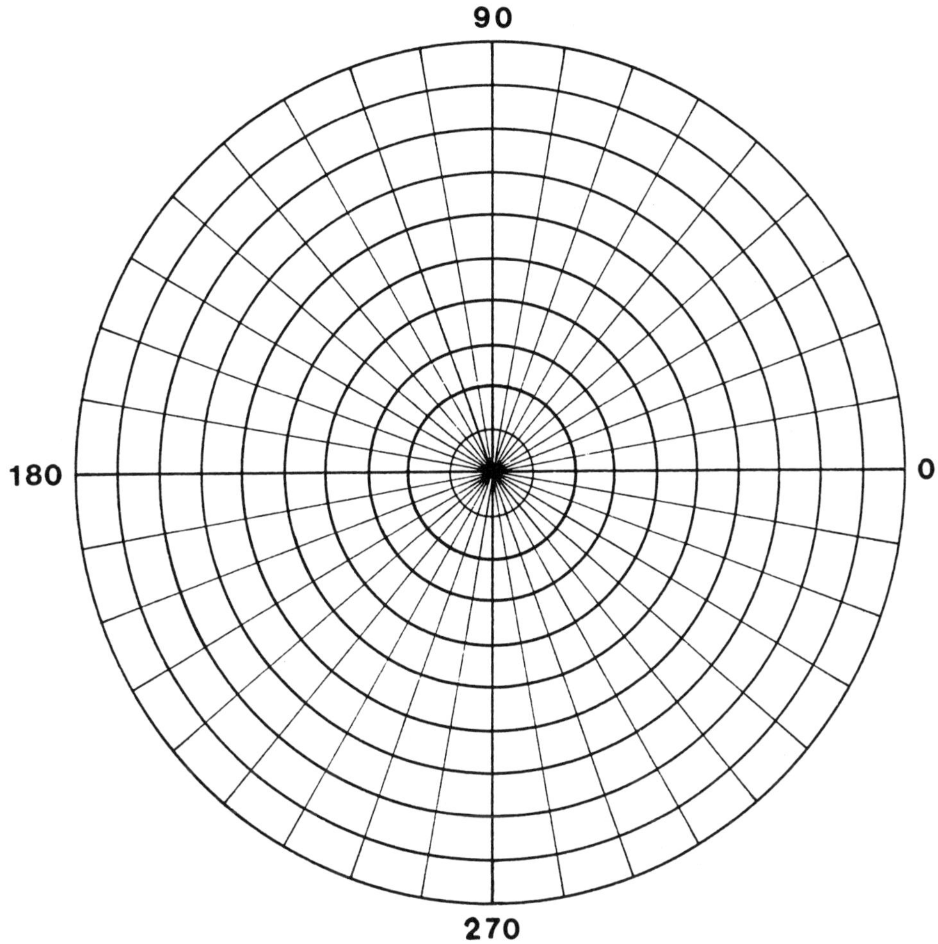

PROBLEM **2**

Construct an angular spiral in each one of the Ø - family rectangles.

PROBLEM **3**

Construct a logarithmic spiral within a Golden Rectangle.

PROBLEM **4**

Construct two different sized Golden Triangles with their respective spirals.

PROBLEM **5**

Use the technique of Construction 40 to develop a "spiral" around a regular hexagon.

PROBLEM **6**

Construct a Baravelle Spiral in a decagon.

PROBLEM **7**

Locate two points at least one inch apart and use them as alternating "poles" to construct a "spiral". You may discover a short cut in the process.

PROJECT **1**

Use the Baravelle Spiral to create an all-over pattern using this square as your inspiration or develop a variation of your own.

PROJECT **2**

Construct a Baravelle or Straight-Line Spiral in any regular polygon and enhance it with a color scheme.

PROJECT **3**

Use the concept of the Spiral to create a three-dimensional art work.

PROJECT **4**

Using the concept of the double helix, create a sculptural form.

PROJECT **5**

Using what you know about the property of similarity, create a group of graduated pyramids based on the Golden Triangle. Join them together to make a spiral form. Color as you choose.

Further Reading

Bain, George. Celtic Art, The Methods of Construction. New York: Dover Publications, Inc. 1973.

Ball, Rouse. A Short Account of the History of Mathematics. New York: Dover Publications, Inc. 1960.

Cook, Theodore A. The Curves of Life. New York: Dover Publications, Inc., 1979.

Edwards, Edward B. Pattern and Design with Dynamic Symmetry: Dover Publications, Inc., 1967.

Huntley, H.E. The Divine Proportion. New York: Dover Publications, Inc., 1970.

Stevens, Peter. Patterns in Nature. Boston: Little, Brown and Co. 1974.

CONCLUSION

In this book we have seen the Golden Ration, ∅, to be the common thread that weaves harmony into the universal patterns. This chart is presented to visually clarify the many connections.

THE GOLDEN RATIO

$$\Phi = 1.61803$$

THE DIVINE PROPORTION

$$\frac{AB}{AC} = \frac{AC}{CB} = \emptyset$$

or

$$\frac{a+b}{a} = \frac{a}{b} = \emptyset$$

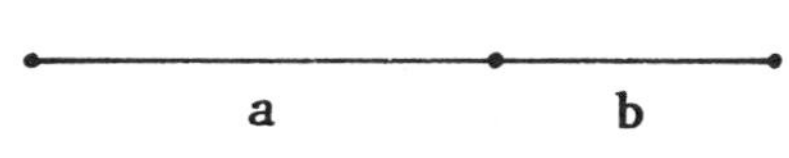

THE FIBONACCI SEQUENCE: THE GOLDEN NUMBERS

n	F_n	n	F_n
1	1	9	34
2	1	10	55
3	2	11	89
4	3	12	144
5	5	13	233
6	8	14	377
7	13	15	610
8	21	16	987

BEYOND F_{15} THE RATIO OF ANY PAIR OF CONSECUTIVE FIBONACCI NUMBERS IS THE SAME AS Ø TO FIVE DECIMAL PLACES.

THE GOLDEN SEQUENCE

$$1,\ \emptyset,\ \emptyset^2,\ \emptyset^3,\ \emptyset^4,\ \emptyset^5,\ \ldots\ldots\ldots$$

We can rewrite this sequence in the following way to show its relationship to the Fibonacci Sequence:

$$1, \emptyset,\ 1+\emptyset,\ 1+2\emptyset,\ 2+3\emptyset,\ 3+5\emptyset,\ 5+8\emptyset,\ \ldots\ldots$$

THE GOLDEN RECTANGLE

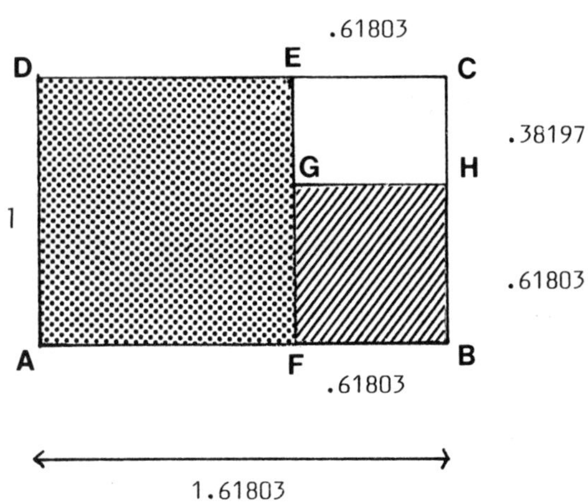

$$\frac{AB}{BC} = \frac{BC}{EC} = \frac{EC}{CH} = \emptyset$$

This is the only rectangle whose gnomen is a square, hence the name "whirling square rectangle".

The Golden Rectangle in relation to the Fibonacci Numbers

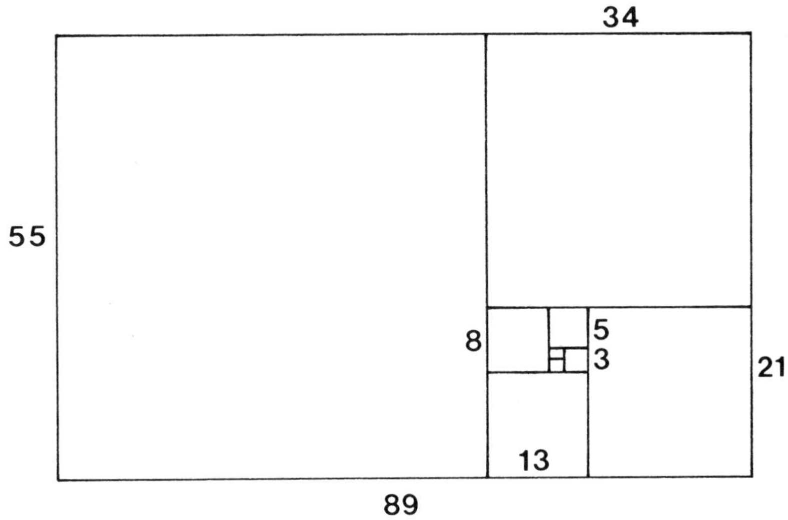

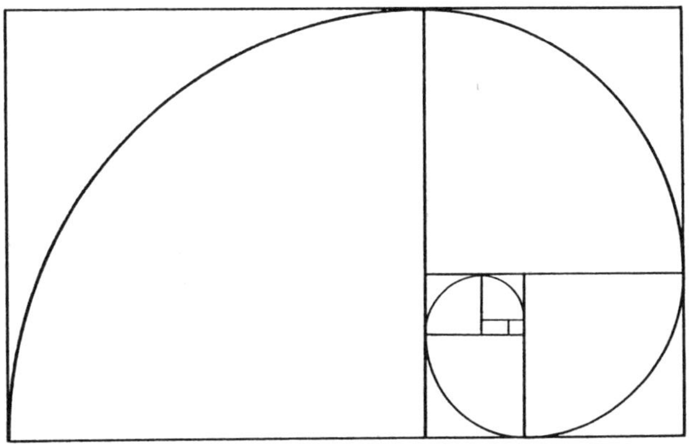

The Golden Rectangle together with the Golden Spiral

THE GOLDEN RATIO IN THE GOLDEN POLYGON: THE PENTAGON WITH PENTAGRAM

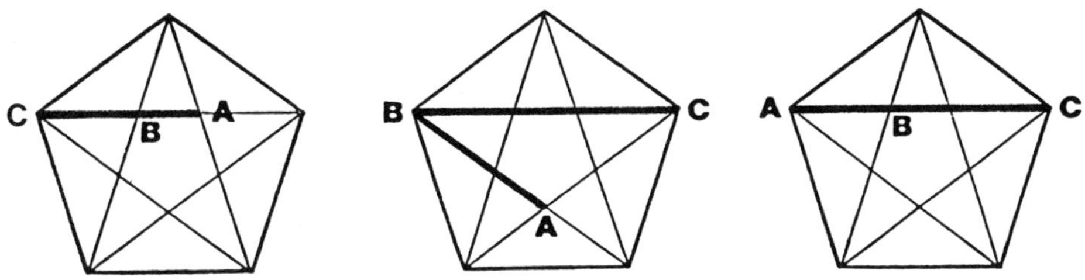

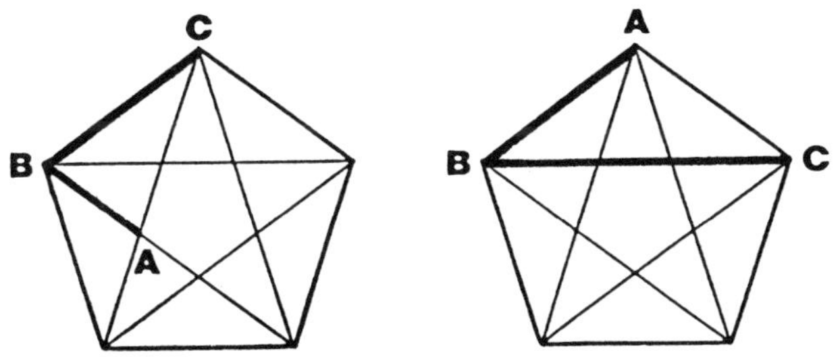

In each case:

$$\frac{AB+BC}{BC} = \frac{BC}{AB} = \phi$$

THE GOLDEN TRIANGLE

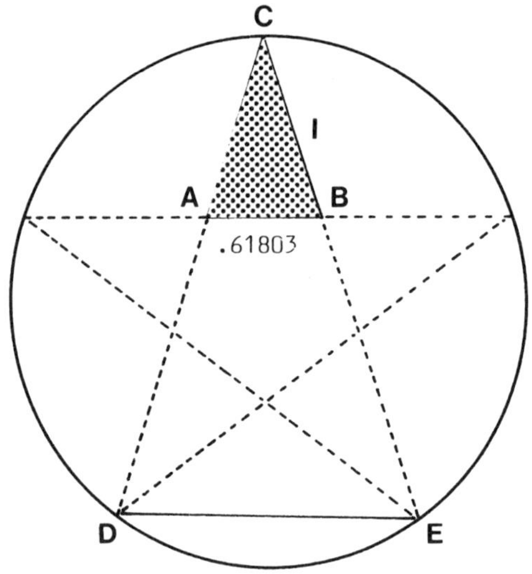

In the Pentagon and Pentagram

$$\frac{CB}{AB} = \frac{CE}{DE} = \phi$$

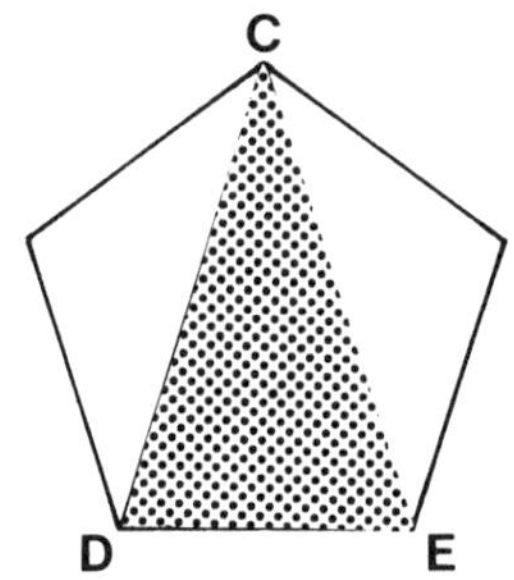

With Its Logarithmic Spiral

$$\frac{AB}{BC} = \frac{BC}{CD} = \frac{CD}{DE} = \frac{DE}{EF} = \frac{EF}{FG} = \phi$$

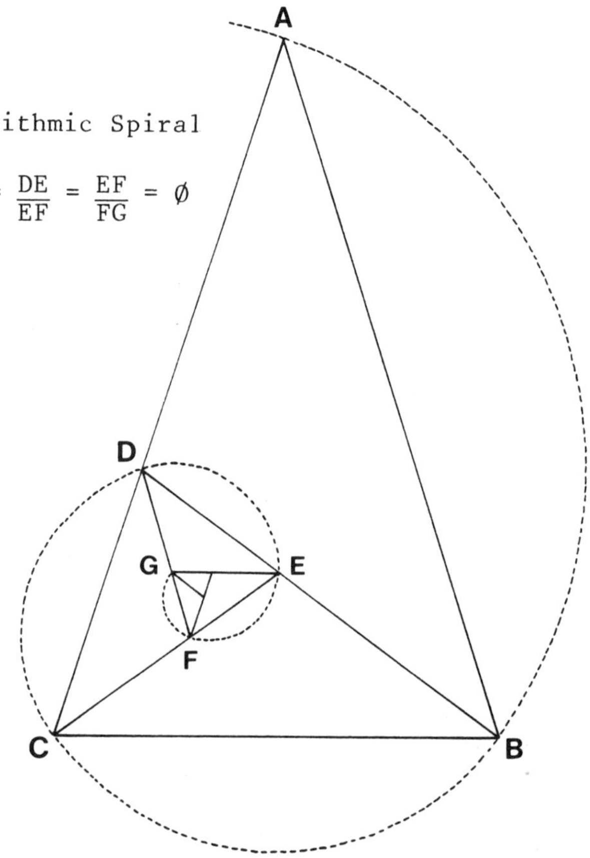

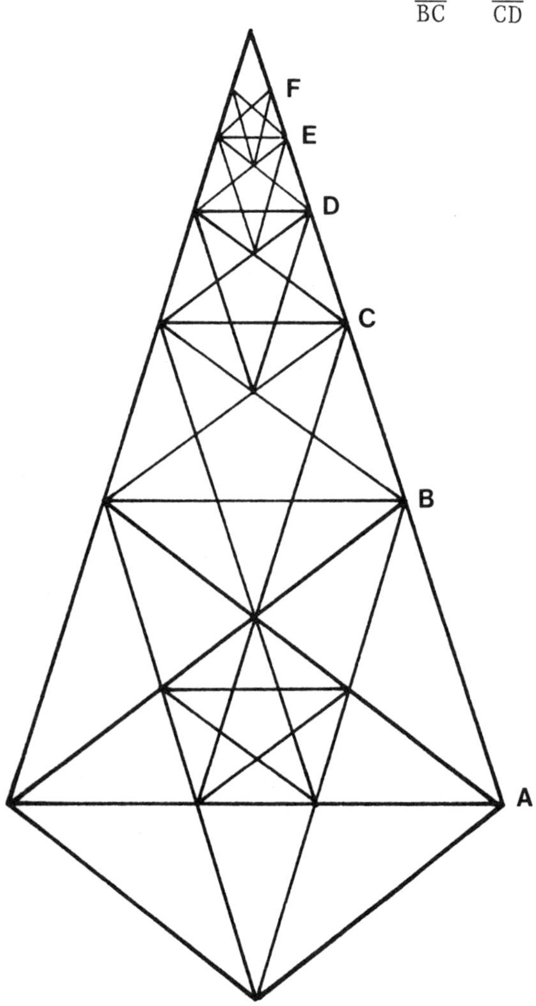

In the Lute of Pythagoras

$$\frac{AB}{BC} = \frac{BC}{CD} = \frac{CD}{DE} = \frac{DE}{EF} = \phi$$

THE PHIDEAN SPIRAL

$$\frac{ED}{DC} = \frac{DC}{CB} = \frac{CB}{BA} = \phi$$

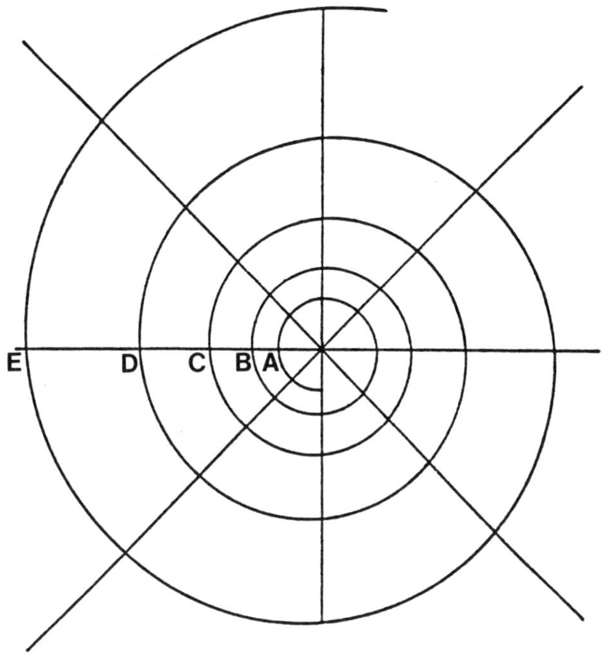

THE PHI-FAMILY RECTANGLES

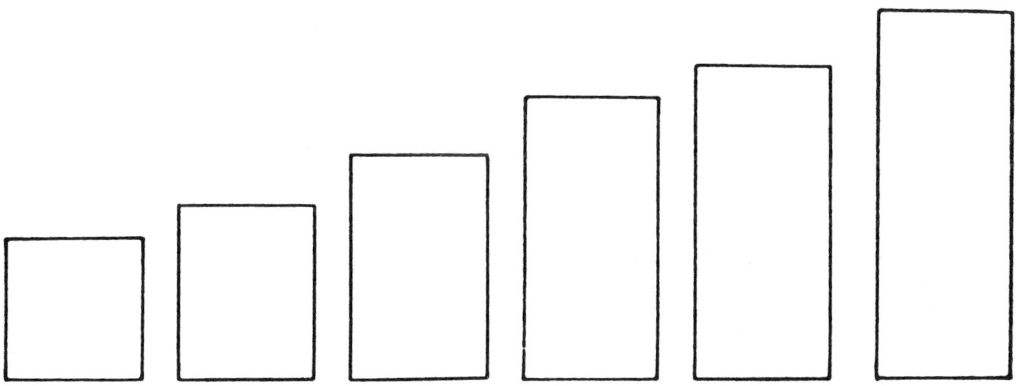

1	$\sqrt{\phi}$	ϕ	$\sqrt{4}$	$\sqrt{5}$	$\phi+1$
square	1.272	1.618	2	2.236	2.618

We have come to the end of Book I. We hope that the reader has gained insight and, like us, has been filled with a deeper appreciation of the omnipresent connections among objects and events of this universe.

SELECTED REFERENCES

Ball, Rouse. A Short Account of the History of Mathematics. New York: Dover, 1960.

Beard, Col. Robert S. Patterns in Space. Palo Alto: Creative Publications, 1973.

Bentley, W.A. and W.J. Humphreys. Snow Crystals. New York: Dover, 1962.

Bezuska, Stanley, Margaret Kenney, Linda Silvey. Designs From Mathematical Patterns. Palo Alto: Creative Publications, 1978.

Bouleau, Charles. The Painter's Secret Geometry. New York: Harcourt, Brace and World, 1963.

Bruni, James. Experiencing Geometry. Belmont: Wadsworth, 1977.

Capra, Fritjof. The Tao of Physics. New York: Bantam Books, 1965.

Capra, Fritjof. The Turning Point. New York: Bantam Books, 1983.

Colman, Samuel. Nature's Harmonic Unity. New York: Putnam's and Sons, 1911.

Cook, Theodore A. The Curves of Life. New York: Dover, 1979.

Coxeter, H.S.M. Introduction to Geometry. New York: John Wiley and Son, 1961.

Critchlow, Keith. Order in Space. New York: The Viking Press, 1970.

Cundy and Rollett. Mathematical Models. New York: Oxford University Press, 1961.

Doczi, Gyorgy. The Power of Limits. Boulder: Shambala Press, 1981.

Edwards, Edward B. Pattern and Design with Dynamic Symmetry. New York: Dover, 1967.

Feininger, Andreas. The Anatomy of Nature. New York: Dover, 1956.

Feininger, Andreas. Nature Close Up: A Fantastic Journey into Reality. New York, 1952.

Fleming, William. Arts and Ideas. New York: Holt, Rinehart and Winston, 1980.

Ghyka, Matila. The Geometry of Art and Life. New York: Dover, 1977.

Gillings, Richard J. Mathematics in the Time of the Pharaohs. New York: Dover, 1972.

Grillo, Paul Jacques. Form, Function and Design. New York, Dover, 1960

Haeckel, Ernst. Art Forms in Nature. New York: Dover, 1974.

Hambidge, Jay. The Elements of Dynamic Symmetry. New York: Dover, 1967.

Hambidge, Jay. The Parthenon and Other Greek Temples, Their Dynamic Symmetry. New Haven: Yale University Press, 1924.

Hambidge, Jay. Practical Applications of Dynamic Symmetry. New Haven: Yale University Press, 1932.

Hogben, Lancelot. The Wonderful World of Mathematics. New York: Garden City Books, 1955.

Hogben, Lancelot. Mathematics for the Millions. New York: W. W. Norton, 1937.

Hogben, Lancelot. Mathematics in the Making. New York: Doubleday, 1960-61.

Holden, Alan. Shapes, Space and Symmetry. New York: Columbia University Press, 1971.

Huntley, H.E. The Divine Proportion. New York: Dover, 1970.

Kline, Morris. Mathematics and the Physical World. New York: Dover, 1959.

Kline, Morris. Mathematics in Western Culture. New York: Oxford University Press, 1953.

Pearce, Peter and Susan Pearce. Polyhedra Primer. New York: Van Nostrand Reinhold, 1978.

Seymour, Dale and Reuben Schadler. Creative Constructions. Palo Alto: Creative Publications, 1974.

Stevens, Peter S. Patterns in Nature. Boston: Little, Brown and Co., 1974.

Thompson, D'Arcy. On Growth and Form. Cambridge: University Press, 1969.

Wenninger, Magnus J. Polyhedron Models. Cambridge: University Press, 1971.

Weyl, Hermann. Symmetry. Princeton: University Press, 1952.

Williams, Christopher. Origins of Form. New York: Visual Communications Books, 1981.

NOTES

NOTES

Appendices

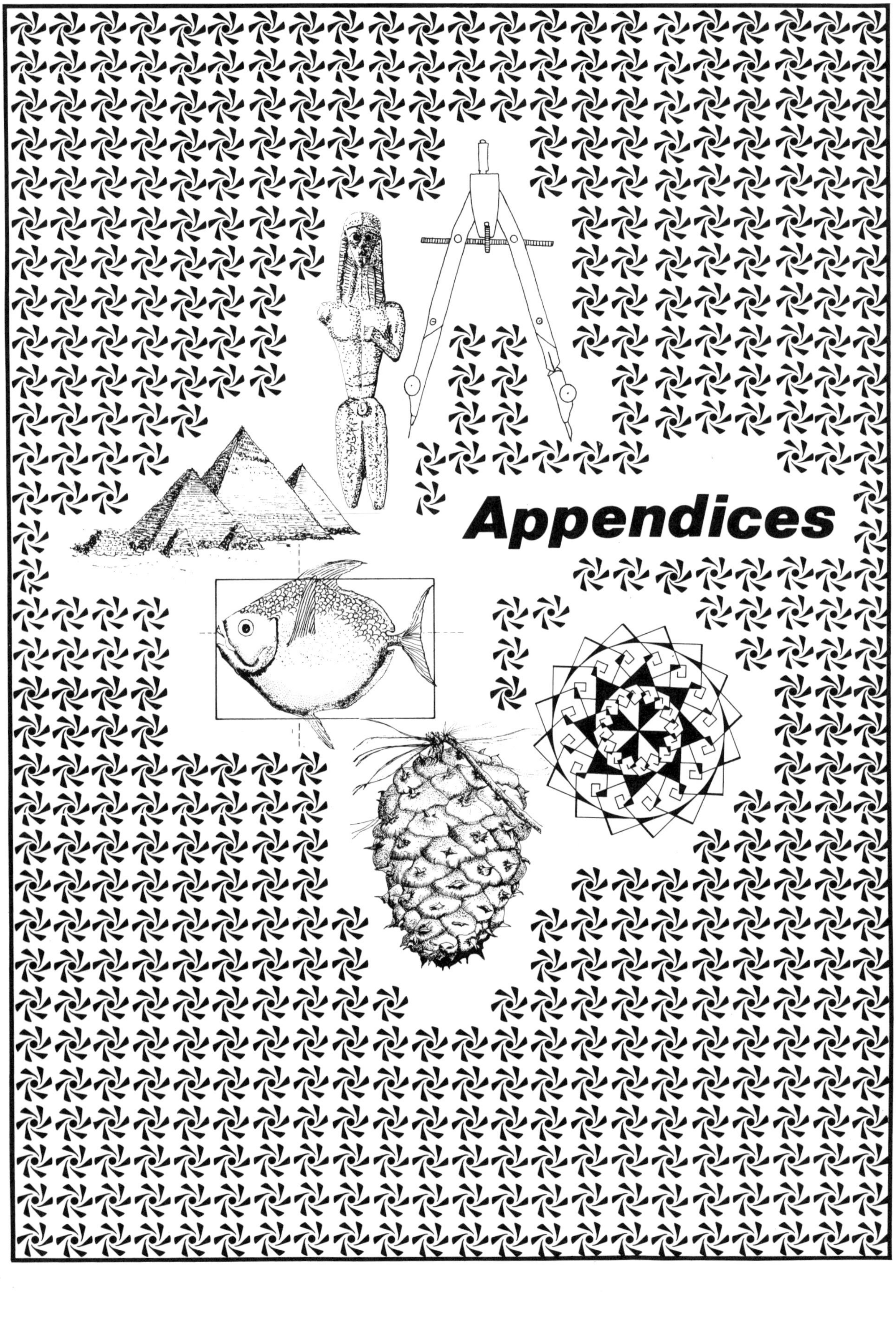

MATHEMATICAL SYMBOLS

$\overline{AB}$	segment AB
$\overleftrightarrow{AB}$	line AB
$\overrightarrow{AB}$	ray AB
$\overset{\frown}{AB}$	arc AB
AB	length of AB (designates a real number)
$\odot P$	circle P
$\triangle ABC$	triangle ABC
$\angle ABC$	angle ABC
$\cong$	is congruent to
$=$	is equal to
$\doteq$	is approximately equal to
$\sim$	is similar to
$\parallel$	is parallel to
$\perp$	is perpendicular to
$>$	is greater than
$\geq$	is greater than or equal to
$<$	is less than
$\leq$	is less than or equal to
$\neq$	is not equal to
$\ngtr$	is not greater than
$\nless$	is not less than
$\pm$	plus or minus (both operations designated by a single symbol)
$\therefore$	therefore
x^{o} / y' / z''	x degrees / y minutes / z seconds — angle measure
x' / y''	x feet / y inches — linear measure
π	pi $\doteq$ 3.14 or 22/7 (the ratio of the circumference of a circle to its diameter)
ϕ	phi $\doteq$ 1.61803 (the Golden Ratio)
x_n	x sub n (a way of naming a variable)
x^n	x to the nth power (multiply x by itself n times)

B

PROPERTIES AND PROOFS

Contents

PROPERTIES OF PROPORTIONS

1. In any proportion the product of the means equals the product of the extremes.

   If $\frac{a}{b} = \frac{c}{d}$, then $ad = bc$.

2. If the product of one pair of numbers equals the product of another pair of numbers, either pair may be made the means and the other pair the extremes of a proportion.

   If $ad = bc$, then $\frac{a}{b} = \frac{c}{d}$ or $\frac{b}{a} = \frac{d}{c}$.

3. In any proportion, the means or the extremes may be interchanged.

   If $\frac{a}{b} = \frac{c}{d}$, then $\frac{d}{b} = \frac{c}{a}$ or $\frac{a}{c} = \frac{b}{d}$.

4. If four quantities are in proportion, they are in proportion by inversion or alternation.

   If $\frac{a}{b} = \frac{c}{d}$, then $\frac{b}{a} = \frac{d}{c}$ or $\frac{c}{a} = \frac{d}{b}$

PROPERTIES OF GEOMETRIC FIGURES

PARALLEL LINES

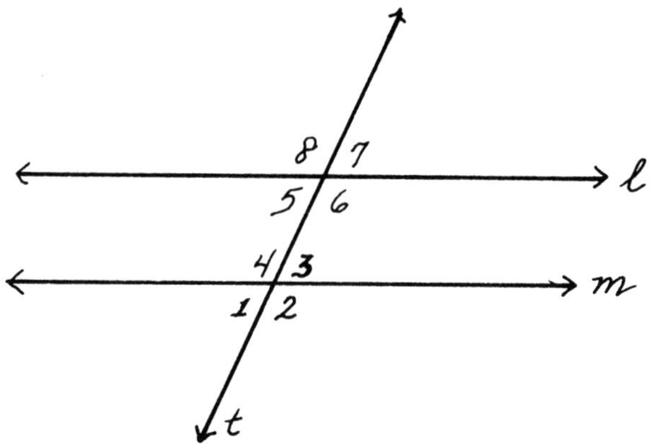

1. If $l \parallel m$ and l and m are cut by transversal t, then:

   $\angle 1 \cong \angle 3 \cong \angle 5 \cong \angle 7$ and

   $\angle 2 \cong \angle 4 \cong \angle 6 \cong \angle 8$

2. Three or more parallels cut proportional segments on two transversals.

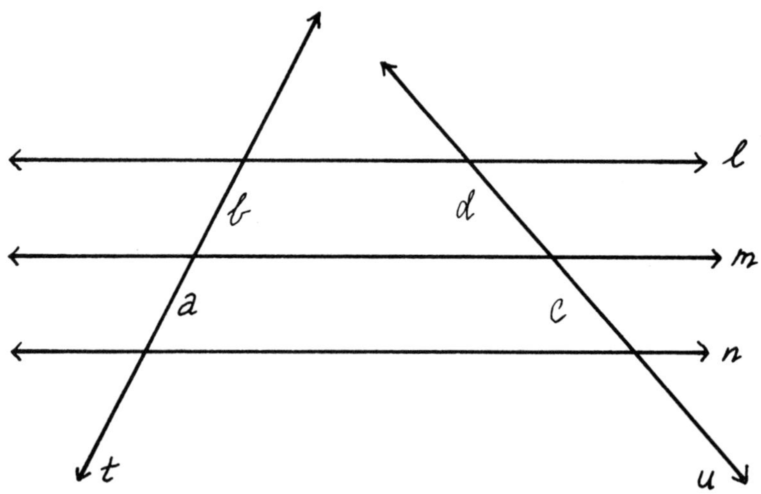

If $l \parallel m \parallel n$ and l, m and n are cut by transversals t and u, then the segments of the transversals are in proportion:

$$\frac{a}{b} = \frac{c}{d}.$$

3. If a line is parallel to one side of a triangle, it divides the other two sides proportionally.

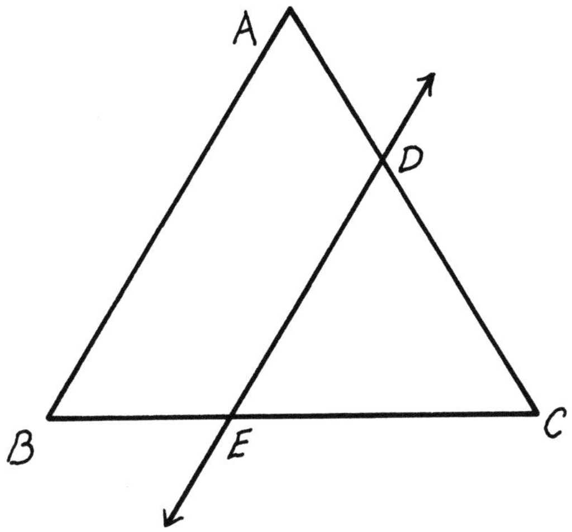

If $\overleftrightarrow{DE} \parallel \overleftrightarrow{AB}$, then $\frac{AD}{BE} = \frac{DC}{EC}$.

POLYGONS

1. The sum of the interior angles of a polygon of n sides is S = (n-2)180.

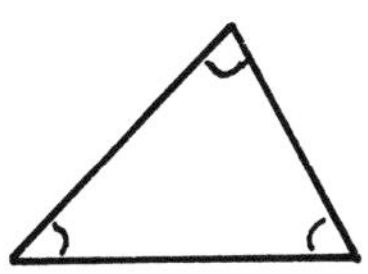

n = 3

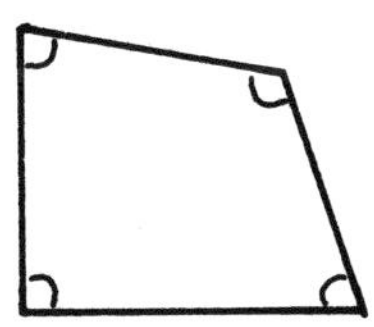

n = 4

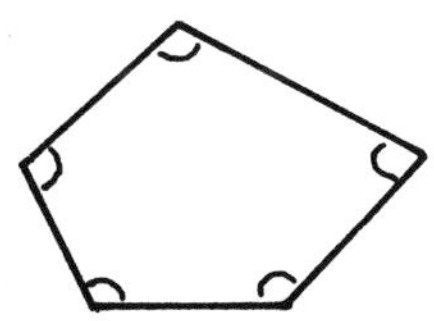

n = 5

S = (3-2)180 = 180°

S = (4-2)180 = 360°

S = (5-2)180 = 540°

2. The perimeters of two similar polygons are proportional to the corresponding sides; but their areas are proportional to the squares of the corresponding sides.

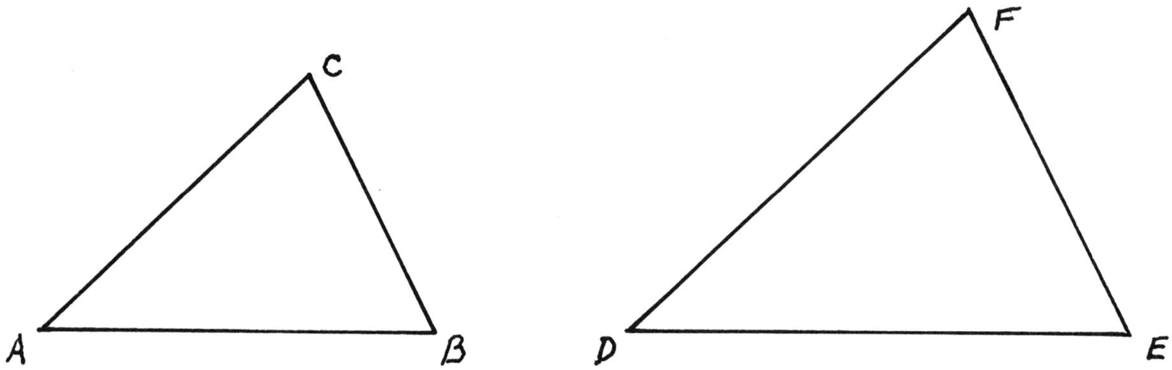

P = perimeter
A = area

P' = perimeter
A' = area

$$\triangle ABC \sim \triangle DEF$$

$$\frac{AB}{DE} = \frac{BC}{EF} = \frac{CA}{FD} = \frac{P}{P'}$$

$$\frac{(AB)^2}{(DE)^2} = \frac{(BC)^2}{(EF)^2} = \frac{(CA)^2}{(FD)^2} = \frac{A}{A'}$$

Triangles

1. Two triangles are congruent if they agree in:
   a. one side and two corresponding angles. (ASA, SAA)
   b. two sides and the included angle. (SAS)
   c. two sides and the right angle opposite one of them. (HL)
   d. three sides. (SSS)

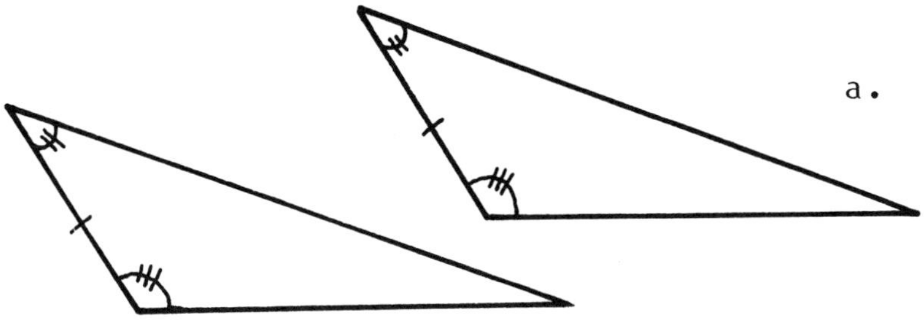

a.

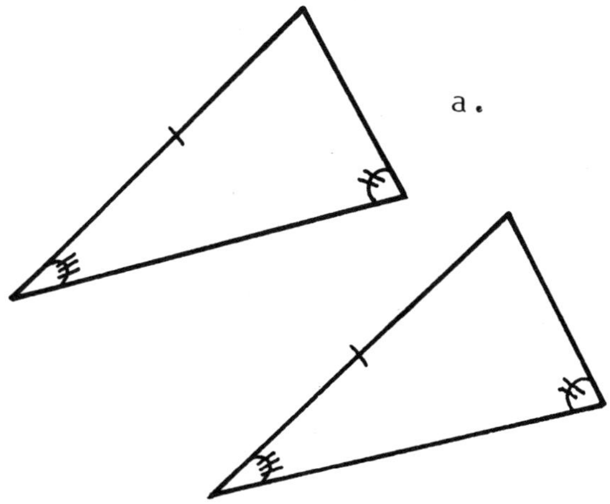

a.

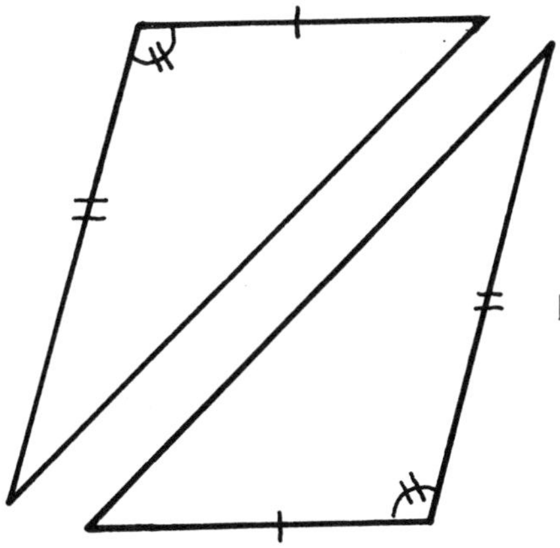

b.

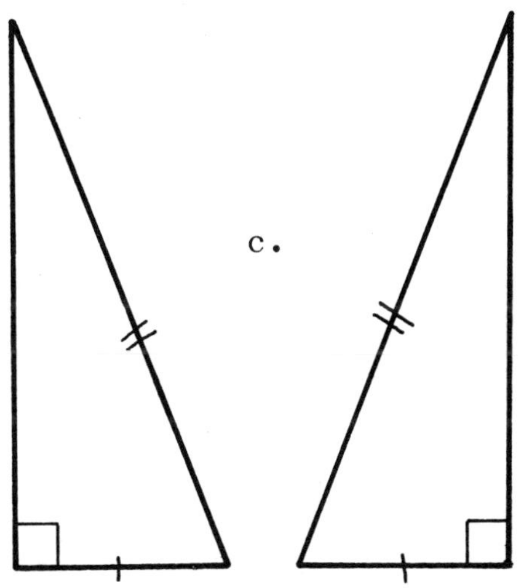

c.

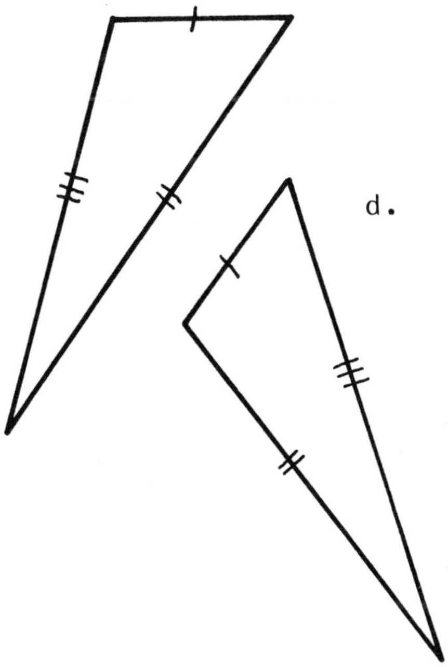

d.

In each case, the corresponding congruent parts of the triangle are sufficient to insure the congruence of the triangles themselves.

2. Two triangles are similar if:

a. two angles in one triangle are congruent to two angles in the other.

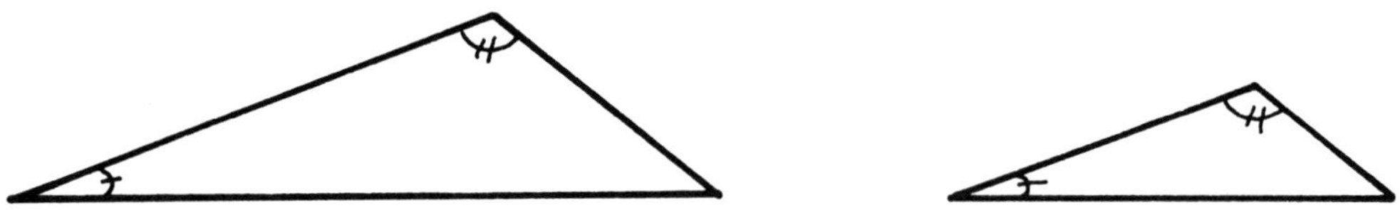

b. an angle of one triangle is congruent to an angle of the other, and the corresponding including sides are proportional.

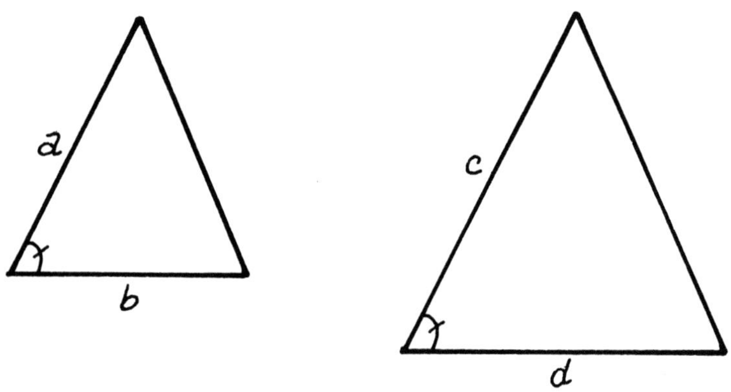

$$\frac{a}{c} = \frac{b}{d}$$

c. three sides of one triangle are in proportion to three sides of the other.

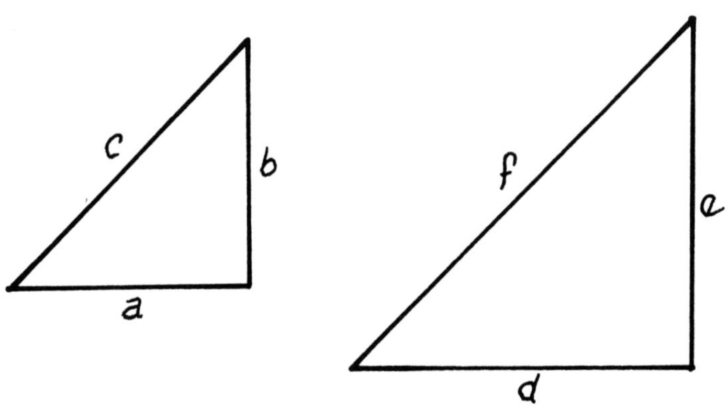

$$\frac{a}{d} = \frac{b}{e} = \frac{c}{f}$$

d. three sides of one triangle are parallel (or perpendicular) to three sides of the other.

The case of parallel sides is illustrated in examples a and b. The case of perpendicular sides is illustrated in example c.

In each case, the listed conditions are sufficient to insure the similarity of the triangles.

3. The altitudes of a triangle meet in a single point. (See **altitude** in glossary.)

4. The medians of a triangle meet in a point that is two-thirds of the distance from each vertex to the midpoint of the opposite side.

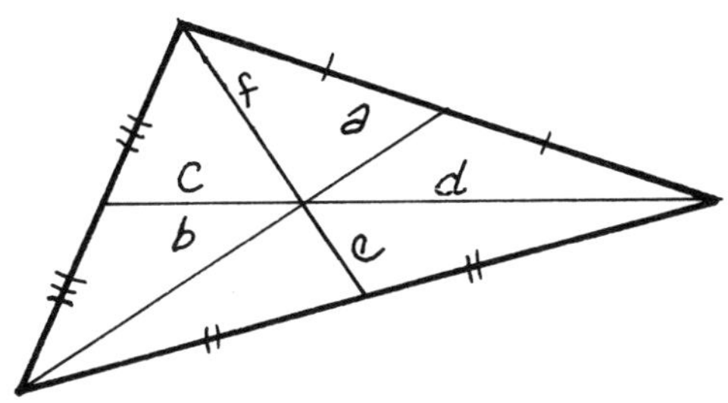

b = 2a
d = 2c
f = 2e

5. The angle bisectors of a triangle meet in a point that is the center of the inscribed circle.

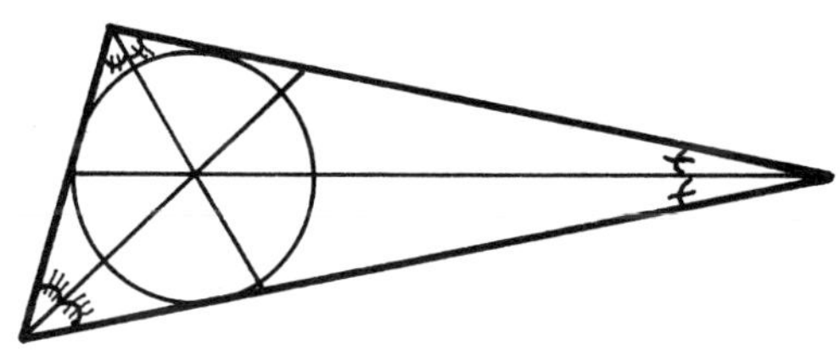

6. The perpendicular bisectors of the sides of a triangle meet in a point that is the center of the circumscribed circle.

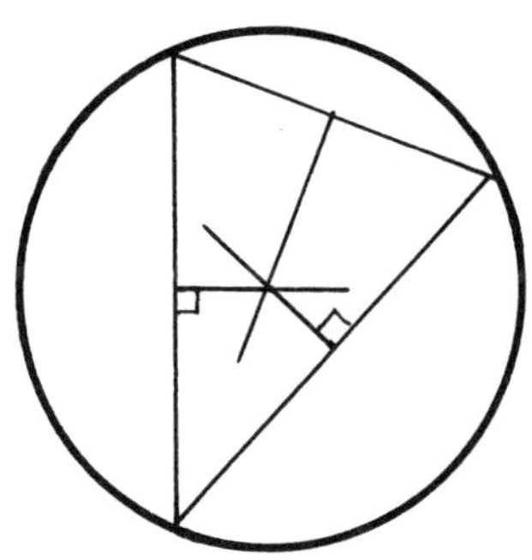

7. The line segment between the midpoints of two sides of a triangle is parallel to the third side and half as long.

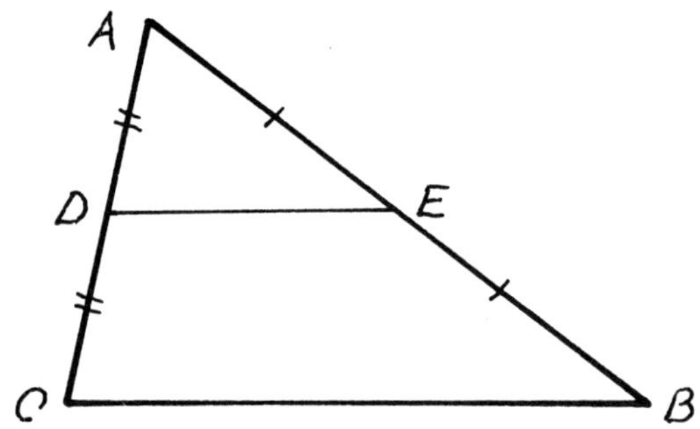

DE = $\frac{1}{2}$CB

8. The median to the hypotenuse of a right triangle is half as long as the hypotenuse.

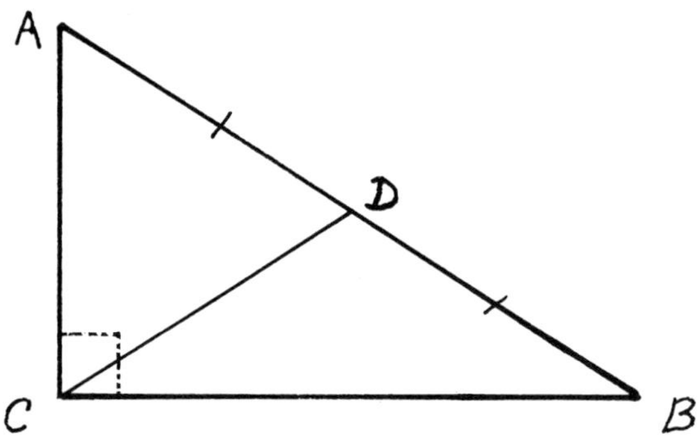

CD = $\frac{1}{2}$AB

9. In a 30-60-90 Triangle, the side opposite the 30° angle is half the length of the hypotenuse.

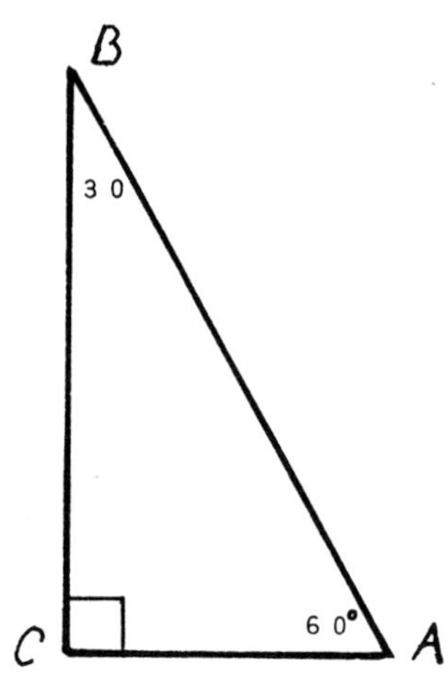

AC = $\frac{1}{2}$AB

10. In an Isosceles Right Triangle the length of the hypotenuse equals the length of the leg times $\sqrt{2}$.

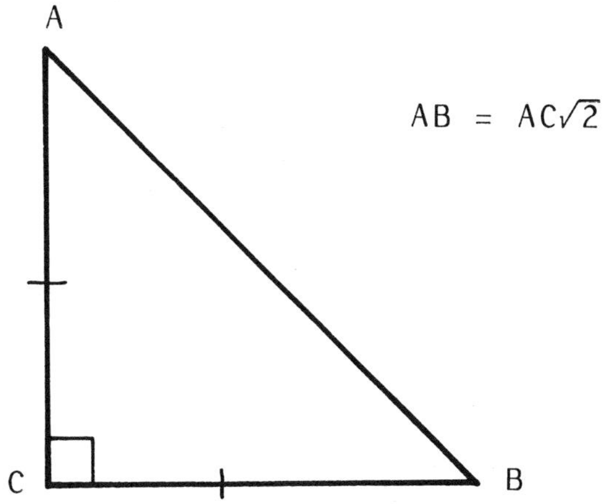

QUADRILATERALS

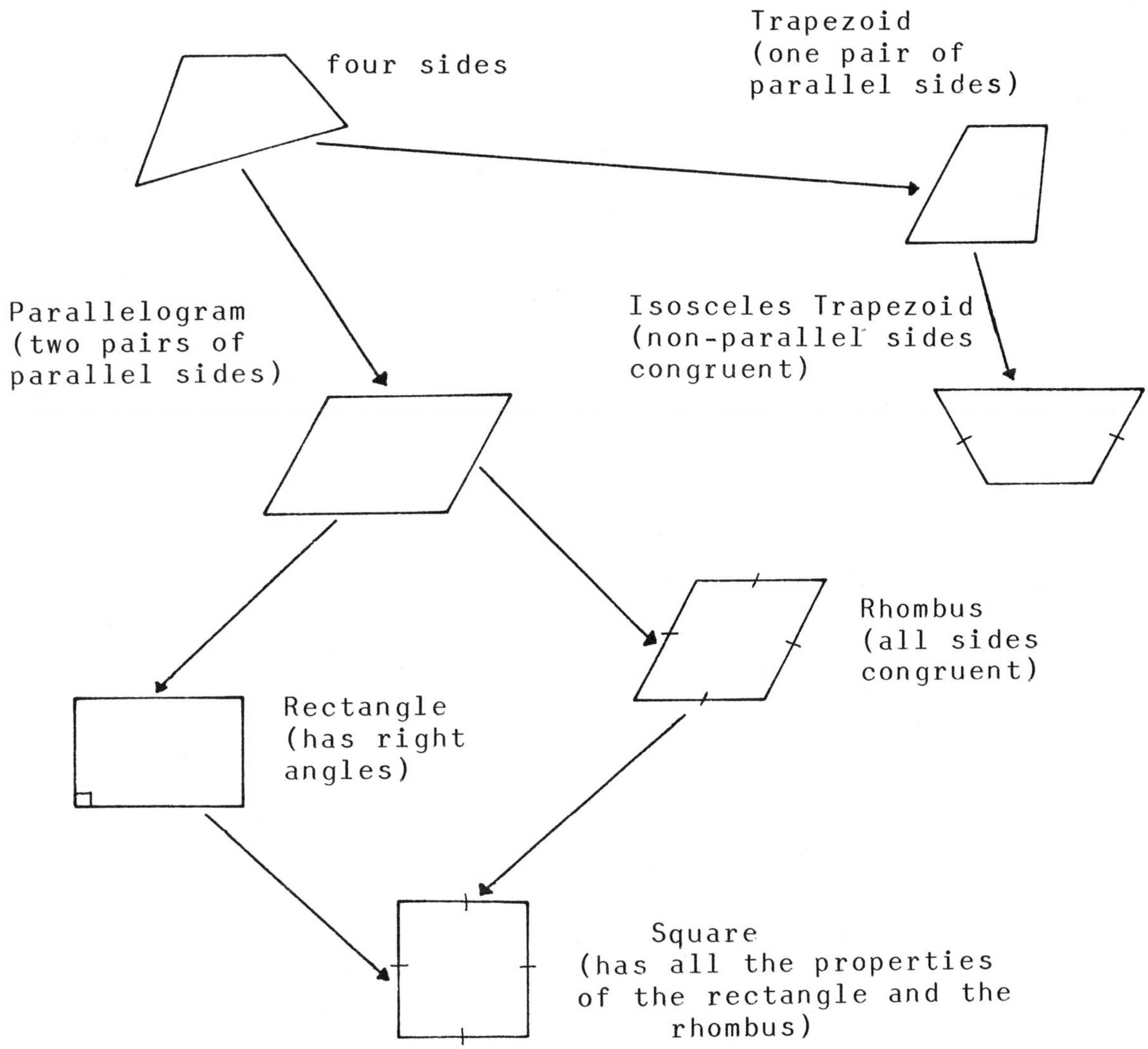

Quadrilaterals

1. In a parallelogram:
   a. opposite sides are congruent,
   b. opposite angles are congruent,
   c. the diagonals bisect each other,
   d. a diagonal divides the parallelogram into two congruent triangles.

a. 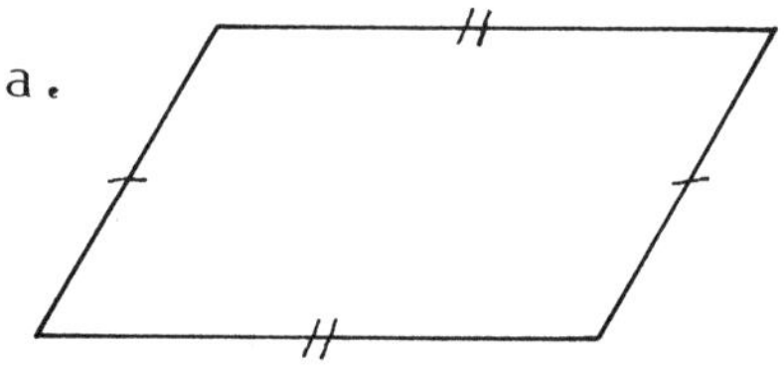

b. 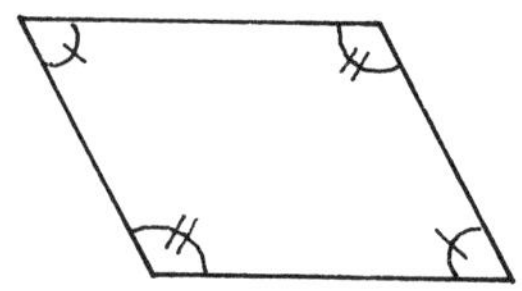

c. 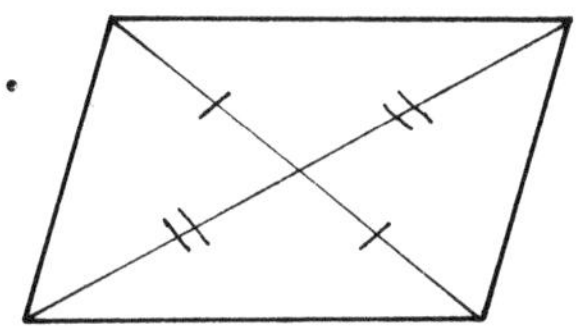

d. 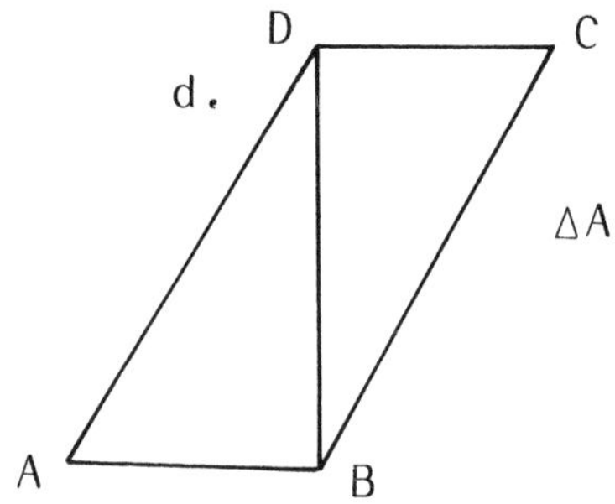

$\triangle ABD \cong \triangle CDB$

2. The diagonals of a rectangle are congruent.

$\overline{AC} \cong \overline{DB}$

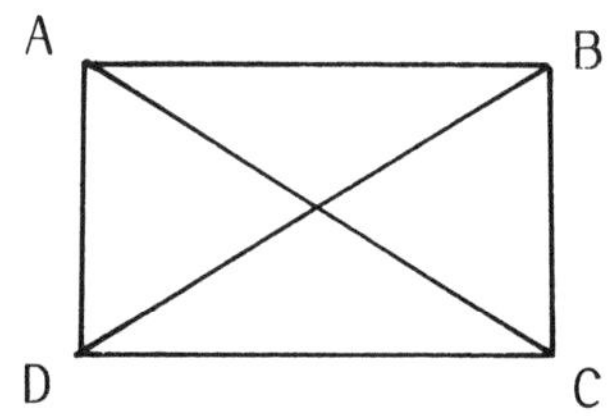

3. The diagonals of a rhombus are perpendicular.

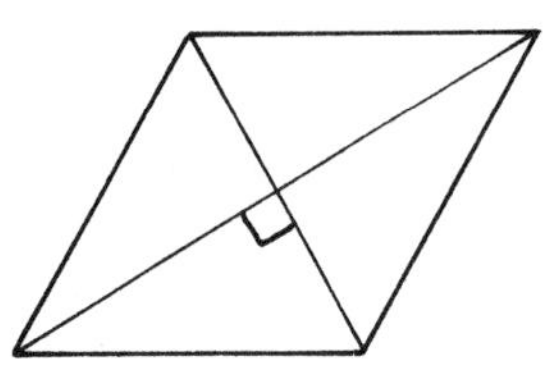

4. The line segment connecting the midpoints of the non-parallel sides of a **trapezoid** is parallel to the bases and half as long as the sum of their lengths.

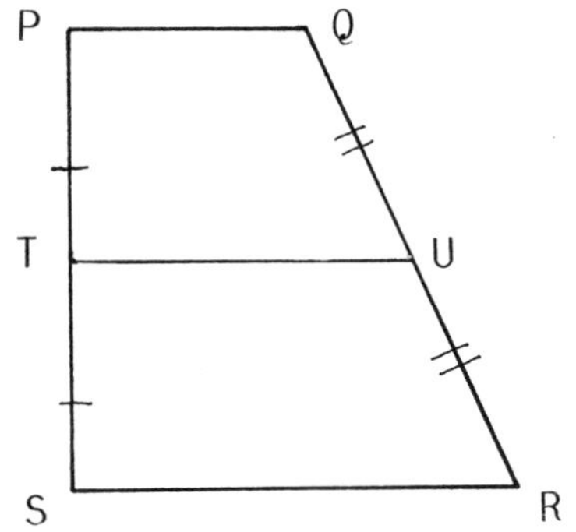

$\overleftrightarrow{PQ} \parallel \overleftrightarrow{TU} \parallel \overleftrightarrow{SR}$

$TU = \frac{1}{2}(PQ+SR)$

5. In an **isosceles trapezoid**:
   a. the base angles are congruent,
   b. the diagonals are congruent.

a.

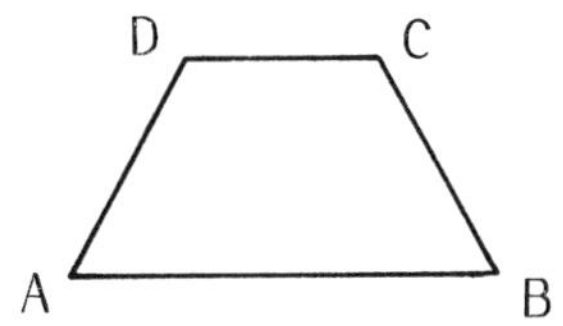

b.

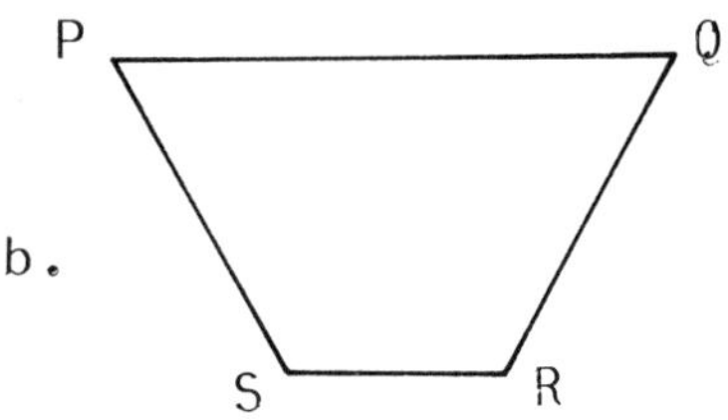

$\overline{PR} \cong \overline{SQ}$

CIRCLES

1. In the same circle or in congruent circles, congruent central angles cut congruent arcs, and conversely.

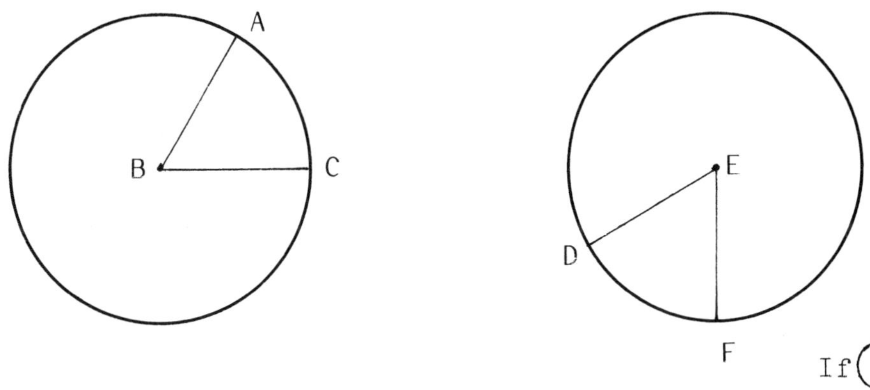

If $\odot B = \odot E$,

If $\triangle ABC \cong \triangle DEF$, then $\widehat{AC} \cong \widehat{DF}$, and

if $\widehat{AC} \cong \widehat{DF}$, then $\triangle ABC \cong \triangle DEF$.

2. In the same circle or in congruent circles, congruent chords cut congruent arcs, and conversely.

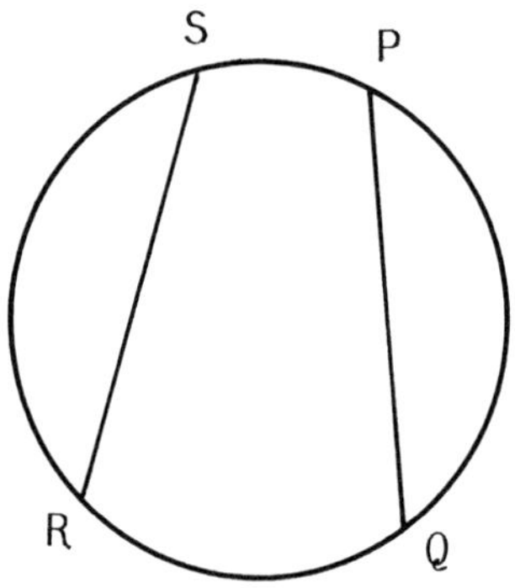

If $\overline{PQ} \cong \overline{RS}$, then $\widehat{PQ} \cong \widehat{RS}$, and

if $\widehat{PQ} \cong \widehat{RS}$, then $\overline{PQ} \cong \overline{RS}$.

3. In the same circle or in congruent circles, congruent chords are equidistant from the center.

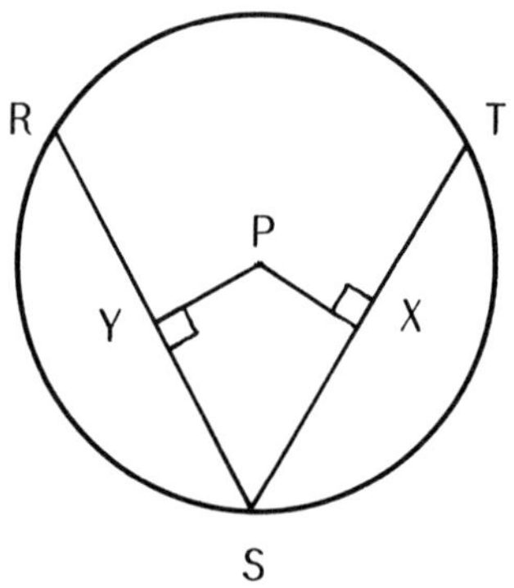

If $\overline{RS} \cong \overline{ST}$, then $\overline{XP} \cong \overline{PY}$.

4. A line perpendicular to a diameter at one endpoint is tangent to the circle, and conversely.

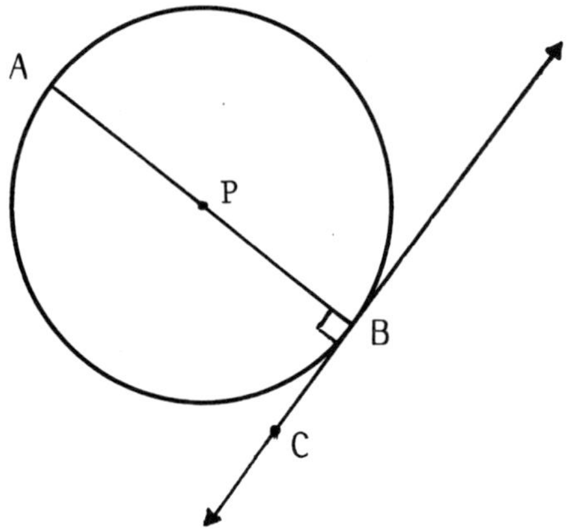

If $\overleftrightarrow{AB} \perp \overleftrightarrow{BC}$, then $\overleftrightarrow{BC}$ is tangent to $\odot$P, and if $\overleftrightarrow{BC}$ is a tangent line, then $\overleftrightarrow{AB} \perp \overleftrightarrow{BC}$.

5. Angles inscribed in the same arc are congruent.

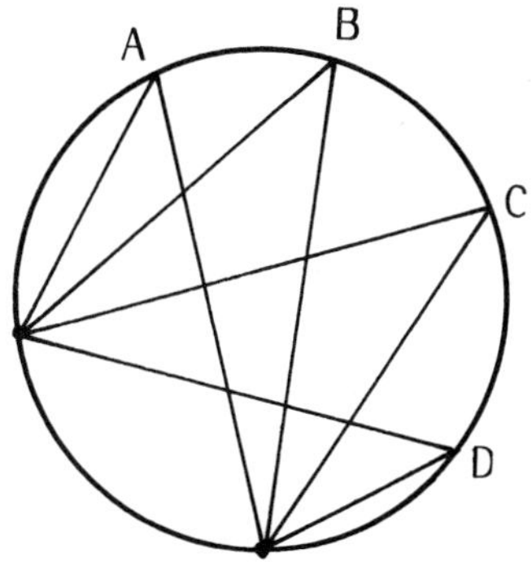

$\angle A = \angle B = \angle C = \angle D$

6. Any angle inscribed in a semicircle is a right angle.

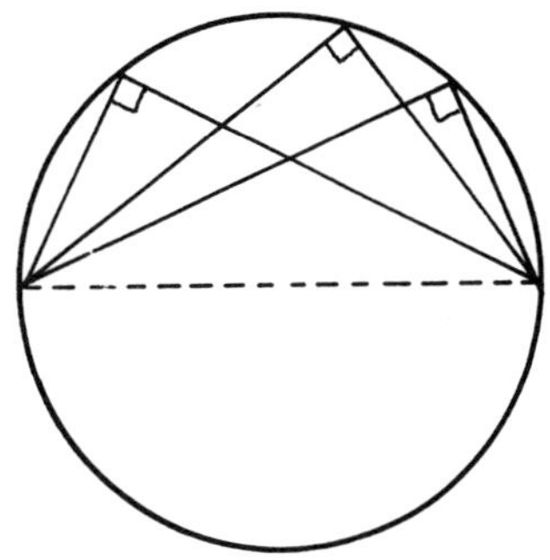

7. A circle, and only one circle, can be circumscribed about or inscribed in any triangle or any regular polygon.

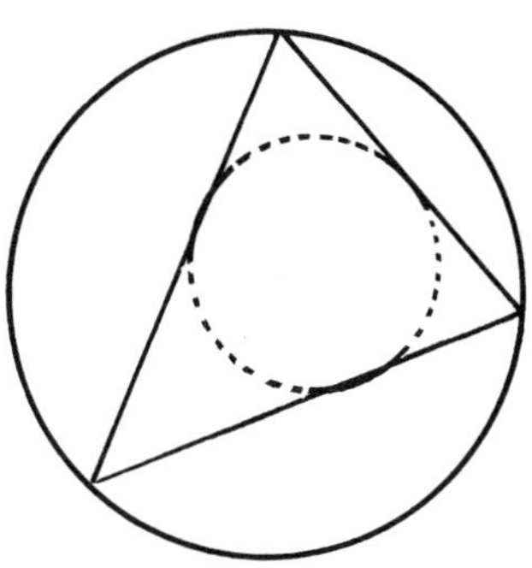

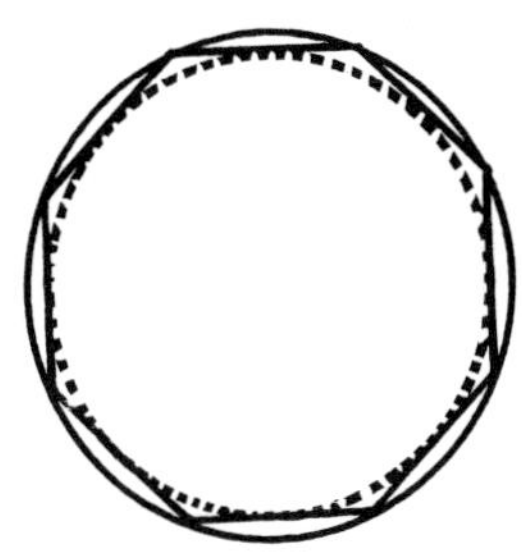

PROPERTIES OF EXPONENTS

1. If n is a positive integer, $a^n = \underbrace{a \cdot a \cdot a \cdots a}_{n \text{ times}}$

   eg. $3^4 = 3 \cdot 3 \cdot 3 \cdot 3 = 81.$

   Note: the exponent only operates on the quantity immediately to its left.

   eg. $2x^3 = 2 \cdot x \cdot x \cdot x$ but $(2x)^3 = 2x \cdot 2x \cdot 2x = 8x^3.$

2. $a^0 = 1$ for all real numbers $a \neq 0$. 0^0 is undefined.

   eg. $1{,}532{,}896^0 = 1.$

3. $a^{-n} = \frac{1}{a^n}$ for all real numbers $a \neq 0$.

   eg. $5^{-3} = \frac{1}{5^3} = \frac{1}{125}.$

4. $(a^n)(a^m) = a^{n+m}.$

   eg. $\phi^2 \cdot \phi^5 = \phi^{2+5} = \phi^7.$

5. $\frac{a^n}{a^m} = a^{n-m}.$

   eg. $\frac{x^5}{x^3} = x^{5-3} = x^2$ or $\frac{s^4}{s^5} = s^{4-5} = s^{-1} = \frac{1}{s}.$

6. $(a^m)^n = a^{mn}.$

   eg. $(n^2)^3 = n^{2 \cdot 3} = n^6.$

7. $a^{p/q} = (\sqrt[q]{a})^p = \sqrt[q]{a^p}.$

   eg. $8^{2/3} = (\sqrt[3]{8})^2 = 2^2 = 4.$

   $8^{2/3} = \sqrt[3]{8^2} = \sqrt[3]{64} = 4.$

PROPERTIES OF PHI

1. $\emptyset - 1 = \frac{1}{\emptyset}$

Proof: We know from the Divine Proportion that $\emptyset^2 - \emptyset - 1 = 0$, or $\emptyset^2 - \emptyset = 1$.

Dividing both sides by $\emptyset$ we get $\frac{\emptyset^2 - \emptyset}{\emptyset} = \frac{1}{\emptyset}$

but $\frac{\emptyset^2 - \emptyset}{\emptyset} = \frac{\emptyset^2}{\emptyset} - \frac{\emptyset}{\emptyset} = \emptyset - 1$.

Therefore, $\emptyset - 1 = \frac{1}{\emptyset}$.

2. $\emptyset \cdot \emptyset' = -1$

Proof: $\emptyset \cdot \emptyset' = \left(\frac{1+\sqrt{5}}{2}\right)\left(\frac{1-\sqrt{5}}{2}\right) = \frac{1-5}{4} = -\frac{4}{4} = -1.$

3. $\emptyset + \emptyset' = 1$

Proof: $\emptyset + \emptyset' = \frac{1+\sqrt{5}}{2} + \frac{1-\sqrt{5}}{2} = \frac{2}{2} = 1.$

4. $\emptyset^n + \emptyset^{n+1} = \emptyset^{n+2}$ for n an integer.

Proof: Let us first consider the case where $n \geq 0$.

Then $\emptyset^n + \emptyset^{n+1} = \emptyset^{n+2}$ for $n \geq 0$ is equivalent to

$\emptyset^{m-1} + \emptyset^m = \emptyset^{m+1}$ for $m \geq 1$.

(a) Using the method of mathematical induction, we first prove the statement true for m=1.

We know that $\emptyset^2 - \emptyset - 1 = 0$

so $1 + \emptyset = \emptyset^2$.

or $\emptyset^0 + \emptyset^1 = \emptyset^2$.

$\therefore$ true for m=1.

(b) Assume true for m=k.

Then $\emptyset^{k-1} + \emptyset^k = \emptyset^{k+1}$.

(c) Prove true for $m = k+1$.

Since $\phi^{k-1} + \phi^{k} = \phi^{k+1}$,

$\phi(\phi^{k-1} + \phi^{k}) = \phi \cdot \phi^{k+1}$

or $\phi^{k} + \phi^{k+1} = \phi^{k+2}$.

$\therefore$ true for $m = k+1$, and by the principle of mathematical induction the statement holds for all positive integral values of m.

Now let us consider the case when $n<0$. If $n<0$, then $\phi^{n} + \phi^{n+1} = \phi^{n+2}$ is equivalent to $\frac{1}{\phi^{m}} + \frac{1}{\phi^{m-1}} = \frac{1}{\phi^{m-2}}$

where $m = -n>0$.

(a') Again using mathematical induction, we first prove the statement true for $m=1$.

We know that ϕ differs from its reciprocal by 1,

that is, $\frac{1}{\phi} + 1 = \phi$.

But this is equivalent to $\frac{1}{\phi} + \frac{1}{1} = \frac{1}{\phi^{-1}}$

which in turn is equivalent to $\frac{1}{\phi^{1}} + \frac{1}{\phi^{0}} = \frac{1}{\phi^{-1}}$,

and the statement is proved for $m=1$.

(b') Assume true for $m=k$.

That is, $\frac{1}{\phi^{k}} + \frac{1}{\phi^{k-1}} = \frac{1}{\phi^{k-2}}$.

(c') Prove true for $m = k+1$.

Since $\frac{1}{\phi^k} + \frac{1}{\phi^{k-1}} = \frac{1}{\phi^{k-2}}$,

$$\frac{1}{\phi}\left(\frac{1}{\phi^k} + \frac{1}{\phi^{k-1}}\right) = \left(\frac{1}{\phi^{k-2}}\right)\frac{1}{\phi}$$

or $\frac{1}{\phi^{k+1}} + \frac{1}{\phi^k} = \frac{1}{\phi^{k-1}}$.

$\therefore$ true for $m = k=1$.

Now by the principle of mathematical induction the statement holds for all positive integral values of m. Since $m = -n$, we have

$$\frac{1}{\phi^{-n}} + \frac{1}{\phi^{-n-1}} = \frac{1}{\phi^{-n-2}}$$

or

$\phi^n + \phi^{n+1} = \phi^{n+2}$, and the theorem is proved for all integral values of n.

THE RELATIONSHIP BETWEEN CUBIT MEASURE AND PHI

Royal Cubit = $.5236 = \phi^2 + \phi + 1$

little cubit = $.4236 = \phi^2 + \phi$

In each case the unit is taken to be .1 meter.

PROOF OF THE CONSTRUCTION OF THE GOLDEN SECTION OF A LINE SEGMENT

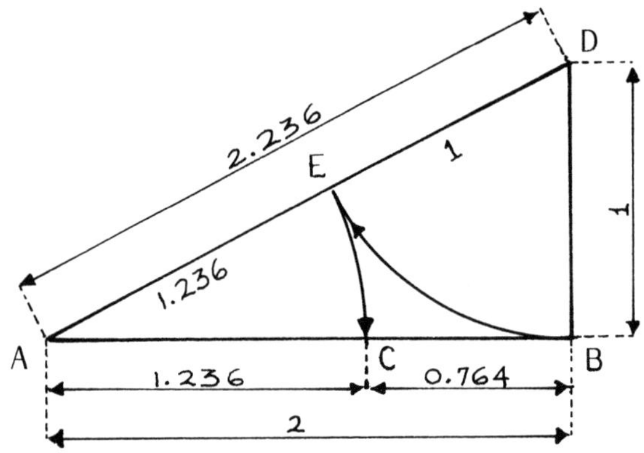

If we assign the length 2 to $\overline{AB}$, then we construct $\overline{BD}$ such that BD = 1.

By the pythagorean Theorem, AD = $\sqrt{5} \doteq 2.236$.

Point E divides $\overline{AD}$ into segments AE and ED having lengths 1.236 and 1 respectively.

Then point C divides $\overline{AB}$ into segments AC and CB having lengths 1.236 and .764 respectively.

$$\frac{1.236}{.764} = 1.618 = \emptyset$$

$\therefore$ C is the Golden Section of $\overline{AB}$.

LENGTHS OF THE DYNAMIC RECTANGLES AND THEIR RECIPROCALS

Name	Length	Length of Reciprocal
$\sqrt{2}$ Rectangle	1.4142	.7071
$\sqrt{3}$ Rectangle	1.7321	.5773
$\sqrt{4}$ Rectangle	2.0000	.5000
$\sqrt{5}$ Rectangle	2.2361	.4472
Golden Rectangle	1.6180	.6180
$\sqrt{\emptyset}$ Rectangle	1.2720	.7862
$\emptyset$+1 Rectangle	2.6180	.3820

PROOF OF THE CONSTRUCTION OF THE DYNAMIC RECTANGLES

Given square ABCD with sides of length 1, prove the resulting rectangles are root rectangles.

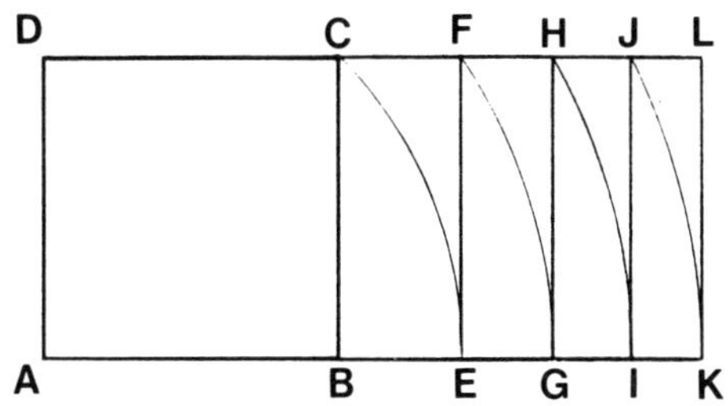

I. $AB=1$, $BC=1$ and, by the Pythagorean Theorem, $AC^2=AB^2+BC^2$ or

$AC^2=1^2+1^2=1+1=2$ or $AC=\sqrt{2}$.

E and C are equidistant from A. Therefore $AE=\sqrt{2}$, and AEFD is a $\sqrt{2}$ Rectangle.

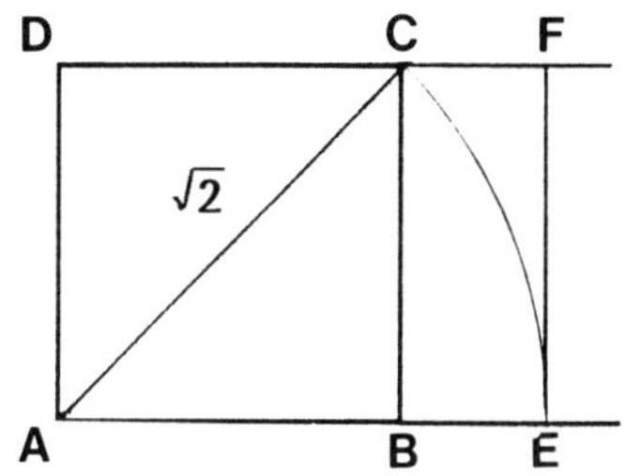

II. $AE=\sqrt{2}$, $EF=1$ and $AF^2=AE^2+EF^2$.

$AF^2=\sqrt{2}^2+1^2=2+1=3$ so $AF=\sqrt{3}$.
F and G are equidistant from A. Therefore $AG=\sqrt{3}$, and AGHD is a $\sqrt{3}$ Rectangle.

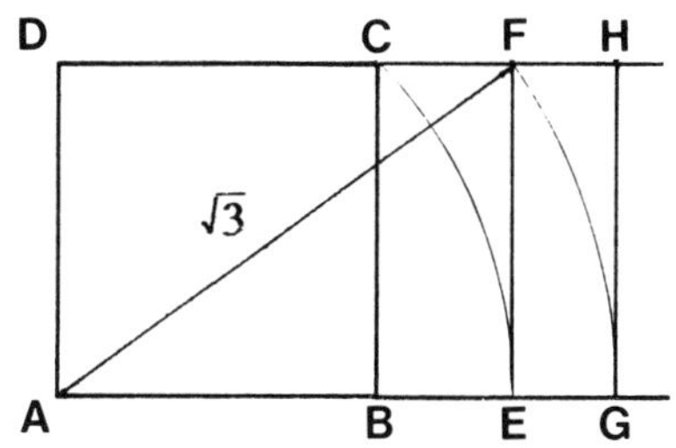

III. $AG=\sqrt{3}$, $GH=1$ and $AH^2=AG^2+GH^2$.

$AH^2=\sqrt{3}^2+1^2=3+1=4$ so $AH=\sqrt{4}$.
H and I are equidistant from A. Therefore $AI=\sqrt{4}$, and AIJD is a $\sqrt{4}$ Rectangle.

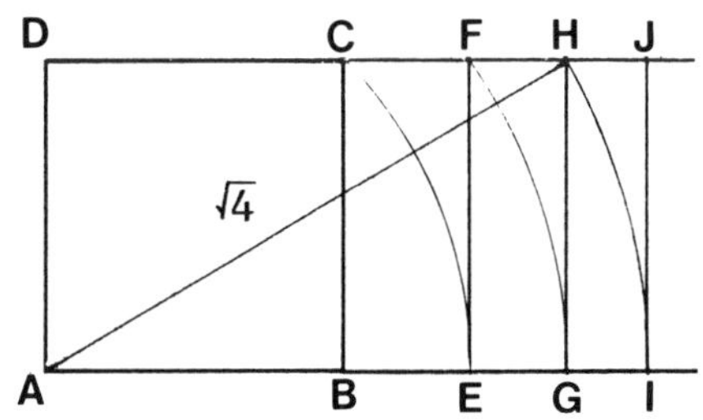

IV. $AI=\sqrt{4}$, $IJ=1$ and $AJ^2=AI^2+IJ^2$.

$AJ =\sqrt{4}^2+1^2=4+1=5$ so $AJ=\sqrt{5}$.
J and K are equidistant from A. Therefore $AK=\sqrt{5}$, and AKLD is a $\sqrt{5}$ Rectangle.

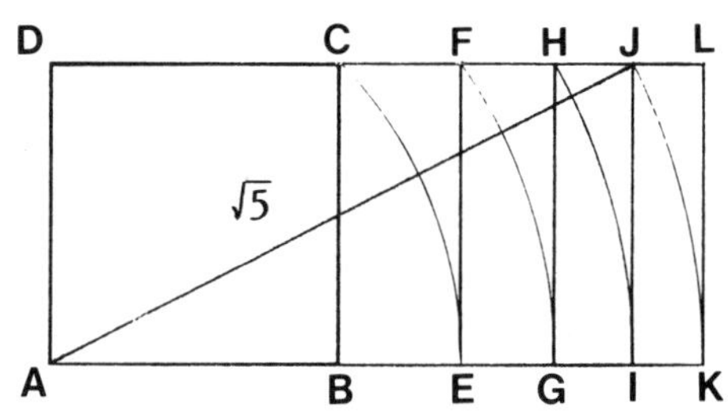

PROOFS OF THE PROPERTIES OF THE FIBONACCI NUMBERS

I. The sum of the first n terms of the Fibonacci sequence is 1 less than the second term beyond F_n, or

$$F_1+F_2+F_3+\cdots+F_n=F_{n+2}-1.$$

Proof: Since the Fibonacci sequence is a summation sequence, $F_n+F_{n+1}=F_{n+2}$ and therefore $F_n=F_{n+2}-F_{n+1}$.

$$F_1=F_3-F_2=F_3-1$$

$$F_2=F_4-F_3$$

$$F_3=F_5-F_4$$

$$\vdots$$

$$F_{n-1}=F_{n+1}-F_n$$

$$F_n=F_{n+2}-F_{n+1}$$

By adding both sides of the equations we have

$$F_1+F_2+F_3+\cdots+F_{n-1}+F_n=F_{n+2}-F_{n+1}+F_{n+1}-\cdots-F_4+F_4-F_3+F_3-1.$$

Or $F_1+F_2+F_3+\cdots+F_n=F_{n+2}-1$.

II. The sum of the even numbered terms of the Fibonacci sequence through F_{2n} is one less than F_{2n+1}, or

$$F_2+F_4+F_6+\cdots+F_{2n}=F_{2n+1}-1.$$

Proof: $F_n+F_{n+1}=F_{n+2}$ or $F_{n+1}=F_{n+2}-F_n$.

$$F_2=F_3-F_1=F_3-1$$

$$F_4=F_5-F_3$$

$$F_6=F_7-F_5$$

$$\vdots$$

$$F_{2n-2}=F_{2n-1}-F_{2n-3}$$

$$F_{2n}=F_{2n+1}-F_{2n-1}$$

Adding we get: $F_2+F_4+F_6+\cdots+F_{2n-2}+F_{2n}=F_{2n+1}-F_{2n-1}+F_{2n-1}-\cdots-F_5+F_5-F_3+F_3-1$. Or $F_2+F_4+F_6+\cdots+F_{2n}=F_{2n+1}-1$.

III. The sum of the odd numbered terms of the Fibonacci sequence through F_{2n-1} is the next term, F_{2n}, or

$$F_1+F_3+F_5+\cdots+F_{2n-1}=F_{2n}.$$

Proof:

$$F_1=1$$

$$F_3=F_4-F_2=F_4-1$$

$$F_5=F_6-F_4$$

$$\vdots$$

$$F_{2n-3}=F_{2n-2}-F_{2n-4}$$

$$F_{2n-1}=F_{2n}-F_{2n-2}$$

Adding we get: $F_1+F_3+F_5+\cdots+F_{2n-3}+F_{2n-1}=F_{2n}-F_{2n-2}+F_{2n-2}-\cdots-F_4+F_4-1+1$, or $F_1+F_3+F_5+\cdots+F_{2n-1}=F_{2n}$.

IV. The square of a Fibonacci number and the product of the Fibonacci numbers that precede and follow it in the sequence differ by one, or

$$(F_n)^2-(F_{n-1})(F_{n+1})=\pm 1.$$

Proof: We can adjust the subscripts without altering the sense of the statement to read $(F_{n+1})^2=(F_n)(F_{n+2})+(-1)^n$ and the exponent n indicates whether the squared term is one more than the product or one less than the product, since $(-1)^n=1$ if n is even and $(-1)^n=-1$ if n is odd.

This property of the Fibonacci sequence can be proved by mathematical induction.

(a) Prove the statement true for n=1.

$$(F_{1+1})^2=(F_2)^2=1^2=1.$$

$$(F_1)(F_{1+2})+(-1)^1=(F_1)(F_3)-1=(1)(2)-1=2-1=1.$$

Therefore, $(F_{1+1})^2=(F_1)(F_{1+2})+(-1)^1$; and the statement is true when n=1.

(b) Assume the statement true for n=k.

Then $(F_{k+1})^2=(F_k)(F_{k+2})+(-1)^k$.

(c) Prove true for n=k+1

By adding $(F_{k+1})(F_{k+2})$ to both sides of the equation in II. we get

$$(F_{k+1})^2+(F_{k+1})(F_{k+2})=(F_k)(F_{k+2})+(F_{k+1})(F_{k+2})+(-1)^k$$

or, by factoring $F_{k+1}(F_{k+1}+F_{k+2})=F_{k+2}(F_k+F_{k+1})+(-1)^k$

But $F_{k+1}+F_{k+2}=F_{k+3}$ and $F_k+F_{k+1}=F_{k+2}$, so we can substitute

to get

$$(F_{k+1})(F_{k+3})=(F_{k+2})^2+(-1)^k \text{ or } (F_{k+2})^2=(F_{k+1})(F_{k+3})+(-1)^{k+1}$$

Therefore, the statement is true for n=k+1 and by the principle of mathematical induction is then true for all values of n.

Photo by Richard Newman

J. Michael Pope
Chapter 5, Project 1.

Student Work

GEOMETRIC FORMULAS

TWO DIMENSIONAL FIGURES

Triangle

Area: $A=\frac{1}{2}bh$

where b is the measure of any side and h is the measure of the altitude to that side.

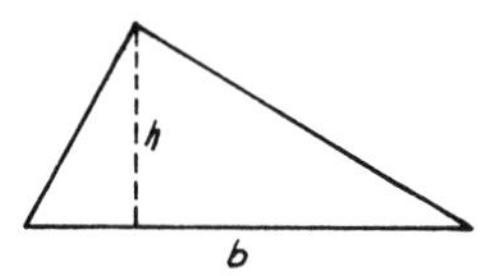

$A=\sqrt{s(s-a)(s-b)(s-c)}$

where a, b and c are the lengths of the sides and $s=\frac{1}{2}(a+b+c)$.

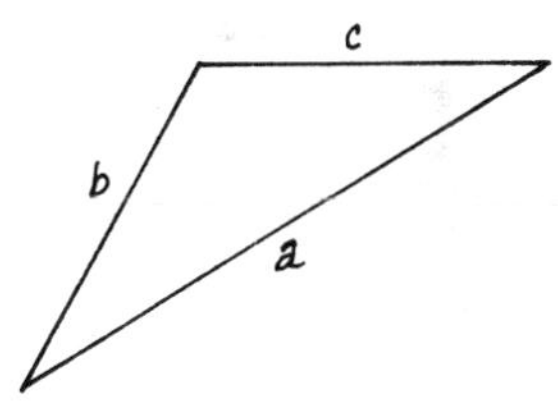

Right Triangle

Pythagorean Theorem:

$a^2+b^2=c^2$

where a and b are the measures of the legs and c is the measure of the hypotenuse.

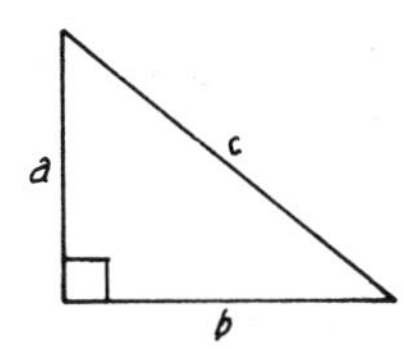

Square

Area: $A= s^2$

Perimeter: $P=4s$

where s is the measure of a side.

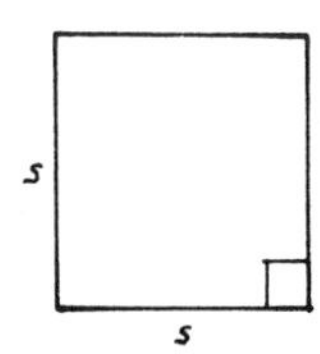

Rectangle

Area: $A=lw$

Perimeter: $P=2l+2w$

where l is the measure of the length and w is the measure of the width.

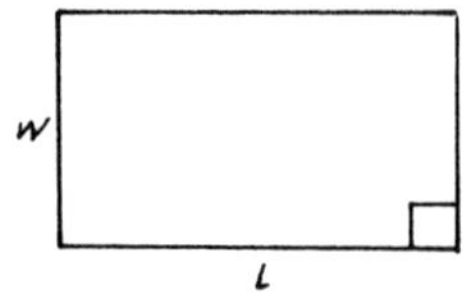

Parallelogram

Area: $A=bh$

where b is the measure of a side and h is the measure of an altitude to that side.

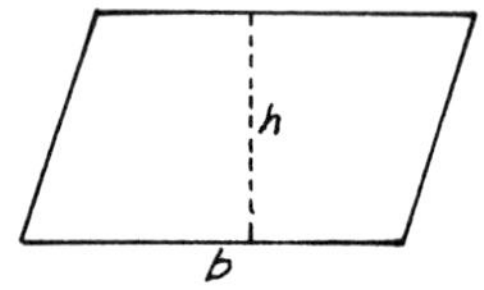

Rhombus

Area: $A=\frac{1}{2}ab$

where a and b are the lengths of the diagonals.

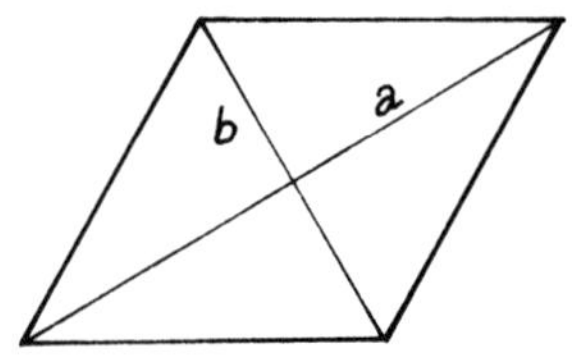

Trapezoid

Area: $A=\frac{1}{2}h(a+b)$

where a and b are the measures of the parallel sides and h is the distance between those sides.

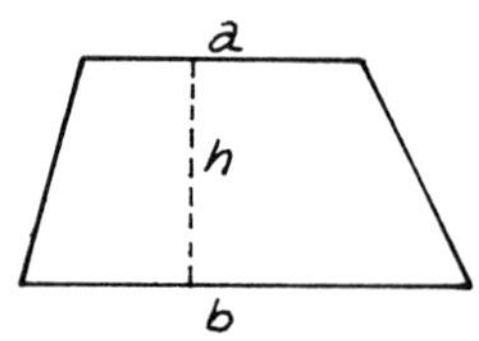

Circle

Area: $A=\pi r^2$

Circumference: $C=\pi d=2\pi r$

where r and d are the radius and diameter respectively, and π may be approximated by 3.14 or $\frac{22}{7}$.

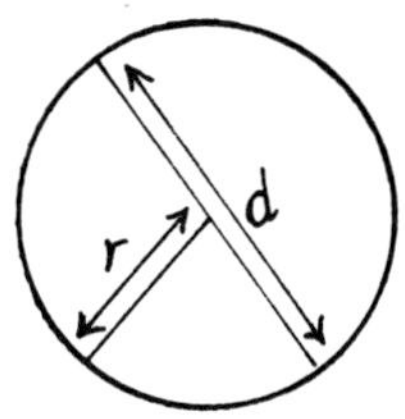

Regular Polygon

Area: $A=\frac{1}{2}asn$

Perimeter: $P=sn$

where a is the measure of the apothem, s is the measure of a side, and n is the number of sides.

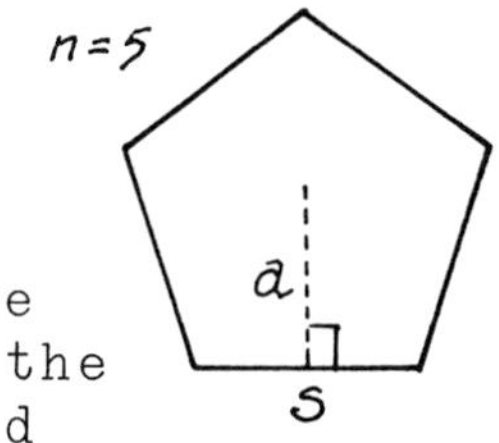

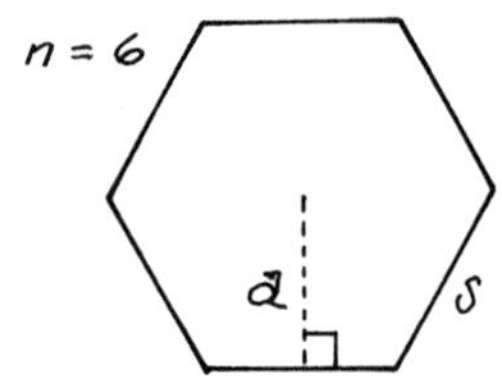

THREE DIMENSIONAL FIGURES

Cube

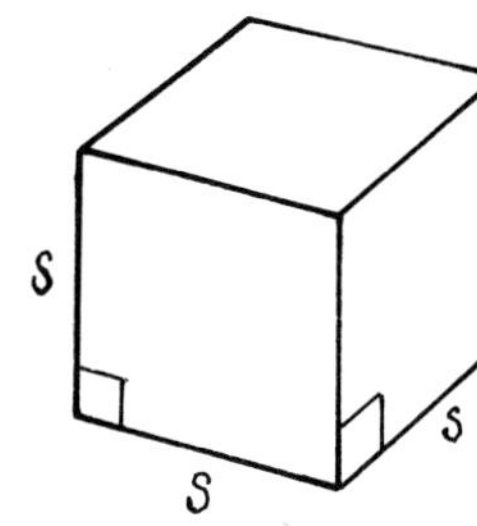

Volume: $V=s^3$
Surface Area: $S=6s^2$
where s is the measure of a side.

Rectangular Solid

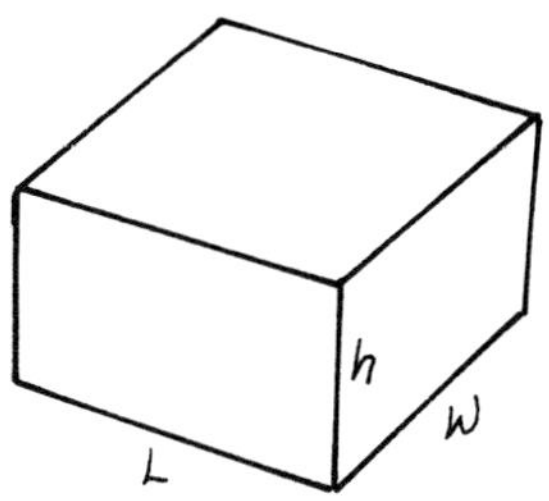

Volume: $V=lwh$
Surface Area: $S=2lw+2wh+2lh$
where l is the length, w is the width, and h is the height.

Sphere

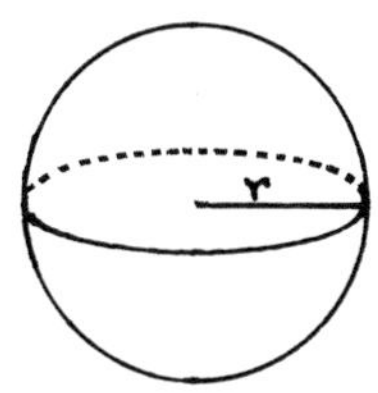

Volume: $V=\frac{4}{3}\pi r^3$
Surface Area: $S=4\pi r^2$
where r is the radius and π may be approximated by 3.14 or $\frac{22}{7}$.

Right Circular Cylinder

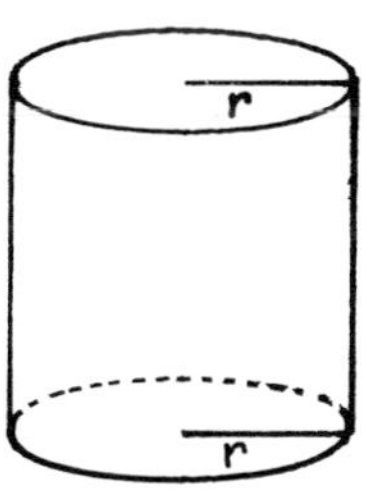

Volume: $V=\pi r^2 h$
Lateral Surface Area: $L=2\pi rh$
(doesn't include the circular ends)
Total Surface Area: $S=2\pi rh+2\pi r^2$
where r is the radius and h is the height, and π may be approximated as above.

Right Circular Cone

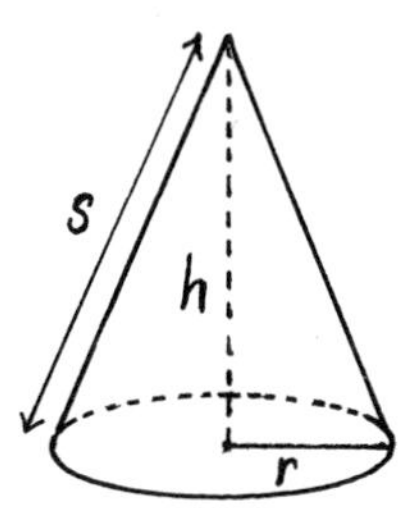

Volume: $V=\frac{1}{3}\pi r^2 h$
Lateral Surface Area: $L=\pi rs$
(doesn't include the circular base)
Total Surface Area: $S=\pi r^2+\pi rs$
Slant Height: $s=\sqrt{r^2+h^2}$
where r is the radius, h is the height, s is the slant height, and π is as above.

Right Pyramid

Volume: $V=\frac{1}{3}hA$

Lateral Surface Area: $L=\frac{1}{2}sP$

(doesn't include the area of the base)

Total Surface Area: $S=\frac{1}{2}sP+A$

where A is the area of the base (choose the appropriate formula for the shape of the base), P is the perimeter of the base (again choose the appropriate formula), h is the height, and s is the slant height (the altitude of one of the triangular faces).

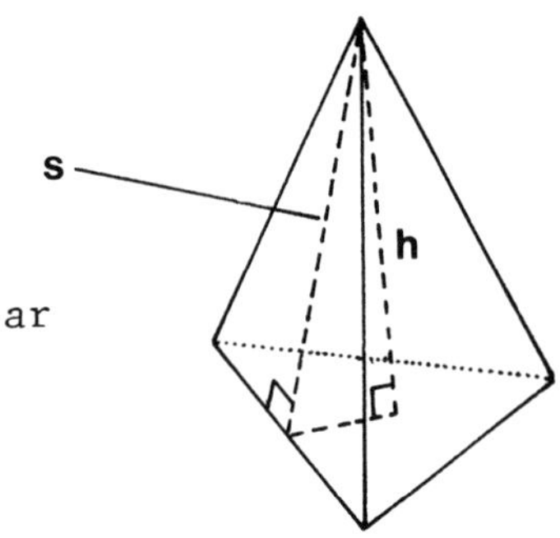

(a) Triangular Pyramid

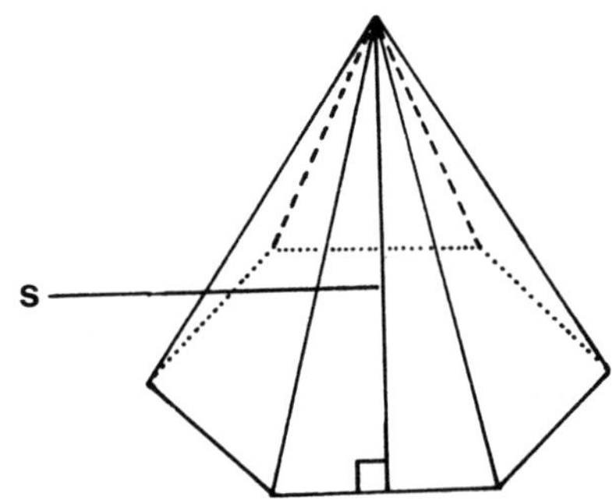

(c) Hexagonal Pyramid

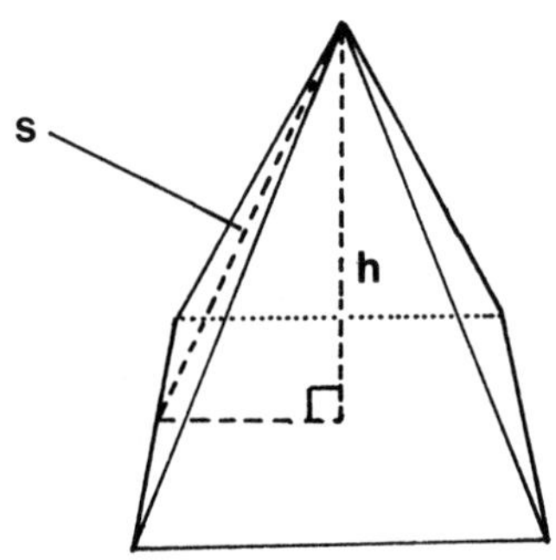

(b) Square Pyramid

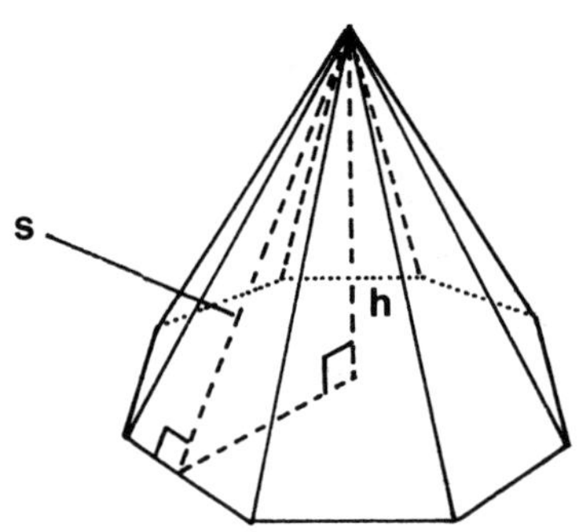

(d) Octagonal Pyramid

Student Work
Chapter 1, Project 1

Carol Gagnon

D

ART MATERIALS AND TECHNIQUES

Contents

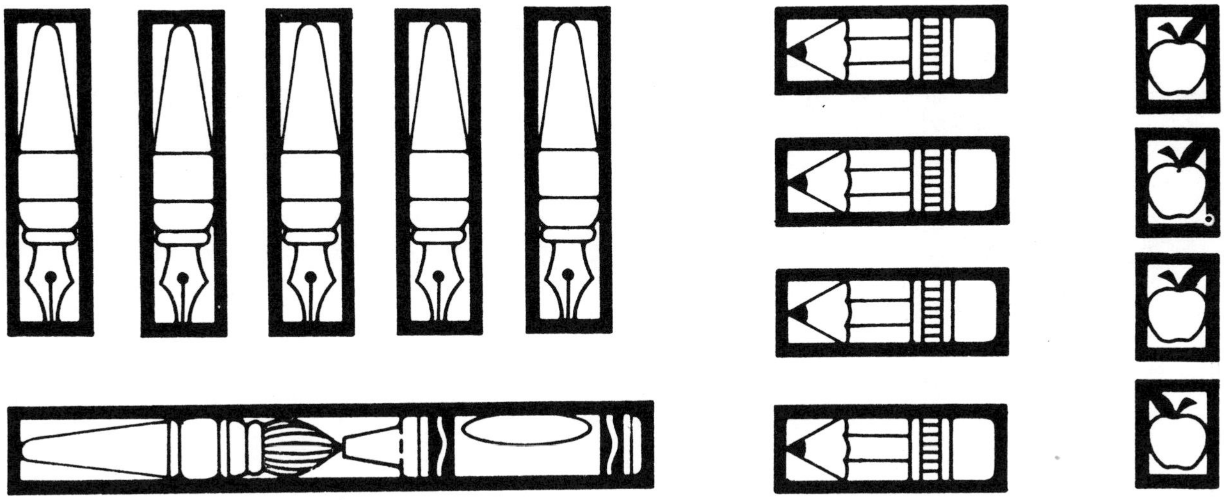

BASIC MATERIALS

The following materials are suggested for the reader who is just beginning his or her investigation of art. This list is by no means complete as the modern market abounds in art supplies of every description, but rather is a list of basic supplies, relatively easy to use, offered with the hope that the first attempts at artistic endeavors will be successful and will encourage the reader to continue.

Adhesives

1. Glue sticks for use on paper
2. Sobo craft glue (heavier than ordinary white glue)
3. Contact cement or epoxy (two part) Devcon #731
   (choice of adhesive depends on material to be glued)

Brushes

1. Pentel color brush - black - with a couple of refills. These come in colors other than black. They are a cross between a marker and a Japanese calligraphy brush.
2. Sable type for use with acrylic polymer paints. At least three different sizes ranging from fine for detail work to ½" for fill-in work.

Construction Tools

1. Metal straightedge
2. Compass - with opening wide enough for good sized circles
3. Scissors
4. Utility knife with extra blades
5. T-square and triangles

Fabric Crayons

These can be used to draw original designs on paper which can then be heat transferred by iron to fabric. The design is reversed when transferred.

Markers

These come in a great many varieties from water base to permanent; from transparent to opaque; from flourescent to metallic.
Do not buy in quantity before testing a particular type of marker on a particular kind of paper.

Paints

1. Gouache - water base opaque paints - not necessarily permanent.
2. Acrylic - water base but permanent when dry.
   Include: (2 oz. tubes)

white	yellow
black	blue
red	violet or purple
matte or gloss medium	

   These can be used transparently or opaquely depending upon the desired effect. Good coverage may require between three and five coats. Brushes should be kept in water to prevent the paint from hardening and ruining them. This can happen very quickly.

3. Alkyd house paint - oil base - permanent.

Paper

1. Pad of white all-purpose drawing paper 18"x24"
2. Package of colored construction paper
3. Bristol type paper - comes singly or in pads - very good for markers
4. Illustration board - cold pressed has smooth surface
   hot pressed has textured surface

Pencils

1. Graphite pencils

H = hard	The number and the letter on the side
B = soft	of the pencil indicates degree of hardness or softness. HB is a middle range.

   $H^{4-3-2-1}HB^{1-2-3-4-5}B$

2. Colored pencils
   These can be purchased individually or in sets, in hard or soft types, and can be blended by drawing on top of one another. To obtain clean, precise lines and reduce smudging, it would be best to avoid crayons, oil pastels, pastels, and charcoal.

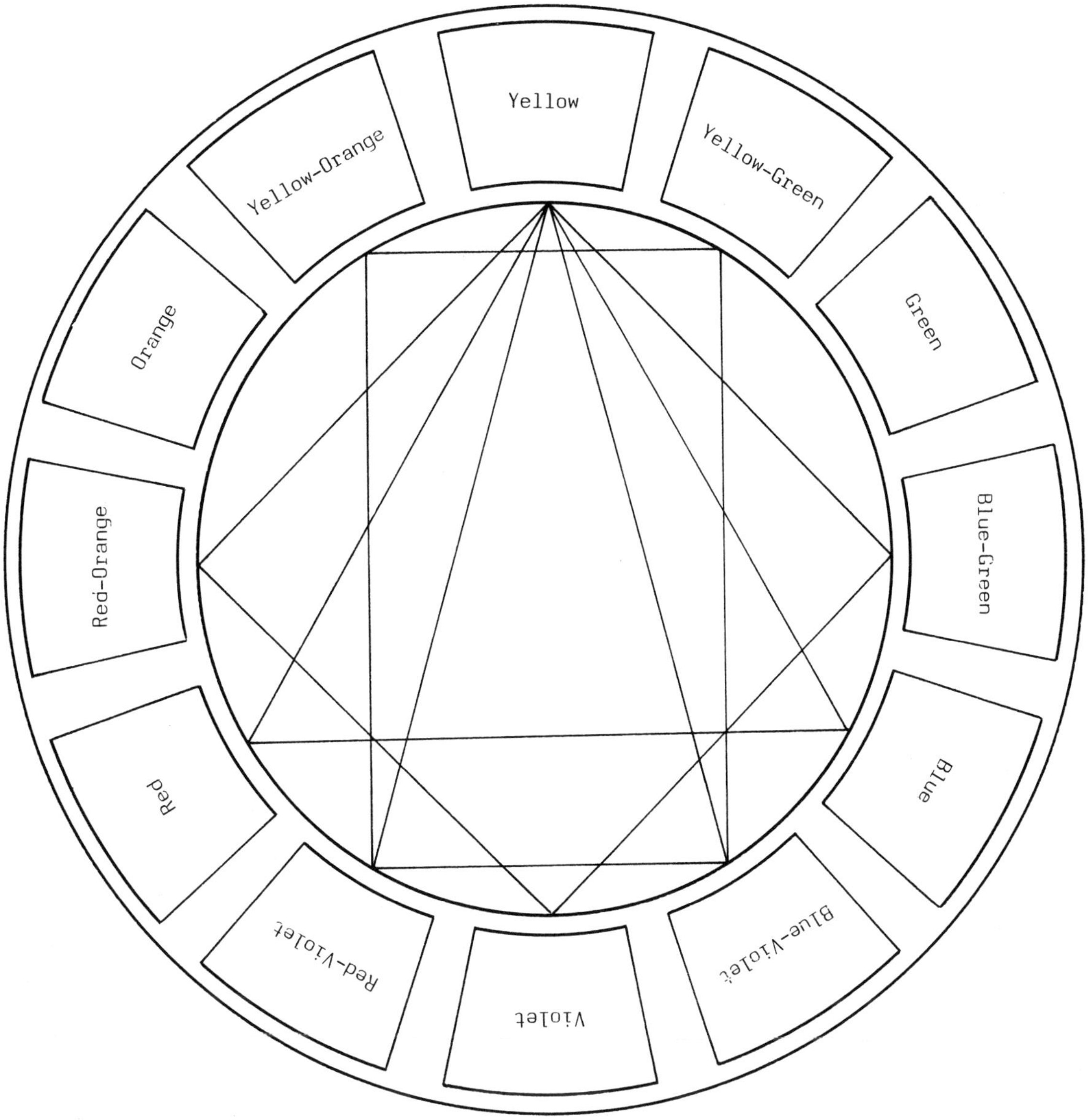

COLOR WHEEL

Yellow, blue and red are the primary colors. All the other colors on the wheel can be mixed from these three colors.

A secondary color is obtained by mixing a pair of primary colors. The secondary colors are orange, green and violet.

A mixture of a primary and a secondary color yields one of the tertiary colors, which are the remaining colors on the wheel.

For all practical purposes, all colors can be obtained by using the primary colors and black and white.

COLOR HARMONIES

In order for a color scheme to be harmonious, a few basic guidelines should be followed in limiting the number of colors used. This can be done by using one of the following schemes:

a. **Monochromatic:** Uses one color and light and dark variations thereof.

b. **Analogous:** Uses three adjacent colors and light and dark variations thereof.

c. **Complementary:** Uses a pair of colors directly opposite one another on the color wheel and light and dark variations thereof.

d. **Triad:** Uses three colors that form the vertices of either an isosceles or an equilateral triangle on the wheel and light and dark variations thereof.

e. **Tetrad:** Uses four colors that form the vertices of a rectangle (including the square) and light and dark variations thereof.

PAPERS AND THEIR USES

TYPE	WEIGHT	CHARACTERISTICS	TEXTURE AND COLOR	USES
Albanene	Light	100% rag, transparent, easily erasable, will not yellow or crack	White, smooth	Tracing
Bristol board	Light to heavy	Tough, many thicknesses	Very smooth, white	Acrylic studies, sculpture
Bond	Light	Somewhat fragile	White, smooth	Pencil studies
Carbon transfer	Very light	Carbon on one side	Several colors	Transfer
Construction	Medium	Inexpensive, fades easily	Full range of color	All-purpose, temporary
Graphite	Light	One side coated with graphite	Gray	Transfers to board and paper
Illustration board	Light to heavy	Lies flat, consistent working surface	White, grays, blacks smooth and rough surfaces	Polyhedra models, acrylic studies
Craft	Medium	Strong	Brown	All purpose
Metal foil	Light	Paper backed with foil	Varied color, shiny and matte	Sculpture
Oak tag	Heavy	Strong, folds without cracking	Full range of colors, smooth	Polyhedra models, paper sculpture
Press-apply	Light	Adhesive backed	Smooth, intense rainbow colors	Surface decoration
Scorasculpture (Dennison Co.)	Medium	Easily scored, very workable	Smooth, white	Polyhedra models, paper sculpture
Tissue	Very Light	Translucent	Full range of color, plain or variegated	Surface decoration
Velour	Medium	Flocked, velvet-like surface	Full range of color	Surface decoration

ADHESIVES AND THEIR USES

General Comments:

* Labels should be checked for proper use of the product.

* When gluing dissimilar materials, their porosity should be determined. When combining two types of materials, the type recommended for the non-porous material should be used.

* Duco Cement or LePage's Household Cement are good all-purpose adhesives. LePage's Glue is good for general paper work.

* Shoe Patch, purchased at a sporting goods store, is a very strong adhesive that might be tried when others fail.

* Polymer medium is useful for preserving, glazing and sealing. It comes in both matte and gloss finish.

* With the exception of polymer medium, available at art supply stores, all other adhesives can be found in hardware stores.

* Adhesives should be tested on a small sample first.

*	Contact cement	Epoxy cement	Epoxy metal cement	Plastic resin urea base	Polymer medium (acrylic emulsion)	Resin glue	Rubber cement	Water-proof glue	White glue	Materials to be glued
N		Ø								China and Glass, ceramic tiles
P				Ø				Ø		Collagraph items for printmaking
P					Ø	Ø			Ø	Cloth with wood
P	Ø				Ø	Ø			Ø	Leather and wood
N	Ø	Ø	Ø							Metal to metal
N	Ø	Ø	Ø			Ø			Ø	Metal to wood
P					Ø	Ø			Ø	Paper and cloth
P					Ø	Ø	Ø		Ø	Paper to paper and cardboard**
P	Ø			Ø	Ø	Ø				Plastic foams (styrofoam)
N				Ø	Ø	Ø				Plastics, vinyls
N	Ø			Ø		Ø		Ø	Ø	Plastic to wood
N							Ø	Ø		Rubber
N	Ø	Ø	Ø							Rubber to wood or metal
N		Ø								Stone and concrete to other items
N				Ø	Ø	Ø			Ø	Sand, stones, beads
P				Ø				Ø		Wood outdoors
P				Ø		Ø		Ø	Ø	Wood to wood

* Porosity P = porous N = nonporous

** **PAPER ADHESIVES**	**CHARACTERISTICS**	**USES**
Glue stick	Very tacky, very strong	Fine for adhering tabs and small areas
Metylan cellulose paste	Inexpensive, dissolves in water	Paper maché and collage
Mucilage	Amber colored, tacky, syrupy	Cardboard
Polyvinyl acetate (Elmer's, Sobo, Duratite, etc.)	White, dries clear, strong	Decoupage, paper maché
Plastic glaze (Mod Podge, Art Podge)	White, dries clear, strong	For protecting paper surfaces
Rubber cement	Transparent, dries quickly, excess easily rubbed away	Excellent for temporary bonding in large areas

MATERIALS AND TECHNIQUES FOR POLYHEDRA CONSTRUCTIONS

Tools

1. Compass
2. Metal Straightedge
3. Utility knife or X-acto Knife with #11 blade
4. Hole punch (optional)
5. Bobby pins or various size paper clips or paper clamps for holding joints together temporarily. Spring clothes pins work for holding joints on larger pieces.
6. Cutting board of either cardboard, chipboard or masonite, or the more expensive self-healing cutting board bought at architectural supply firms.
7. Tweezers

Materials

1. Pencils
2. Choice of surface material - some suggestions follow:
    a. Tag board, card stock or colored poster board (sometimes called railroad board),
    b. Two-ply bristol board or two-ply strathmore paper (both take color well),
    c. Clear and colored acetate or rigid vinyl sheet plastic - transparent or opaque - .030",
    d. Balsa wood in sheet or stick form.
3. Adhesive - see previous materials list
4. Tapes
    a. Scotch Brand Magic Tape #810 ½" for paper models,
    b. Scotch Brand Polyester Tape #853 ½" for plastic models,
    c. Architect's tape - comes in a variety of widths and colors - decorative for joining edges.
5. Color for surfaces
    a. Permanent ink markers,
    b. Acrylic paints or alkyd house paints,
    c. Polymer matte medium and gloss varnish.

Procedure for construction

Read directions through first.

1. Obtain the appropriate materials and supplies. Allow plenty of time and space to work slowly, neatly and accurately.
2. Choose a specific polyhedron and construct the template or plane "net" for the polyhedral form. Take care with this construction or the inaccuracies

will compound with the construction of additional units.

3. A convenient working size is a 2" or 3" edge length. Make sure the edges of all the faces are of equal length. The larger the polyhedron, the sturdier the material must be.
4. Cut the units or net on a cutting board.
5. Some methods of construction require tabs on the individual units or on the nets; some do not. When working with rigid plastic, the units are taped together and do not require tabs.
6. Bond edges or tab faces carefully making sure that joints are accurate and tight.
7. On lightweight paper use a ballpoint pen to score all the necessary fold lines; otherwise use the X-acto knife.
8. Paint may be applied before scoring or bending into three dimensions. Or the form may be constructed before the paint is applied. Use multiple coats of paint for a good surface.
9. Use a matte or gloss varnish over the paint for a final coat.

GLOSSARY

acute angle: An angle that measures between 0° and 90°.

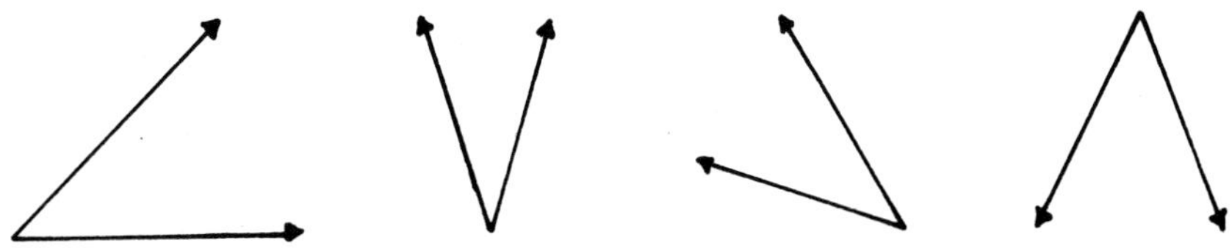

acute angles

acute triangle: A triangle having three acute angles.

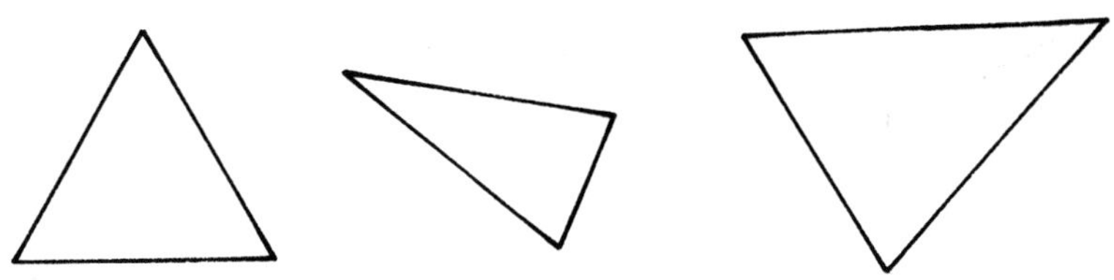

acute triangles

adjacent: Next to.

altitude: In a triangle: the line segment from a vertex perpendicular to the opposite side. Note: every triangle has three altitudes and they might not all be in the interior of the triangle.

In an acute triangle all three altitudes lie in the interior of the triangle.

altitudes: $\overline{AF}$, $\overline{BD}$, $\overline{CE}$

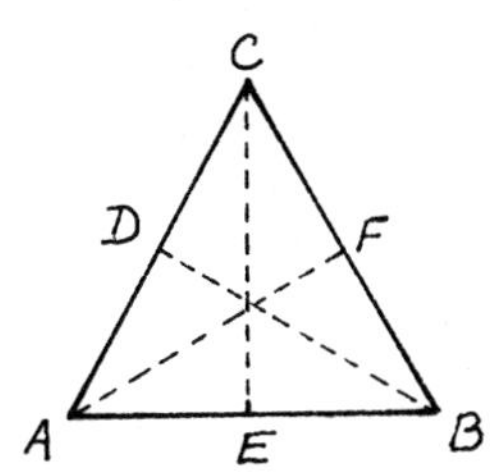

In a right triangle the legs are also altitudes.

altitudes: $\overline{AC}$, $\overline{CB}$, $\overline{CE}$

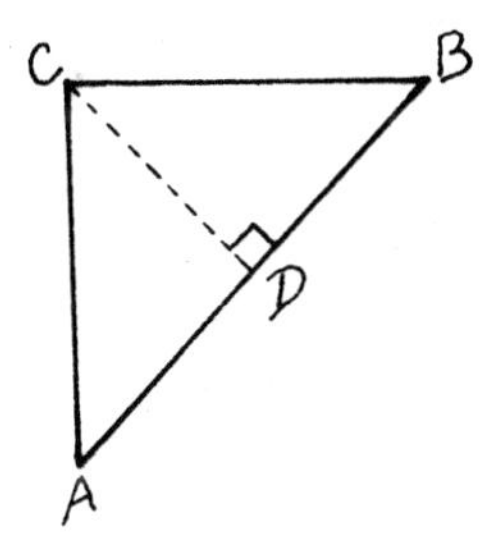

<u>In an obtuse triangle</u> two of <u>the altitudes lie in</u> the exterior of the triangle.

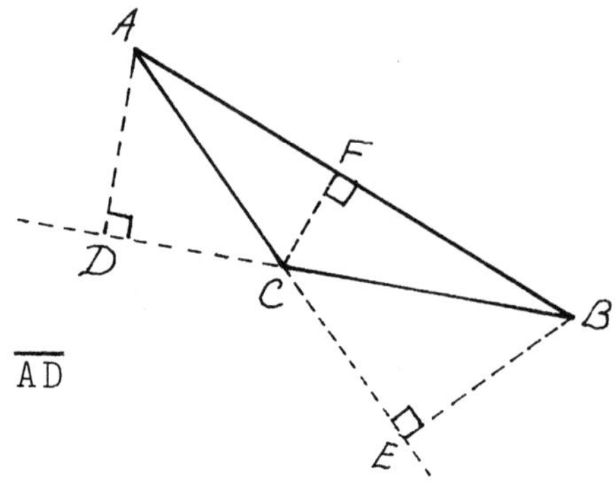

altitudes: $\overline{CF}$, $\overline{BE}$, $\overline{AD}$

<u>In a parallelogram:</u> <u>the</u> altitude is the distance between two parallel sides and any line segment joining two sides and perpendicular to both is <u>an</u> altitude.

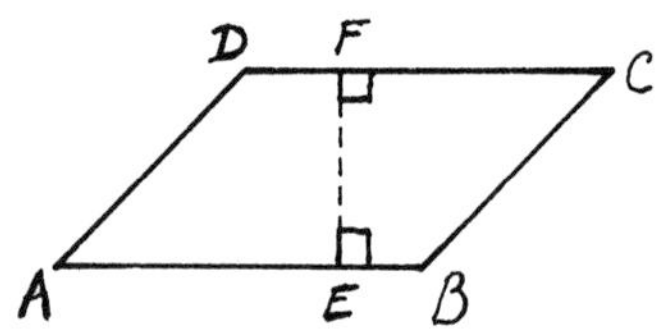

$\overline{EF}$ is an altitude of ABCD.

<u>analogous color scheme</u>: A color scheme that uses hues that are next to each other on the color wheel. Light and dark values of the hues may be used.

<u>angle</u>: The figure formed by two rays having a common endpoint. The common endpoint is called the vertex of the angle, and the rays are called the sides of the angle.

$\angle$ABC names the angle whose vertex is B such that A is a point on one side and C is a point on the other. This angle could also be named $\angle$CBA. The vertex B must be between the other two points named.

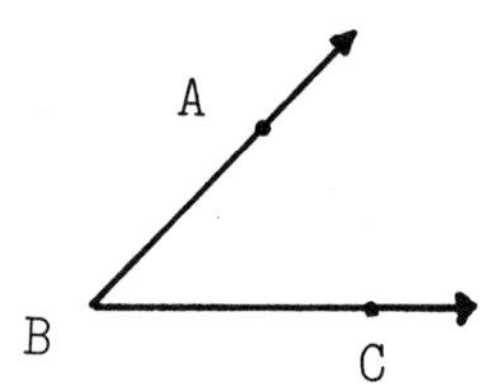

If there is no ambiguity, an angle may be named only by its vertex. ∠A makes sense in figure 1 but not in figure 2 since there are three angles at A, namely ∠BAC, ∠CAD, and ∠BAD.

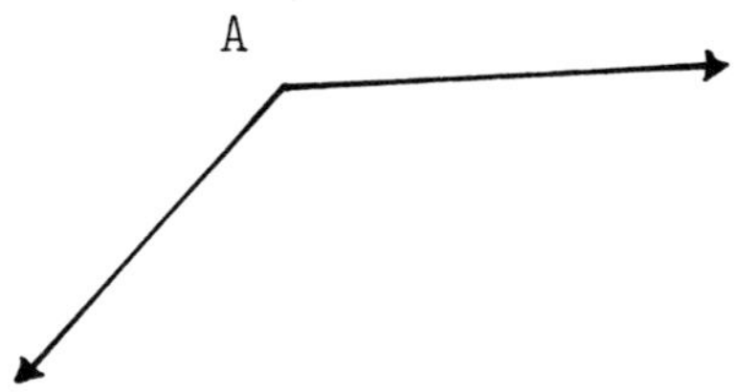

fig. 1

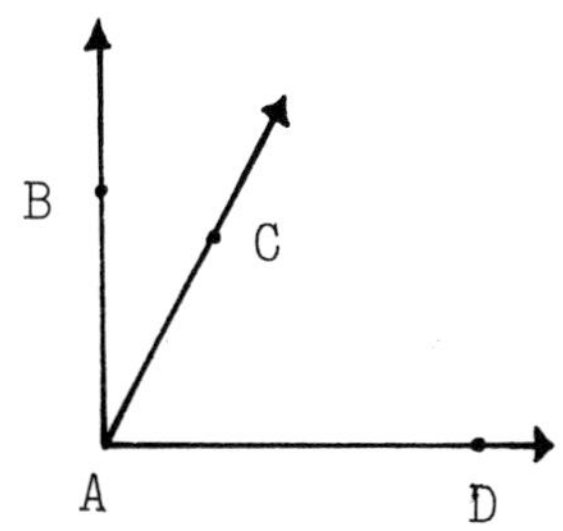

fig. 2

apothem: A line segment from the center of a circle perpendicular to a side of an inscribed regular polygon.

$\overline{OP}$ is an apothem of polygon ABCDEF

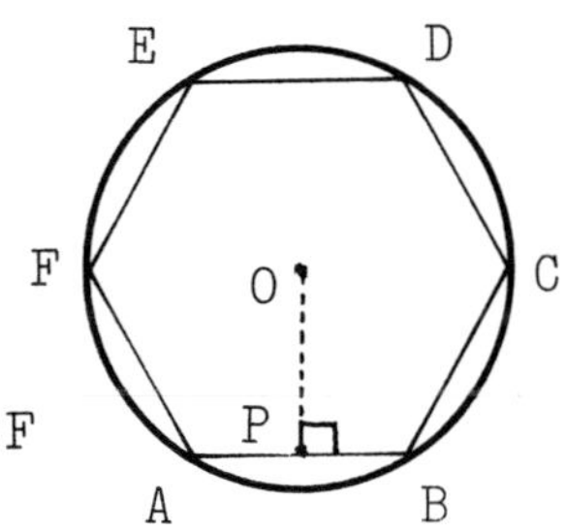

arc: The figure formed by two points and the portion of a circle between those two points. The two points are called the endpoints of the arc, and an arc is named by its endpoints.

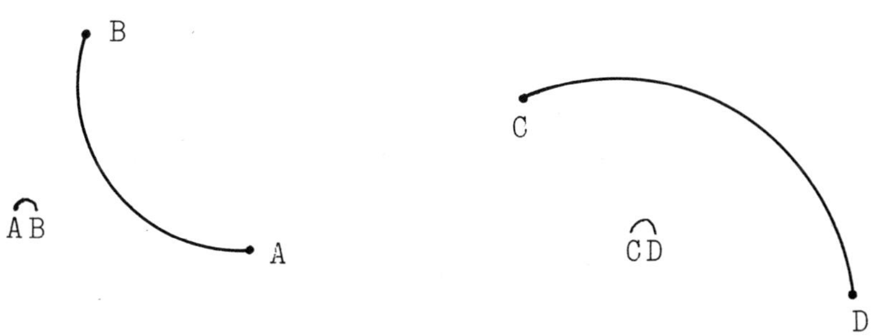

If two points lie on a circle, they determine two different arcs. To avoid confusion, the arcs may be named by the endpoints and any other point on the arc. The third point is listed between the endpoints. In fig. 1 A and B are endpoints of a diameter. Therefore, arc ACB and arc ADB are both semicircles. In fig. 2 A and B are not endpoints of a diameter. Arc ACB is called the minor arc (the smaller one) and arc ADB is called the major arc (the larger one).

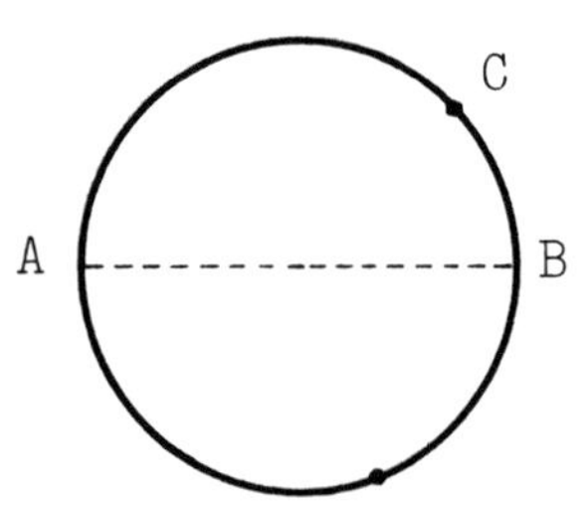

fig. 1

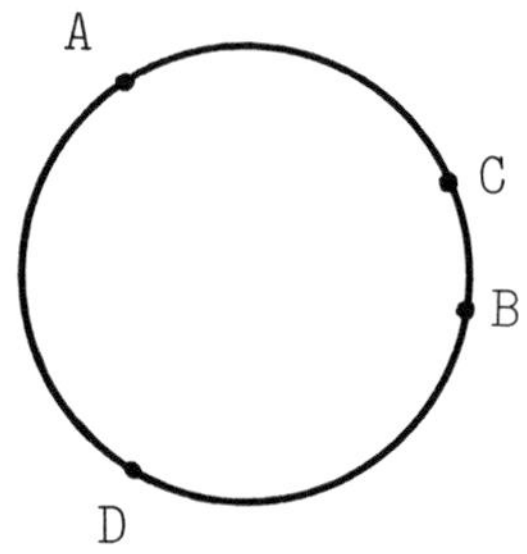

fig. 2

<u>Archimedean spiral</u>: See spiral.

<u>area</u>: The measure in square units of any bounded portion of a plane.

<u>arithmetic sequence</u>: A sequence in which each term is gotten by adding the same number to the preceding term.

eg. 1, 5, 9, 13, 17, ... is an arithmetic sequence where 4 is added to each term to produce the following term.

<u>axis</u>: A line used as a reference.

In the Cartesian coordinate plane all points are located with reference to two perpendicular axes: the x-axis and the y-axis.

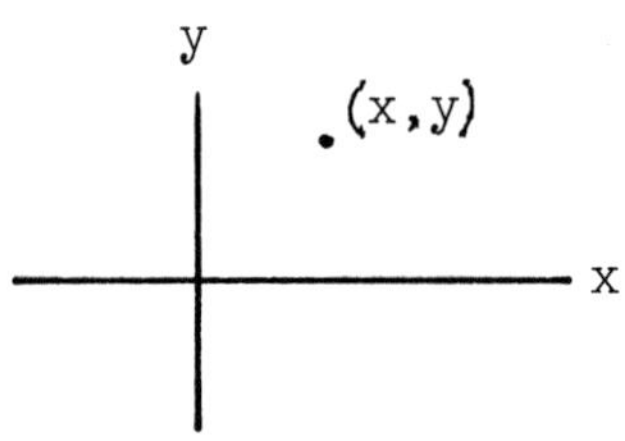

An axis of symmetry is a line which divides the plane into two halves which are mirror images of each other.

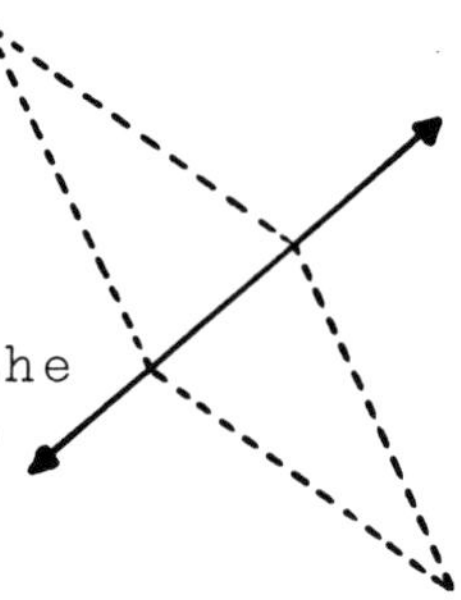

<u>Baravelle Spiral</u>: The straight line design based on repeatedly connecting consecutive midpoints of sides of regular polygons.

<u>base angles</u>: <u>In an isosceles triangle</u>: the two congruent angles.

In the figure, ∠A and ∠B are the base angles of ΔABC, and ∠R and ∠S are the base angles of ΔRST.

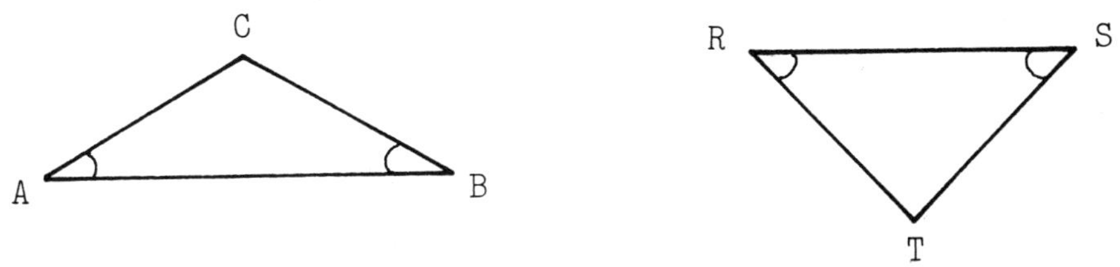

<u>In a trapezoid</u>: a pair of angles with a base as a common side. Each trapezoid has two pairs of base angles.

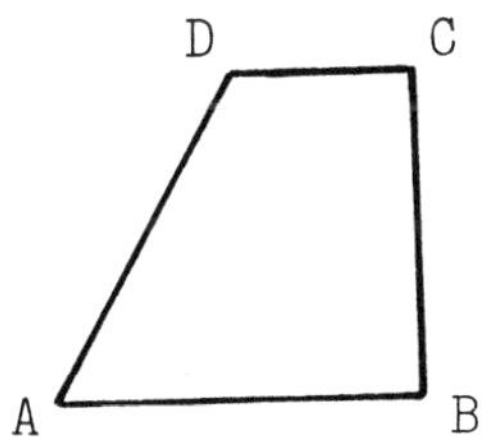

In ABCD, ∠A and ∠B are one pair of base angles, and ∠C and ∠D are the other pair.

<u>bisect</u>: To divide into two congruent figures.

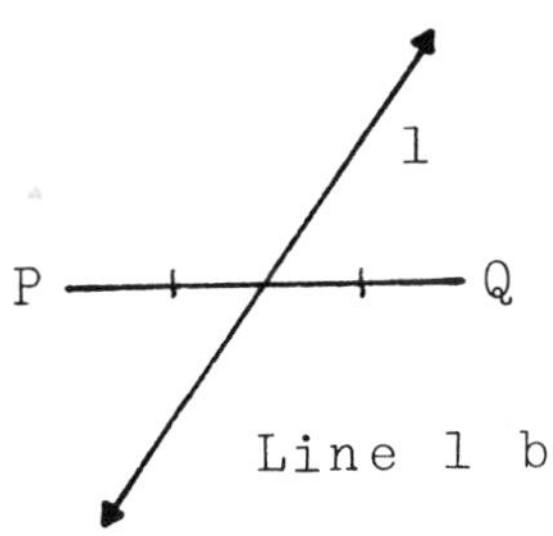

Line l bisects $\overline{PQ}$

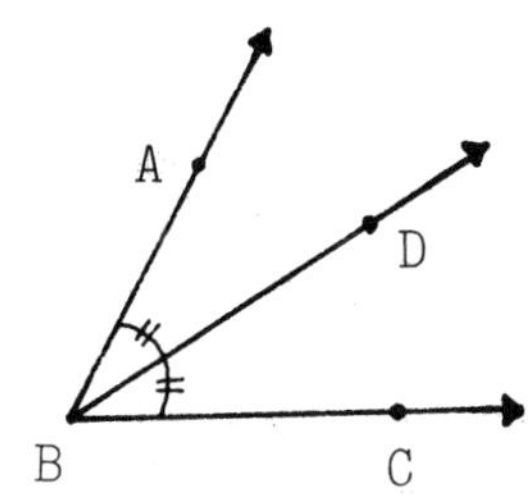

$\overline{BD}$ bisects ∠ABC

bisector: The line or ray that divides a segment or angle into congruent figures.

perpendicular bisector: A line that passes through the midpoint of a segment and is at right angles to the segment.

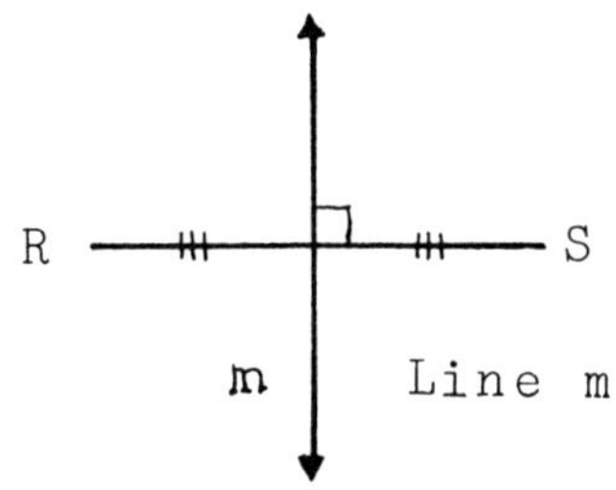

Line m is the perpendicular bisector of $\overline{RS}$.

C

Cartesian coordinate plane:

A rectangular system for locating points in a plane by using a pair of perpendicular axes:

the horizontal line is the x-axis and the vertical line is the y-axis.

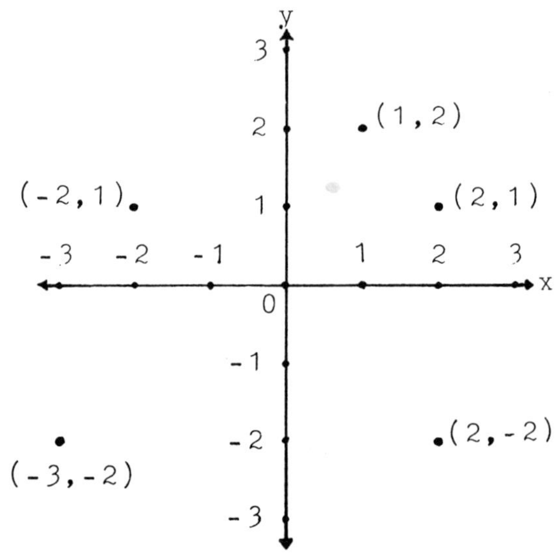

A point is located by an ordered pair of numbers (x,y), which locates it with respect to both axes.

cella: In Greek temples, the rectangular interior space which housed the cult statue.

central angle: The angle formed by two radii of a circle.

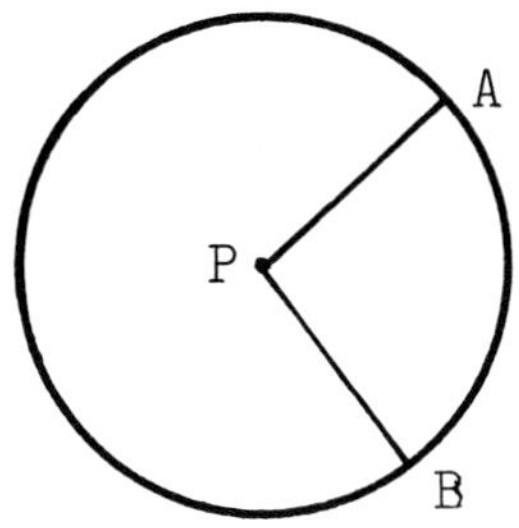

In ⊙P, ∠APB is a central angle.

chord: A line segment whose endpoints lie on a circle.

In ⊙P, $\overline{RS}$ is a chord.

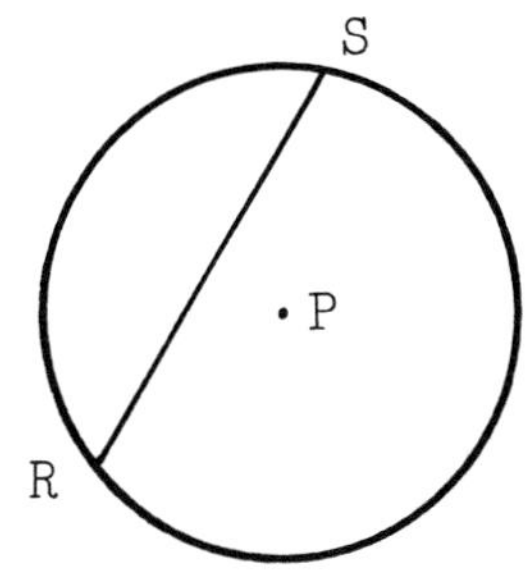

circle: The set of all points in a plane equidistant from a given point. That point is called the center of the circle, although it is not part of the circle. A circle is named by its center.

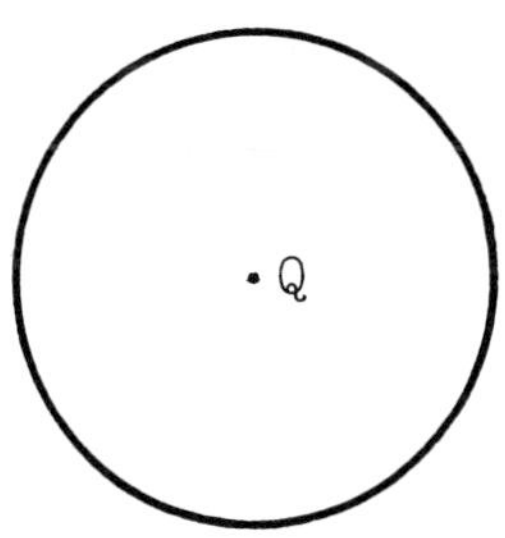

⊙Q means "circle Q" or the circle whose center is Q.

circumference: The distance around a circle. The circumference of a circle is given by the formula $C = \pi d$, where d is the diameter of the circle.

coefficient: A factor in a product generally taken to mean a number multiplied by a variable quantity.

In the expressions $2x$, $-3a^2b$, $-pq$:
2 is the coefficient of x,
-3 is the coefficient of a^2b, and
-1 is the coefficient of pq.

collinear: Lying on the same line.

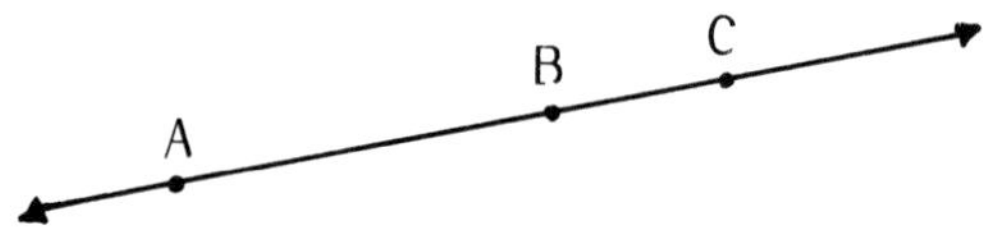

eg. In the figure points A, B and C are collinear.

commensurable: Able to be measured by a common unit.

complementary color scheme: A color scheme that uses two hues that are directly opposite each other on the color wheel. It can include light and dark values of the hues.

complementary angles: Two angles the sum of whose measures is 90°. The angles are said to be complements of each other.

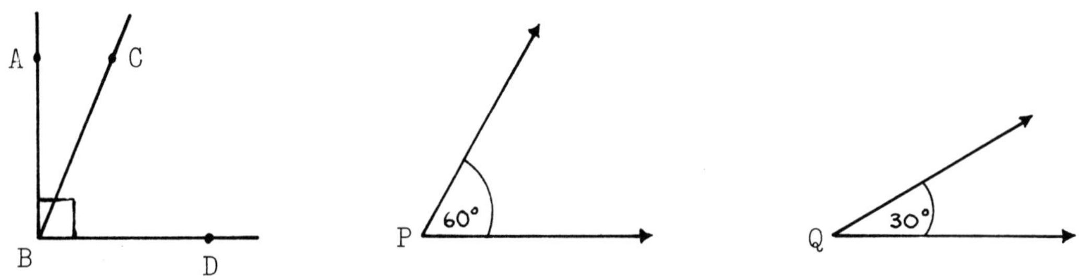

∠ABC is complementary to ∠CBD.
∠P and ∠Q are complementary angles.

compass: An instrument used for drawing circles.

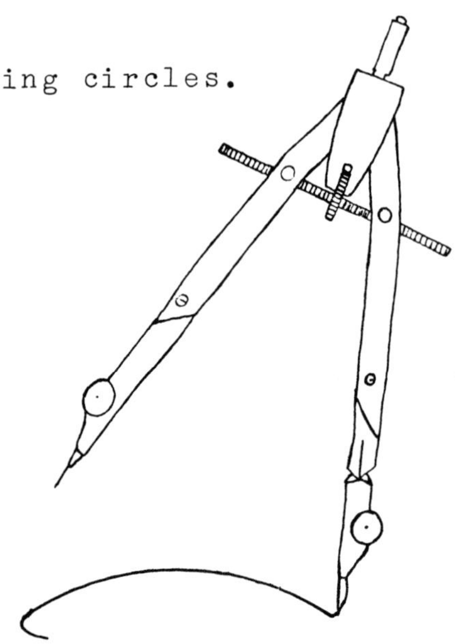

composite number: A natural number, different from one, that is not prime. The number one is neither prime nor composite.

concave polygon: A polygon that has at least one interior angle that measures more than 180°.

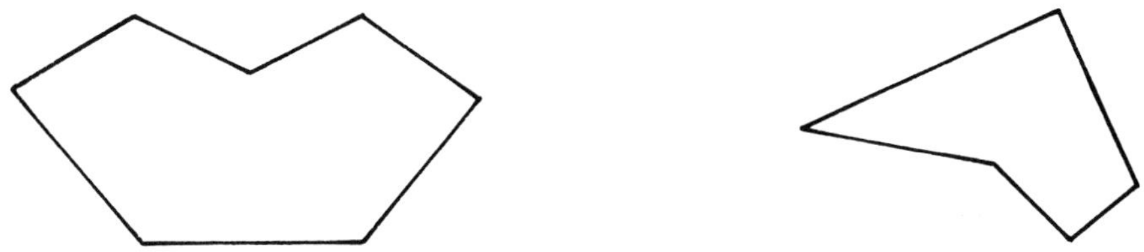

concave polygons

concentric: Having the same center.

The two circles with center Q are concentric circles.

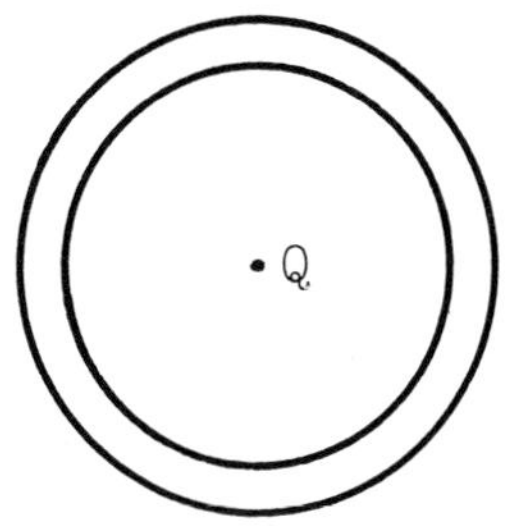

concho spiral: The three dimensional equivalent of the logarithmic spiral. This is the spiral found on the surfaces of sea shells and pine cones.

concurrent: Occurring at the same time.

cone: A three dimensional figure with a circular base and a surface generated by the lines passing through a point outside the plane of the circle and the points on the circle. See Appendix C for formulas related to the cone.

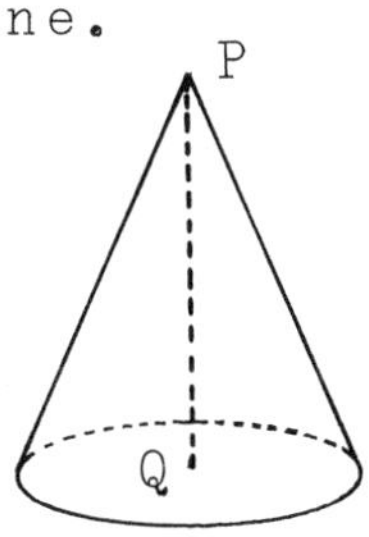

Q is the base of the cone.
P is called the vertex of the cone.

If $\overleftrightarrow{PQ}$ is perpendicular to the plane of the circle, the cone is a right circular cone.

congruent: Having the same size and shape.

conical helix: The spiral like curve formed by the uniform wrapping of a line around a cone from its vertex to its base.

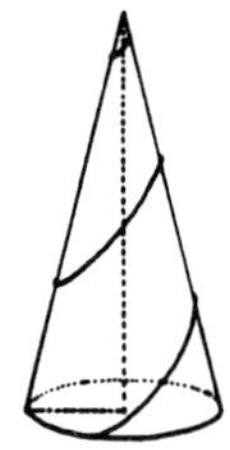

consecutive: In order or in sequence.

eg. 2,3,4 are consecutive integers.
5 and 7 are consecutive odd integers.

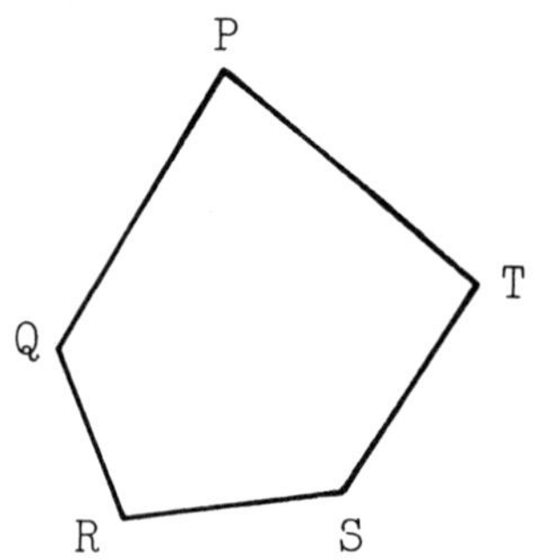

In a polygon, consecutive sides or consecutive vertices are those that appear in order as you travel around the polygon in either direction starting at any vertex.

In PQRST, a pair of consecutive vertices might be Q and R, and a pair of consecutive sides might be $\overline{TP}$ and $\overline{PQ}$.

convex polygon: A polygon whose interior angles each measure less than 180°.

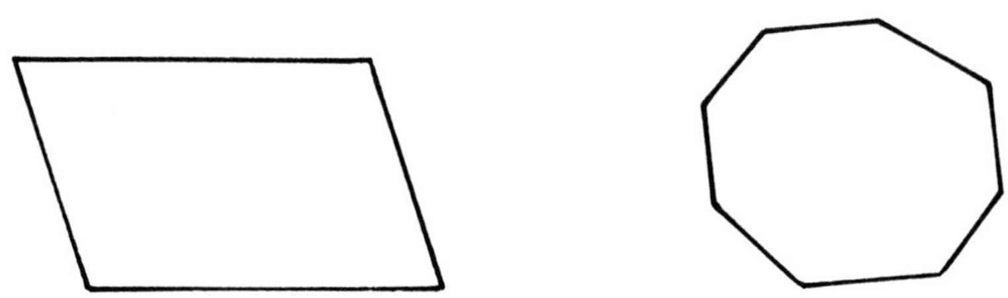

convex polygons

cube: As a noun:
A polyhedron having six square faces.

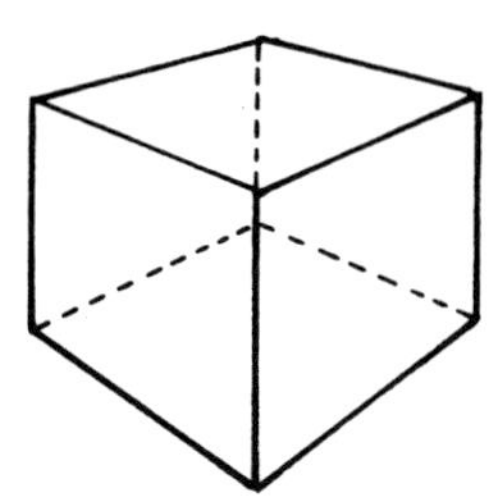

As a verb:

To raise to the third power.

$$\text{eg.} \quad 4^3 = 4 \text{ cubed} = 4 \cdot 4 \cdot 4 = 64$$

cubit: An ancient Egyptian unit of linear measure.
The little cubit measured .4236m.
The Royal Cubit measured .5236m.

curve: A continuous set of points having only one dimension, namely length.

Note: by this definition, a straight line is a curve.

cylinder: A three dimensional figure generated by a circle whose center moves along a line segment. See Appendix C for formulas related to the cylinder.

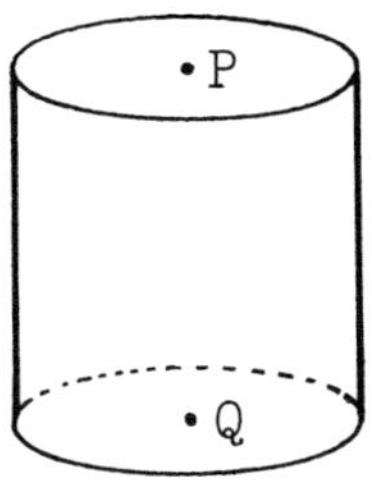

⊙P and ⊙Q are the bases of the cylinder. If $\overleftrightarrow{PQ}$ is perpendicular to the planes of ⊙P and ⊙Q, then the cylinder is a right circular cylinder.

D

decagon: A polygon having ten sides.

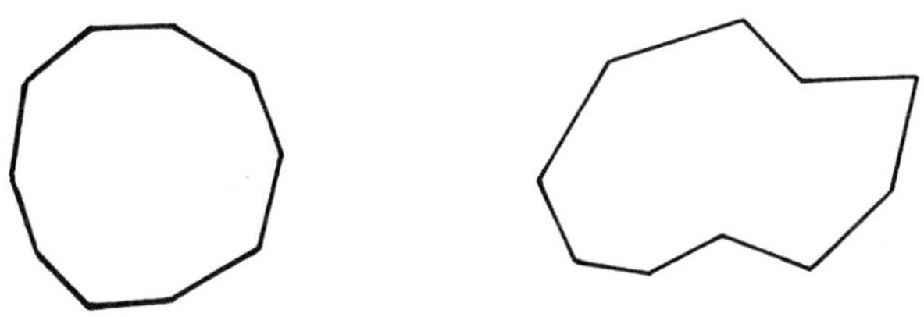

decagons

__deduction__: A method of reasoning that allows specific conclusions to be drawn from general truths. This is the form of reasoning used in mathematical proof.

__denominator__: In a fraction, the number that is written below the line.

__diagonal__: A line segment that joins two nonconsecutive vertices of a polygon. The number of diagonals a polygon has is dependent on the number of sides.

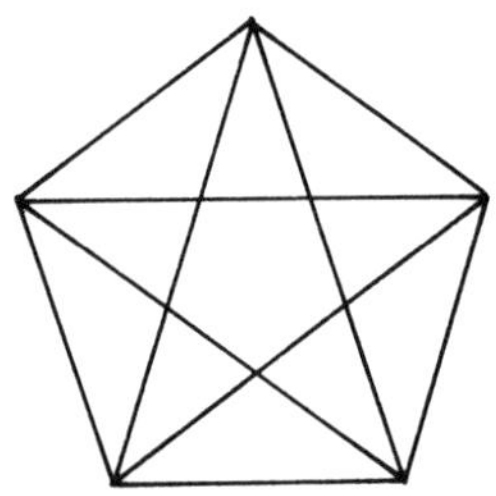

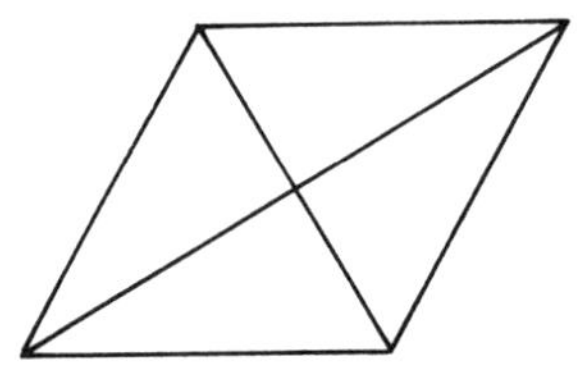

Each of the line segments in the interior of the figures is a diagonal.

__diameter__: A chord that passes through the center of a circle. A circle has infinitely many diameters - all the same length. The measure of any one of them is called __the__ diameter of the circle. Note: a diameter cuts the circle into two semicircles.

$\overline{QR}$ is __a__ diameter of ⊙P.
QR is __the__ diameter of ⊙P.

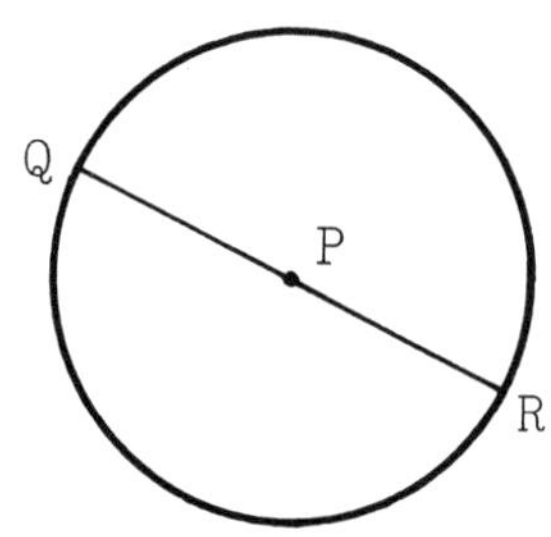

__dihedral angle__: The angle formed by two adjacent faces of a polyhedron.

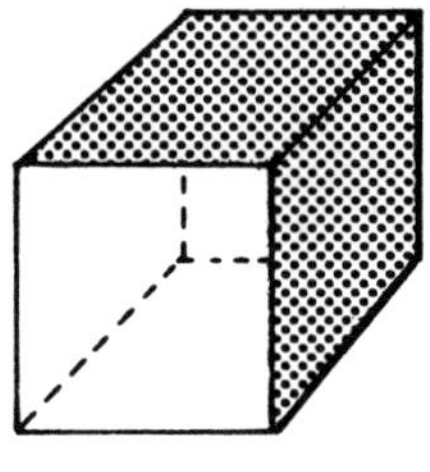

In this figure the faces are perpendicular giving a dihedral angle of 90°.

distance: 1. Between points: the length of the line segment joining the points.

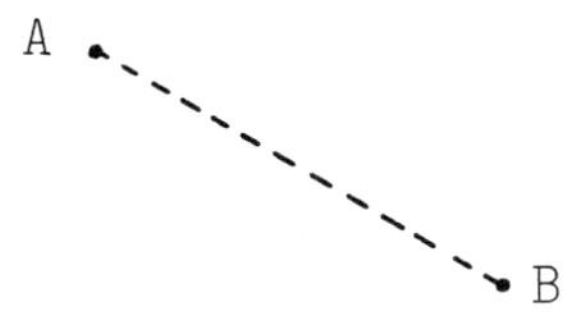

AB is the distance from A to B.

2. From a point to a line: the length of the line segment from the point to the line and perpendicular to the line.

PQ is the distance from P to line m.

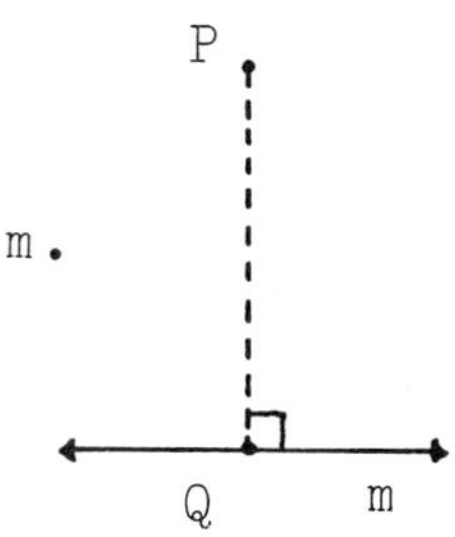

3. Between parallel lines: the length of any segment having one endpoint on each line and perpendicular to both.

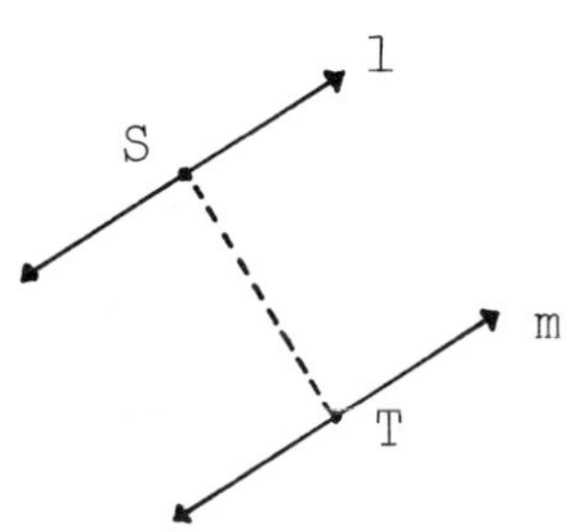

ST is the distance from line l to line m.

divergency: If we choose a leaf on a stem and travel up the stem in a spiral fashion, we will eventually encounter another leaf situated directly over the first. If n is the number of leaves passed on the way, and p the number of complete turns made around the stem, then p/n is the divergency constant in the plant.

Divine Proportion: The proportion derived from the division of a line segment into two segments such that the ratio of the whole segment to the longer part is the same as the ratio of the longer part to the shorter part.

$$\frac{AB}{AC} = \frac{AC}{CB}$$

$$\frac{a+b}{a} = \frac{a}{b}$$

dodecagon: A polygon having twelve sides.

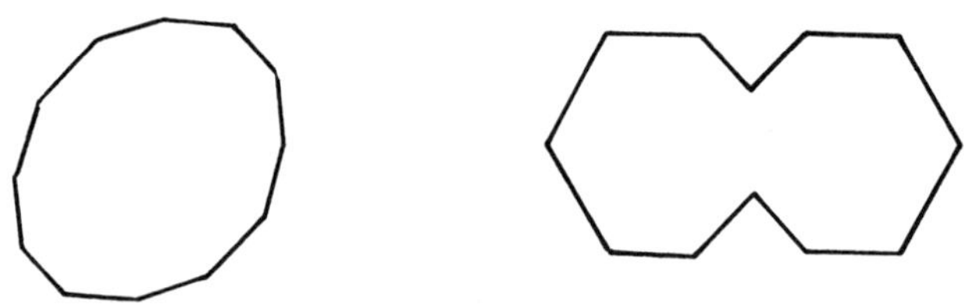

dodecagons

dodecahedron: A polyhedron having twelve faces.

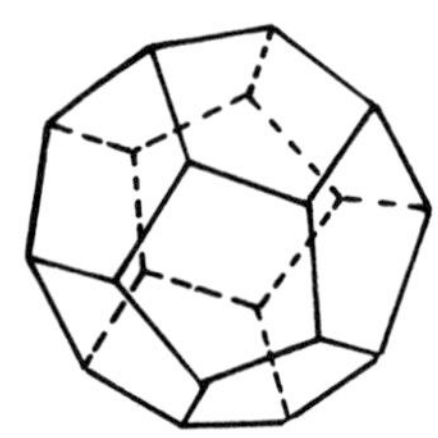

Dynamic Rectangles: A family of rectangles all derived from the square that give harmonious proportions. Included in the family are the root rectangles and the Phi-family Rectangles. The Dynamic Rectangles are sometimes called the Euclidean Series.

Dynamic Parallelogram: Any parallelogram whose sides are in the same ratio as one of the Dynamic Rectangles.

Dynamic Spiral: Any logarithmic spiral which uses as its framework one of the Dynamic Rectangles. A Dynamic Spiral is named by its related Dynamic Rectangle.

eg. the spiral created within a $\sqrt{2}$ Rectangle would be named a $\sqrt{2}$ Spiral.

E

<u>edge</u>: In a polyhedron, the common side between two faces.

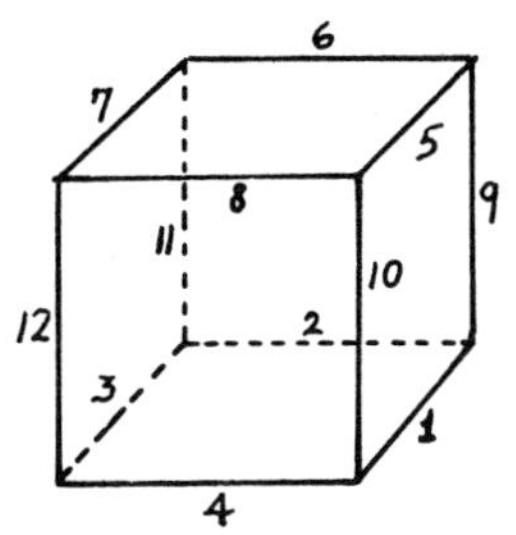

eg. In the cube there are twelve edges joining the eight vertices.

<u>Egyptian Triangle</u>: See Triangle of Price.

<u>equiangular</u>: Having angles of the same measure.

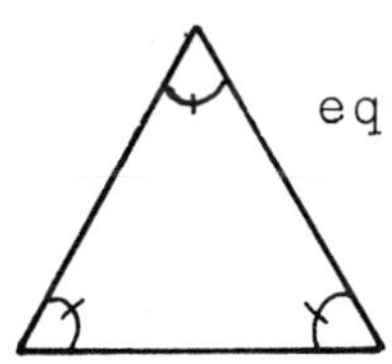

equiangular triangle

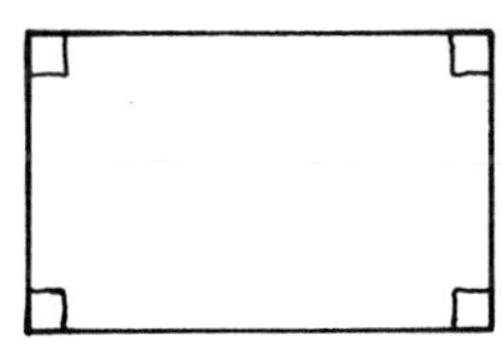

equiangular quadrilateral

<u>equiangular spiral</u>: See spiral (logarithmic spiral).

<u>equilateral</u>: Having sides of the same measure.

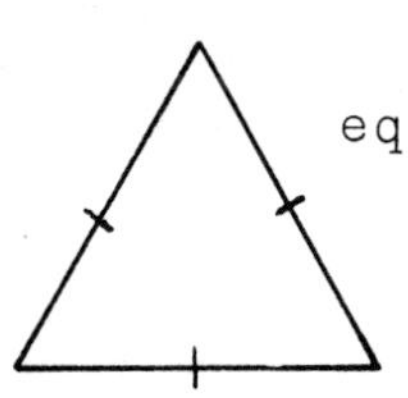

equilateral triangle

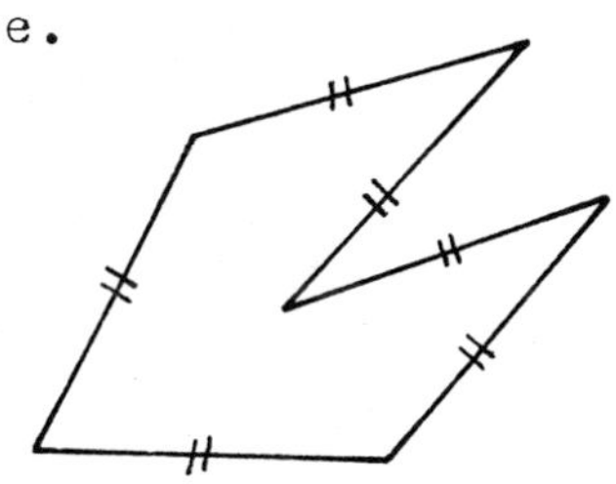

equilateral hexagon

escribe: To draw around.

Euclidean Series: See Dynamic Rectangles.

even number: A whole number that is divisible by two.

$2, 4, 6, 8, 10, \cdots$

If n is any whole number, $2n$ will always designate an even number.

exponent: A number that indicates the number of times a quantity is to be used as a factor.

In the expression a^4, 4 is the exponent.

$a^4 = a \cdot a \cdot a \cdot a$

See Appendix B for properties of exponents.

exterior angle of a polygon: An angle formed by a side and the extension of an adjacent side in a polygon.

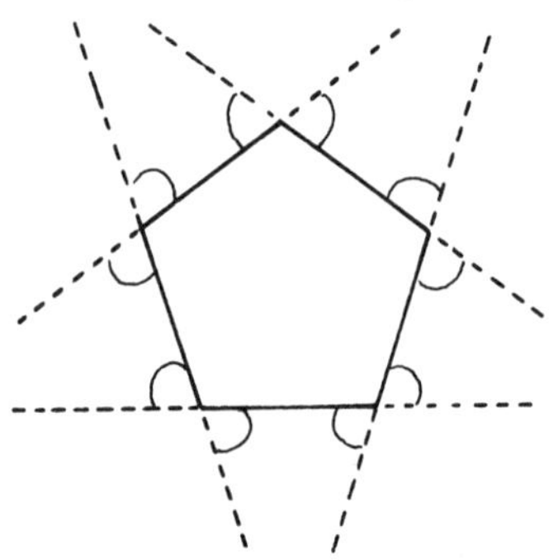

Note: there are two exterior angles at each vertex.

exterior (of an angle or a polygon): The set of all points in the plane that do not lie on the figure or in the interior of the figure.

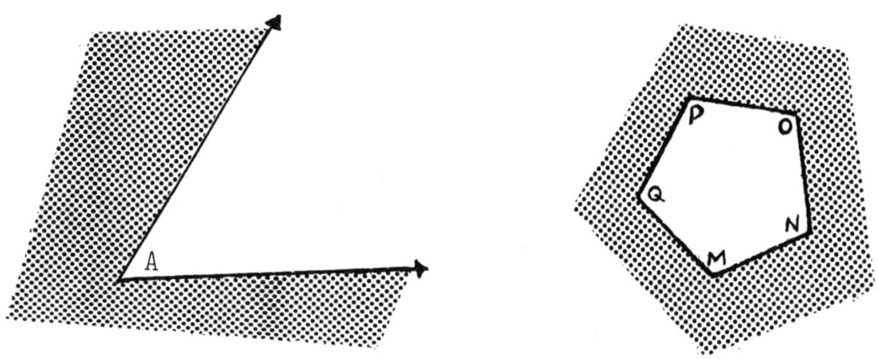

The shaded portions of the plane represent the exteriors of $\angle A$ and polygon MNOPQ respectively.

__extremes__: See proportion.

F

__face__: A polygonal shape, together with its interior, that forms part of the surface of a polyhedron.

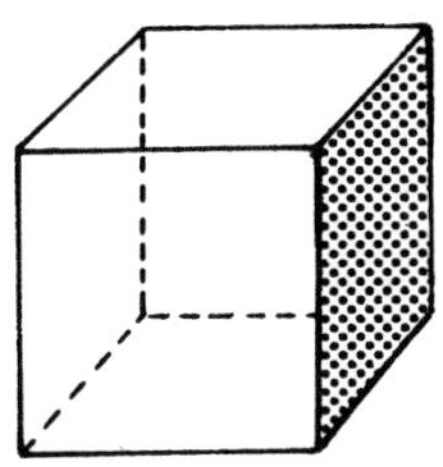

The shaded portion of the diagram represents one of the six faces of the cube.

__Fibonacci Angle__: See Ideal Angle.

__Fibonacci Sequence__: The summation sequence whose terms are as follows:

1, 1, 2, 3, 5, 8, 13, • • •

G

__geodesic__: The curve on a surface that is the shortest distance between two points on that surface.

<u>geometric sequence</u>: A sequence in which each term is gotten by multiplying the preceding term by a constant quantity.

If r is the constant, the nth term, s_n, is equal to rs_{n-1}.

eg. 2, 4, 8, 16, 32, • • • is a geometric sequence where each term is multiplied by 2 to produce the following term.

<u>glide reflection</u>: A symmetry operation that involves both a translation and a reflection.

<u>gnomon</u>: Any figure that when added to a given figure results in a figure similar to the original.

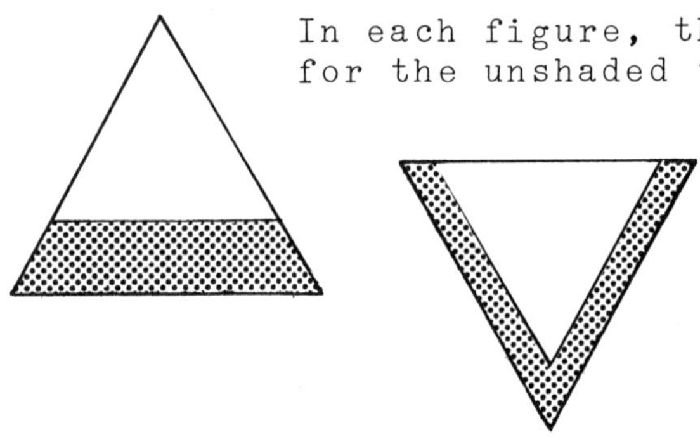

In each figure, the shaded portion is a gnomon for the unshaded triangle.

<u>Golden Cut</u>: See Golden Section.

<u>Golden Mean</u>: See Golden Section.

<u>Golden Ratio</u>: The ratio of 1.61803 to 1, derived from the Divine Proportion.

<u>Golden Rectangle</u>: Any rectangle whose sides are in the ratio of 1.61803 to 1.

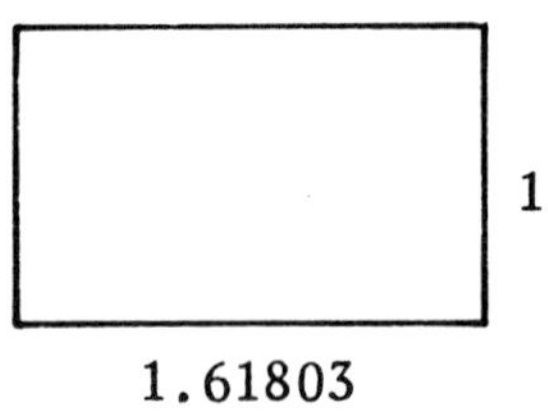

<u>Golden Section</u>: The point on a line segment that divides the segment into two segments whose lengths are in the Golden Ratio.

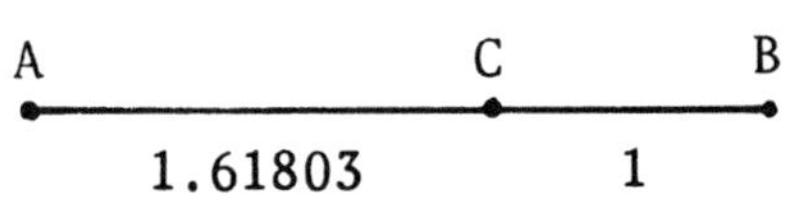

Golden Sequence: The sequence whose terms are the successive powers of ϕ.

$$\phi,\ \phi^2,\ \phi^3,\ \phi^4,\ \ldots,\phi^n,\ldots$$

Golden Spiral: The logarithmic spiral formed within the pattern of whirling squares in a Golden Rectangle.

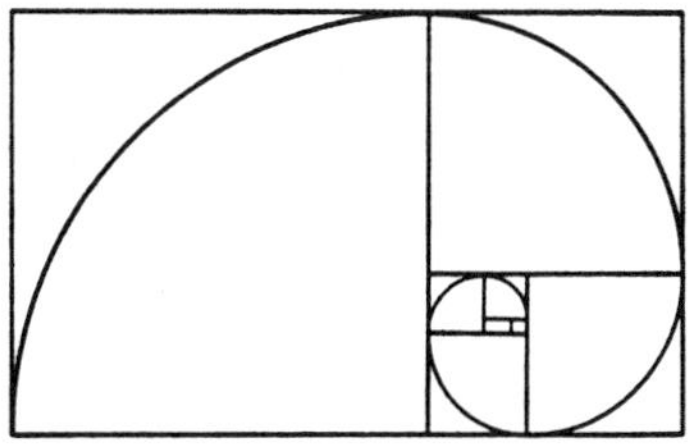

Golden Triangle: Any isosceles triangle in which the ratio of a leg to the base is 1.61803 to 1. Also called the Triangle of the Pentalpha or the Sublime Triangle.

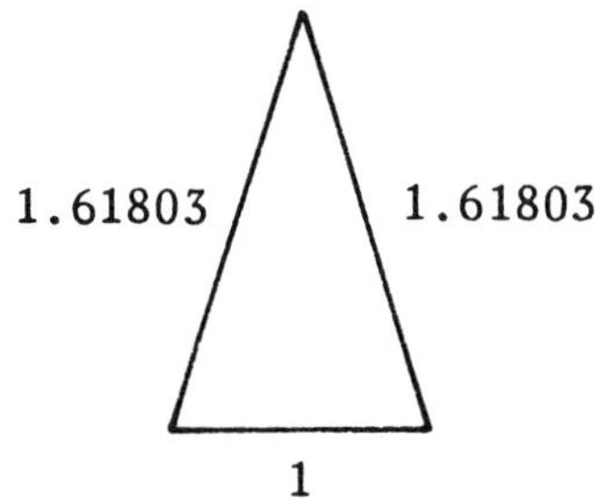

grid: A plane rectangular coordinate system.

H

Harmonic Triangle: The right triangle derived from the radius and apothem of a regular pentagon.

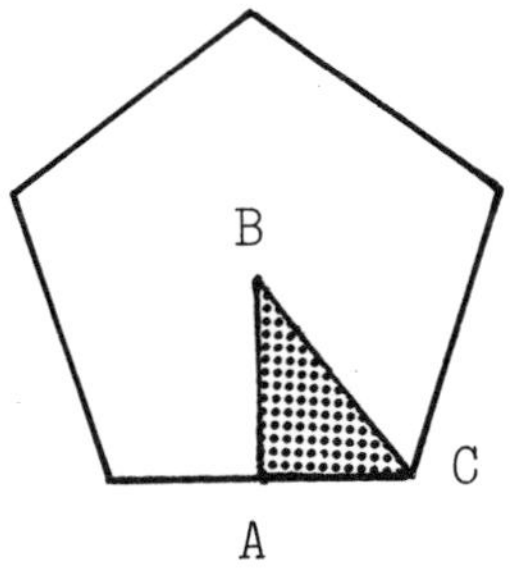

ΔABC is an Harmonic Triangle.

helix: A spiral-like curve formed by wrapping a cylinder uniformly in a single layer.

eg. the thread of a screw is a helical curve.

heptagon: See septagon.

hexagon: A polygon having six sides.

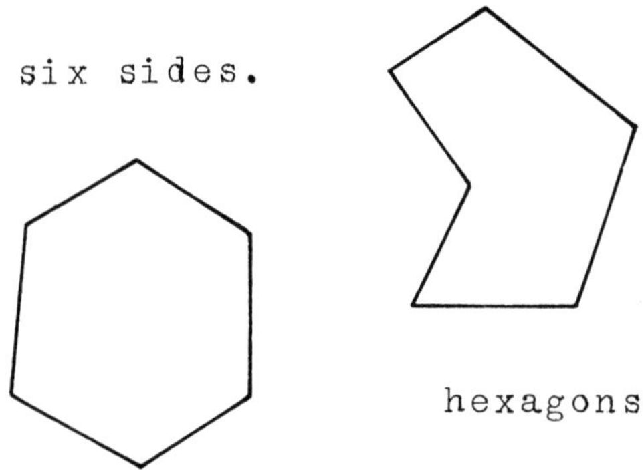

hexagons

hypotenuse: In a right triangle, the side opposite the right angle.

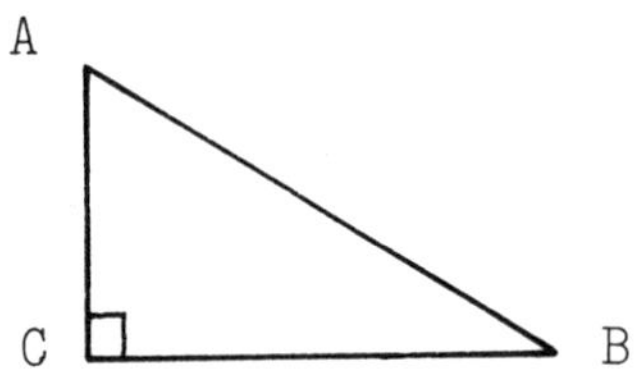

In $\triangle ABC$, $\overline{AB}$ is the hypotenuse.

hypothesis: A conjecture that requires testing for verification.

I

icosagon: A polygon having 20 sides.

icosagon

icosahedron: A polyhedron having 20 faces.

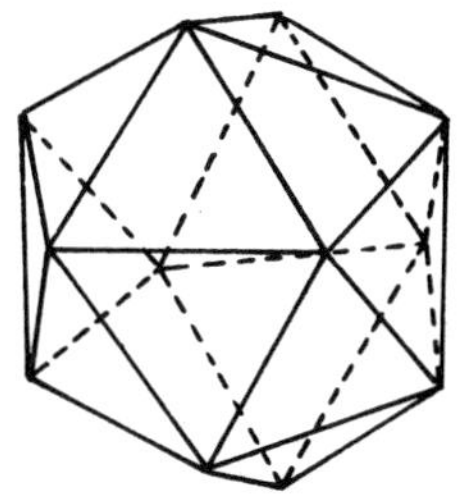

Ideal Angle: The angle measuring about 137.5° related to plant divergency which would provide maximum exposure to sunlight as well as minimum overlapping of leaves.

incommensurable: Not able to be measured by a common unit.

included angle: The angle formed by two specified adjacent sides of a polygon.

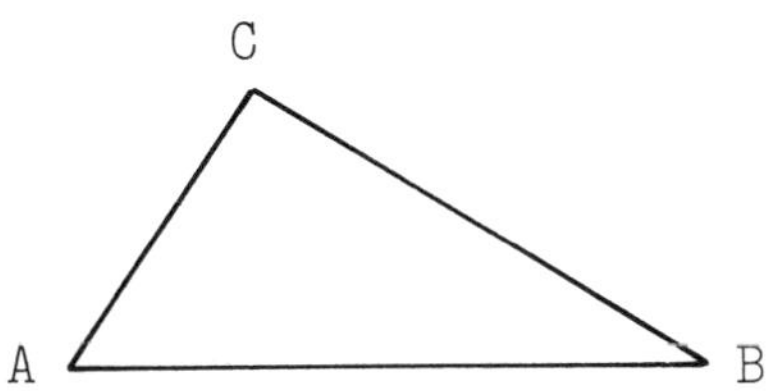

$\angle A$ is included between $\overline{AB}$ and $\overline{AC}$.
$\angle B$ is included between $\overline{AB}$ and $\overline{BC}$.
$\angle C$ is included between $\overline{AC}$ and $\overline{BC}$.

included side: In a polygon, the common side between two specified angles.

$\overline{AC}$ is included between $\angle A$ and $\angle C$.
$\overline{AB}$ is included between $\angle A$ and $\angle B$.
$\overline{BC}$ is included between $\angle B$ and $\angle C$.

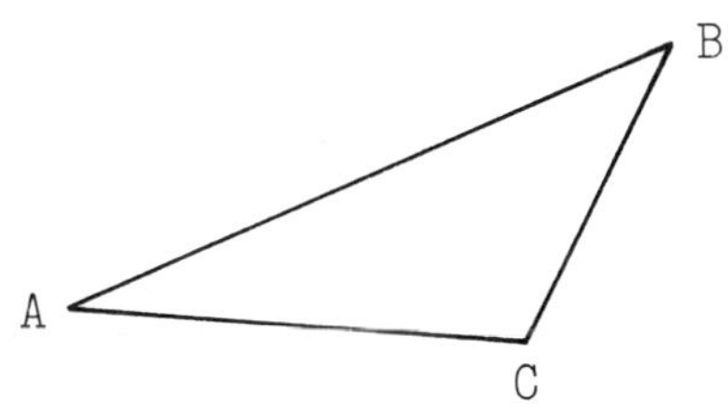

inductive reasoning: A method of reasoning that allows a generalization to be made from observing specific cases. This method of reasoning can be likened to making educated guesses. Unlike deductive reasoning, it can lead to a false conclusion. This is the form of reasoning used in the Scientific Method.

induction: See inductive reasoning.

inscribed angle: An angle whose vertex lies on a circle and each of whose sides intersects the circle.

$\angle CAB$ is inscribed in $\odot P$.

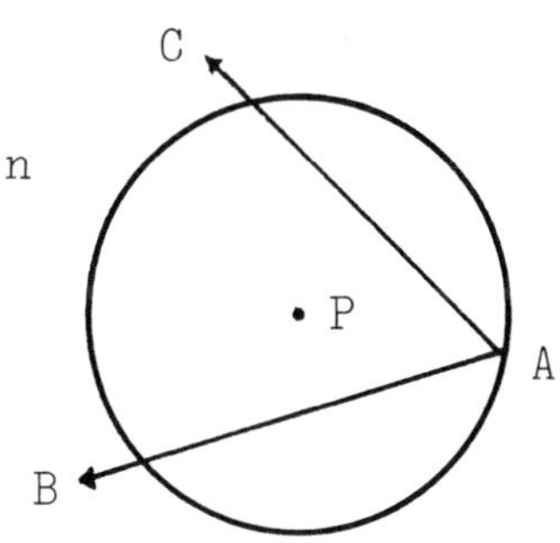

Inscribed circle: A circle lying in the interior of a polygon such that each side of the polygon is tangent to the circle.

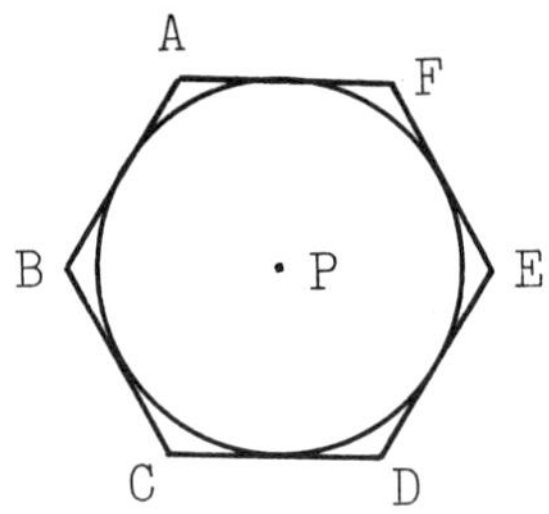

P is inscribed in ABCDEF.

inscribed polygon: A polygon whose vertices lie on a circle.

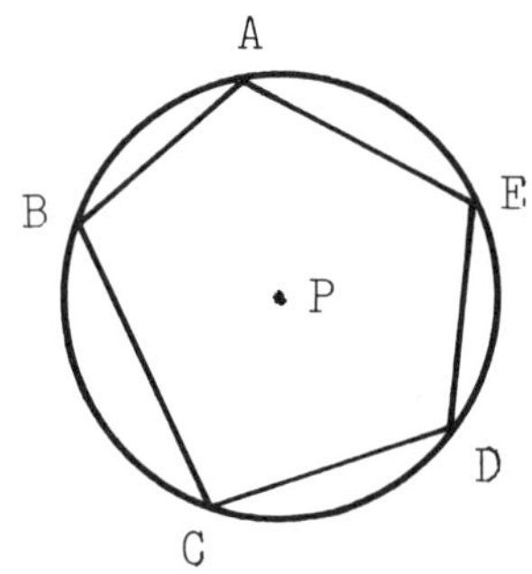

ABCDE is inscribed in P.

integer: Any member of the following infinite set of numbers:

$$\ldots, -3, -2, -1, 0, 1, 2, 3, \ldots$$

interior angle of a polygon: An angle formed by adjacent sides of a polygon. Note: the number of interior angles is the same as the number of sides or vertices.

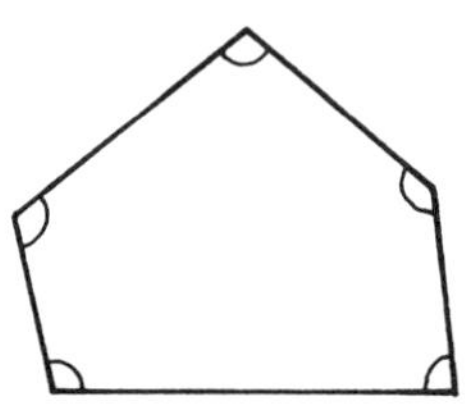

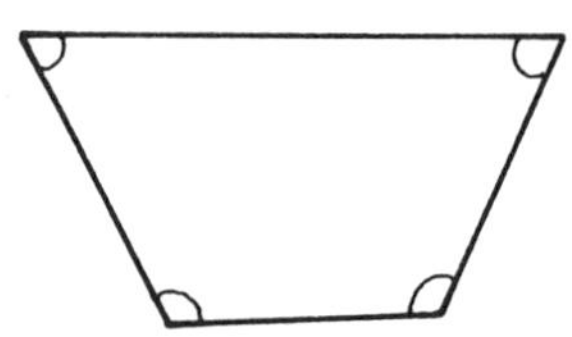

interior of an angle: The set of all points not on an angle having the property that if any two are joined by a line segment, the segment will not intersect the angle.

The shaded portions represent the in teriors of angles ABC and DEF respectively.

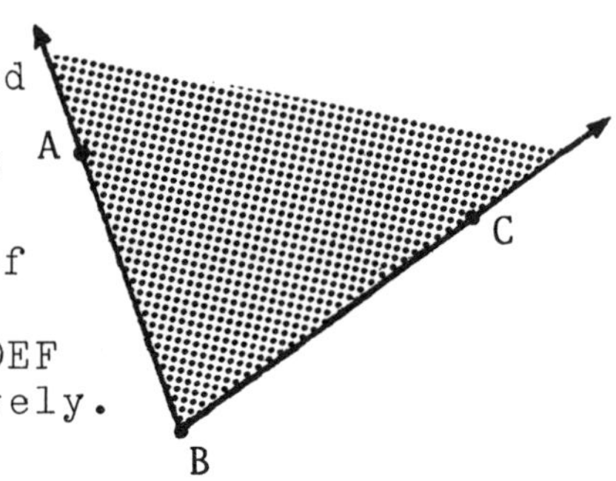

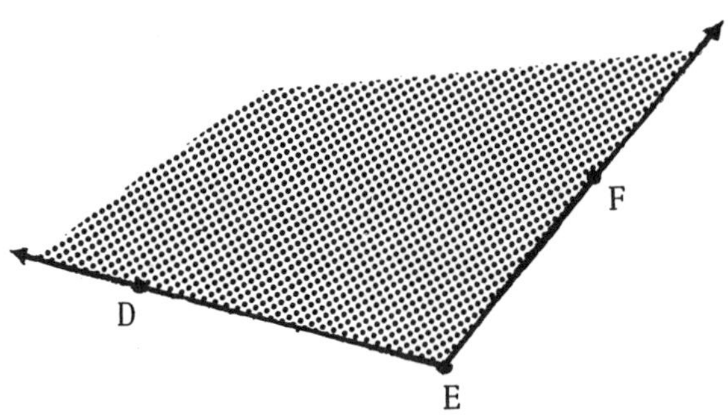

interior of a closed curve: The set of all points in the plane that are enclosed by the curve.

The shaded portion represents the interior of polygon PQRSTU.

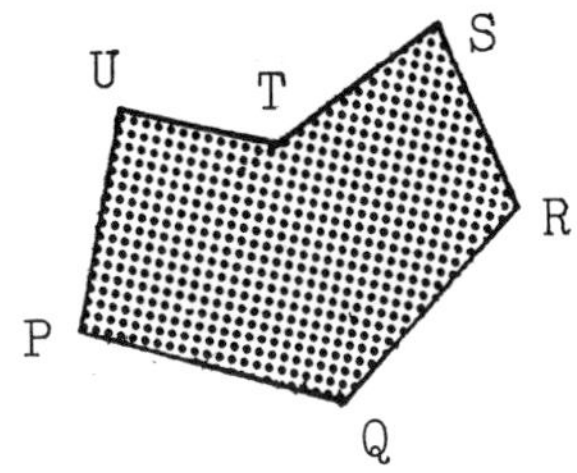

intersect: To have at least one point in common.

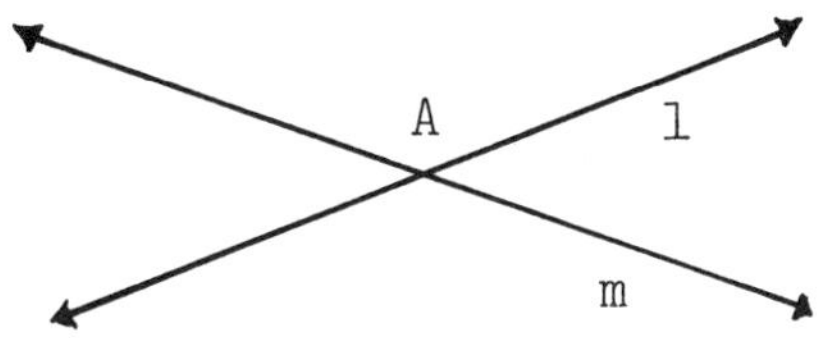

Lines l and m intersect at point A, and A is called the point of intersection.

irrational number: Any number that cannot be expressed as the ratio of two integers.

isosceles triangle: A triangle having a pair of congruent sides.

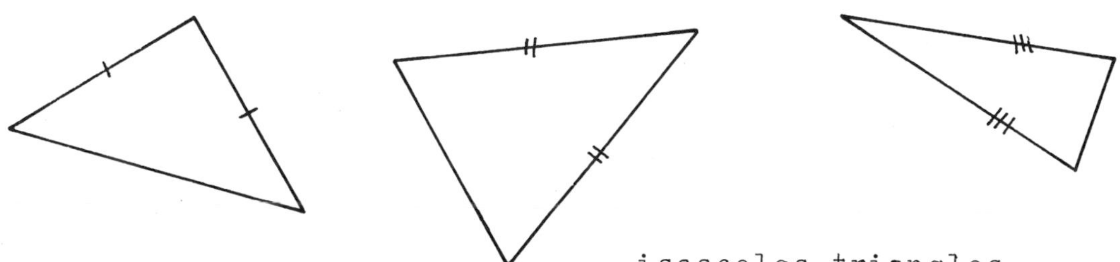

isosceles triangles

isosceles trapezoid: A trapezoid in which the nonparallel sides are congruent. These sides are called the legs of the trapezoid.

ABCD is an isosceles trapezoid with legs $\overline{AB}$ and $\overline{CD}$.

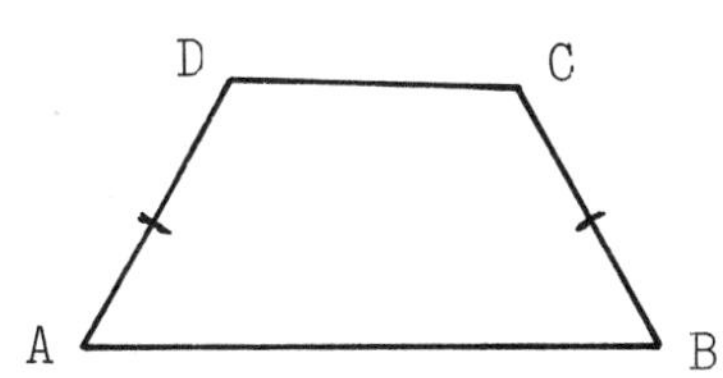

isometry: Equality in measured parts.

L

<u>leg(s)</u>: In a right triangle, the sides that form the right angle.

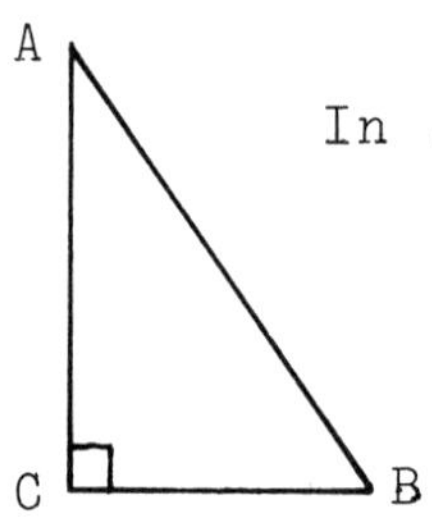

In $\triangle ABC$, $\overline{AC}$ and $\overline{CB}$ are the legs.

In an isosceles triangle, the congruent sides.

In $\triangle PQR$, $\overline{PQ}$ and $\overline{PR}$ are the legs.

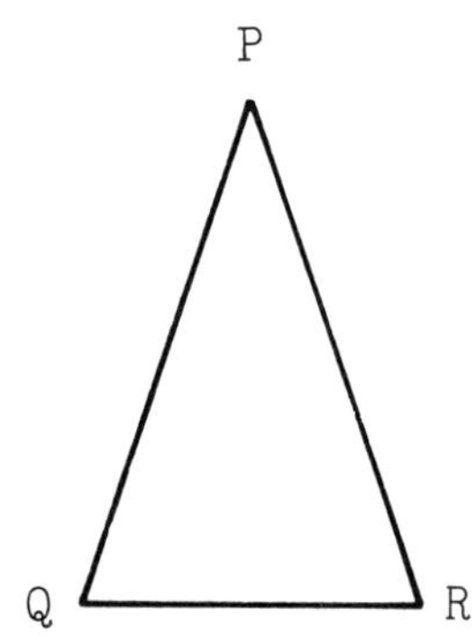

In an isosceles trapezoid, the nonparallel sides.

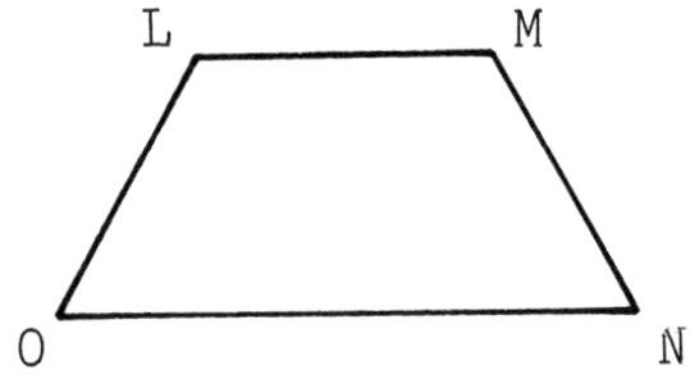

In LMNO, $\overline{LO}$ and $\overline{MN}$ are the legs.

<u>line</u>: An undefined term in geometry. It has the properties of infinite length, continuity, and no width. It is a straight curve.

$\overleftrightarrow{AB} = \overleftrightarrow{BC} = \overleftrightarrow{AC}$

line ℓ

A line is named by any two of its points or by a lower case letter.

<u>line segment</u>: The figure formed by two points on a line and all the points in between those two points. The two points are called the endpoints of the segment, and a line segment is named by its endpoints.

$\overline{AB} = \overline{BA}$

<u>logarithmic spiral</u>: See spiral.

M

<u>mean proportion</u>: A proportion in which the means are equal.

eg. $\frac{a}{b} = \frac{b}{c}$

<u>means</u>: See proportion.

median: A line segment that joins a vertex of a triangle to the midpoint of the opposite side. Every triangle has three medians.

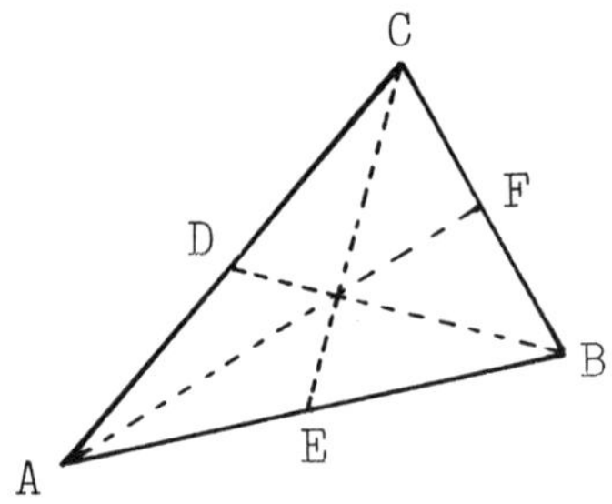

medians $\overline{AF}$, $\overline{CE}$, $\overline{BD}$

midpoint: The point that divides a line segment into two congruent segments.

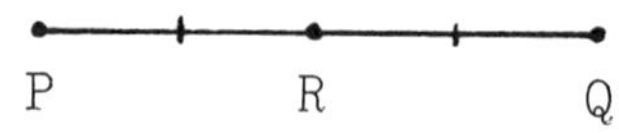

R is the midpoint of $\overline{PQ}$.

minute: One-sixtieth of a degree in angle measure. $60' = 1^\circ$.

$46^\circ 17'$ reads 46 degrees, 17 minutes.

monochromatic color scheme: A color scheme that uses one hue and the light and dark values thereof.

morphology: The study of form.

natural numbers: The counting numbers. The set of natural numbers is named by the letter N.

$$N = \{1, 2, 3, \cdots\}$$

negative reciprocal: The number that when multiplied by a given number produces a product of -1.

If a is any real number, then $-\frac{1}{a}$ is its negative reciprocal.

nonagon: A polygon having nine sides.

nonagons

noncollinear: Not lying on the same line.

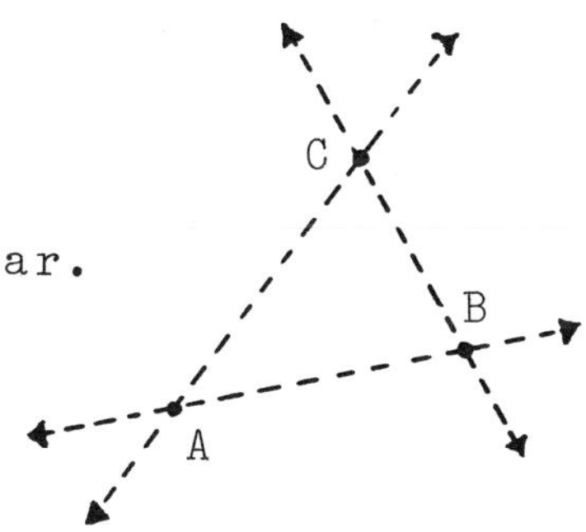

Points A, B and C are noncollinear.

nonconsecutive: Not in order or in sequence. See consecutive for more information.

numerator: In a fraction, the number written above the line.

O

obtuse angle: An angle that measures between 90° and 180°.

obtuse angles

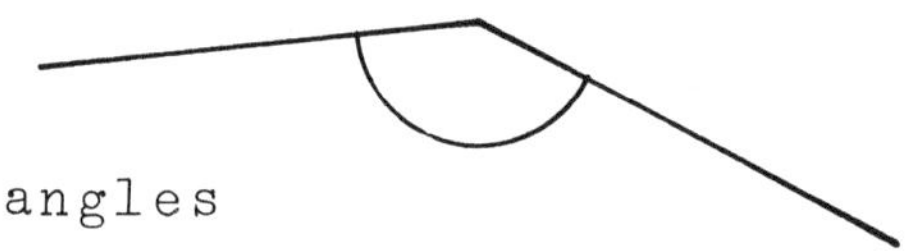

obtuse triangle: A triangle having one obtuse angle.

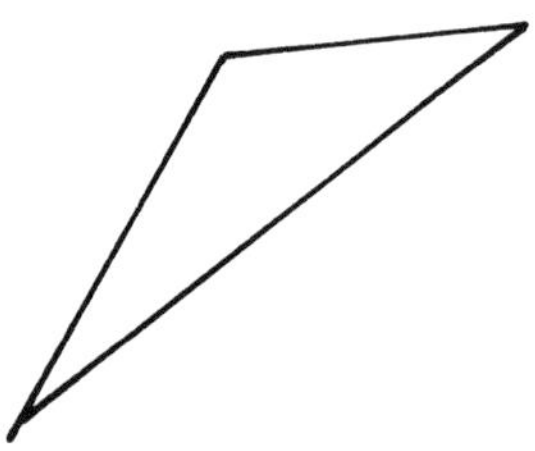

obtuse triangles

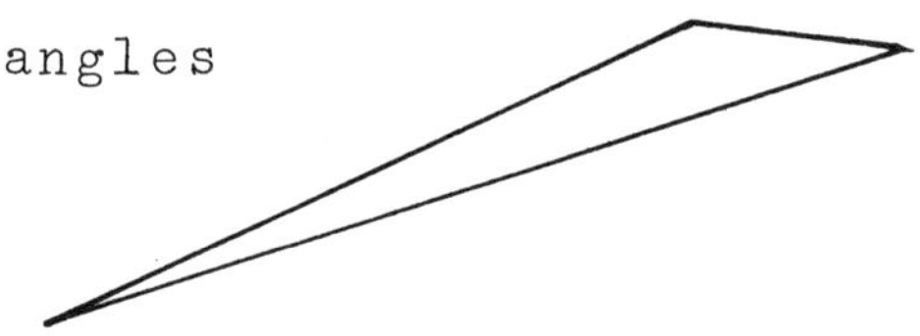

octagon: A polygon having eight sides.

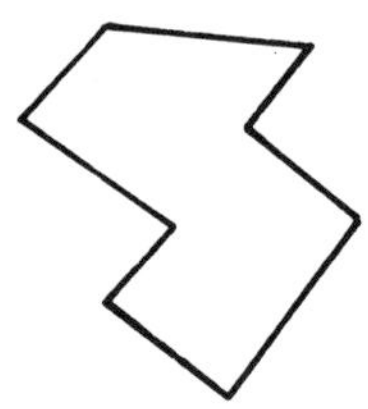

octogons

octahedron: A polyhedron having eight faces.

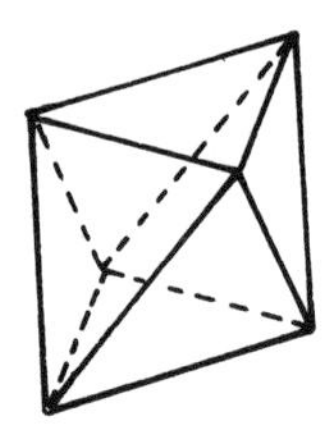

odd number: A whole number that is not divisible by two.
eg. 1,3,5,7,9,···

If n is any whole number, $2n-1$ or $2n+1$ will always designate an odd number.

orthogonal: Perpendicular.

opposite rays: The two rays determined by a single point on a line that travel in different directions. Two rays are opposite rays if they are collinear and have only their endpoint in common.

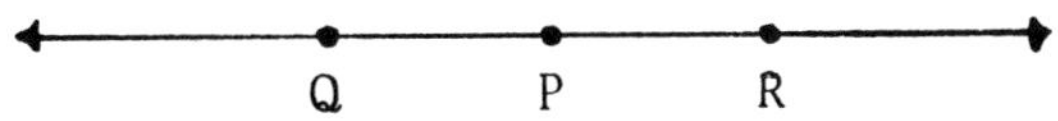

$\overrightarrow{PQ}$ and $\overrightarrow{PR}$ are opposite rays.

P

palindrome: A word or number that reads the same from either end.
eg. 12321 pop

parallel: Two lines are parallel if they lie in the same plane and have no points in common.

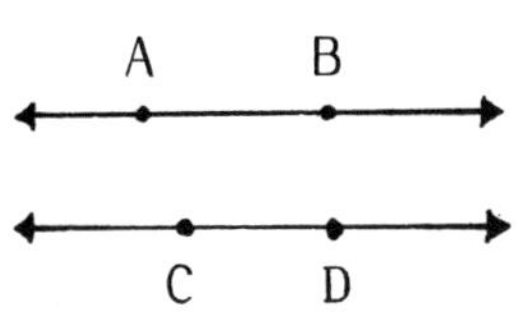

$\overleftrightarrow{AB} \parallel \overleftrightarrow{CD}$

Two segments are parallel if they lie on parallel lines.

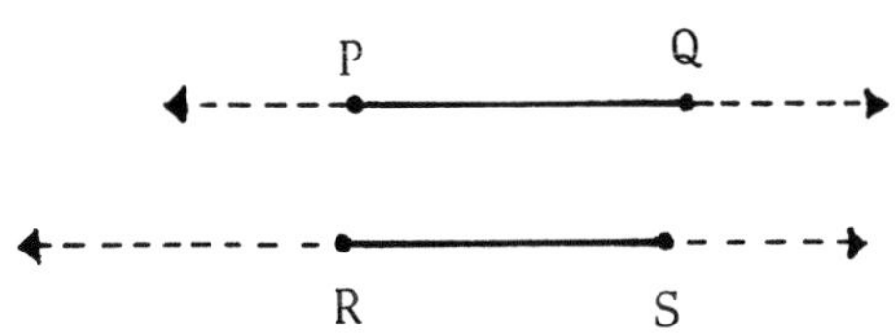

$\overline{PQ} \parallel \overline{RS}$

parallelogram: A quadrilateral in which opposite sides are parallel. See Appendix B for properties of a parallelogram.

ABCD is a parallelogram.

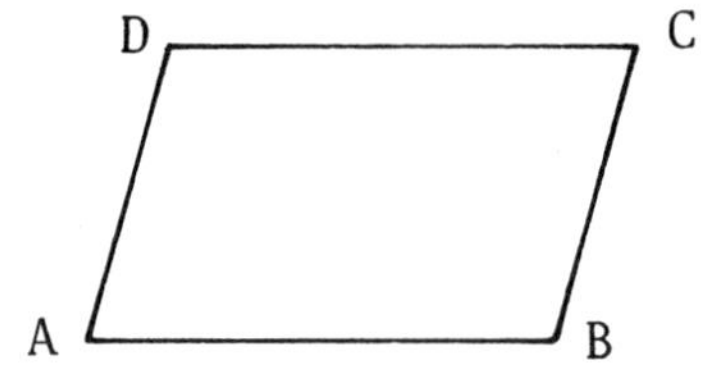

pattern: A repitition of units in a ordered sequence.

pentacle: See pentagram.

pentagon: A polygon having five sides.

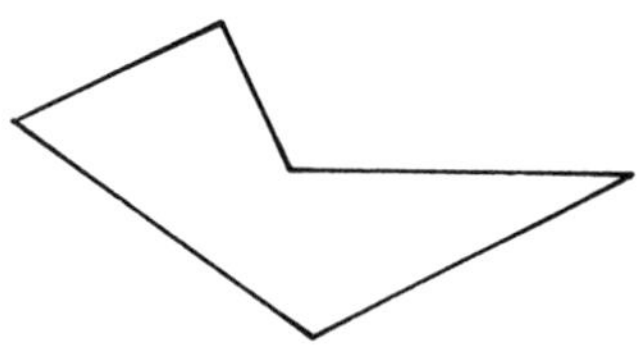

pentagons

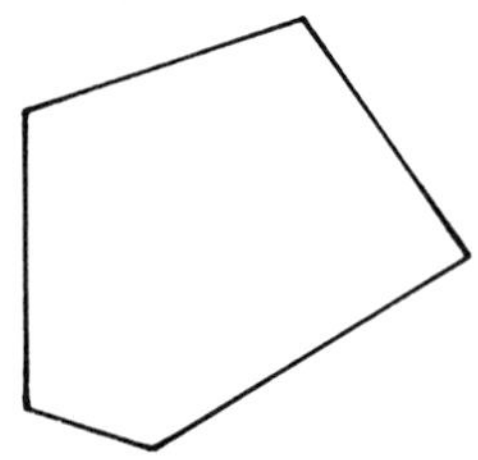

pentagram: The star shaped figure obtained by drawing all the diagonals of a regular pentagon. Also called a pentalpha, pentacle, or pentangle.

The solid segments represent the pentagram. The dotted segments represent the corresponding pentagon.

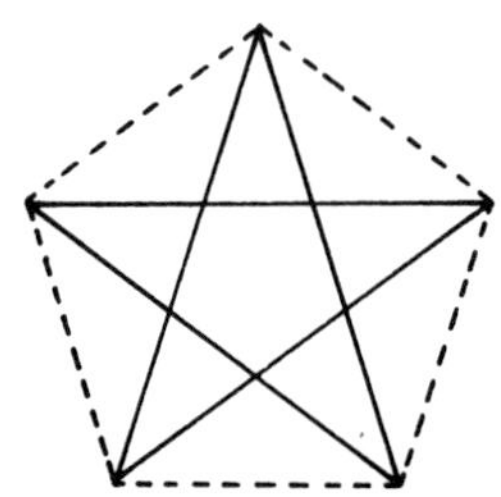

pentalpha: See pentagram.

pentangle: See pentagram.

perimeter: The distance around a figure.

perpendicular: Two lines are perpendicular if they meet to form right angles.

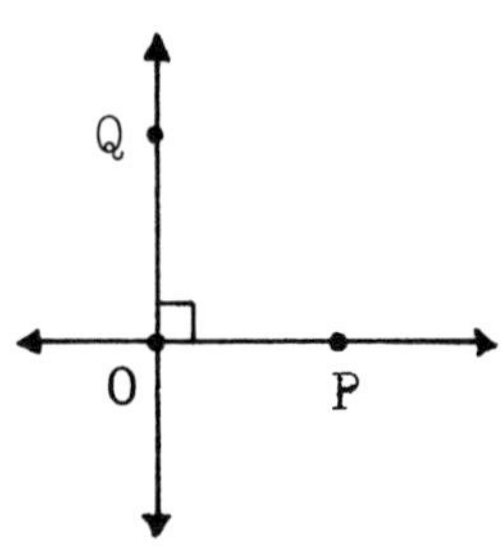

$\overleftrightarrow{OQ} \perp \overleftrightarrow{OP}$

Two segments are perpendicular if they lie on perpendicular lines.

$\overline{AB} \perp \overline{CD}$

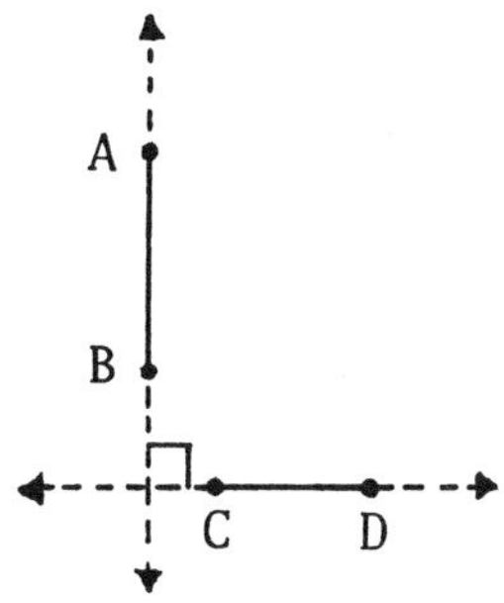

<u>perpendicular bisector</u>: See bisect.

<u>phi</u>: The number 1.61803 naming the ratio that appears in the Divine Proportion. The Greek letter phi (Ø) is used in this text, but the letter tau (τ) is also used by mathematicians to name this ratio.

<u>Phidian spiral</u>: The logarithmic spiral whose successive whorls cut any radius vector into Ø-proportions.

<u>Ø-Family Rectangles</u>: The group of rectangles related to the Golden Rectangle or generated by Ø. They are:

the square, $\sqrt{\emptyset}$ Rectangle, the Golden Rectangle itself, $\sqrt{4}$ Rectangle, $\sqrt{5}$ Rectangle, and the Ø+1 Rectangle.

<u>Ø-Family Spirals</u>: The logarithmic spirals that can be developed within the Ø-Family Rectangles.

Construction methods may be found in Edward B. Edwards' book, <u>Patterns and Design with Dynamic Symmetry</u>.

<u>phyllotaxis</u>: The distribution or arrangement of leaves or buds on a stem or seeds in a flower head.

<u>plane</u>: An undefined term in Geometry. A plane is often thought of as "a slice of space" having infinite length and width but no thickness.

<u>Platonic Solids</u>: The five regular polyhedra: the tetrahedron, the cube, the octahedron, the dodecahedron, and the icosahedron.

<u>point</u>: An undefined term in Geometry. A point is often thought of as a location in space having no dimensions.

<u>pole</u>: The central point of a spiral.

polygon: A simple closed curve composed of line segments each intersecting exactly two others, one at each endpoint. The line segments are called the sides of the polygon and the points of intersection are called the vertices of the polygon. A polygon is named by its vertices read consecutively in either a clockwise or a counter-clockwise direction.

The first polygon pictured could be named ABCDE. It could also be named DCBAE, among others.
The second polygon might be named WRSTUV or VUTSRW, among others.

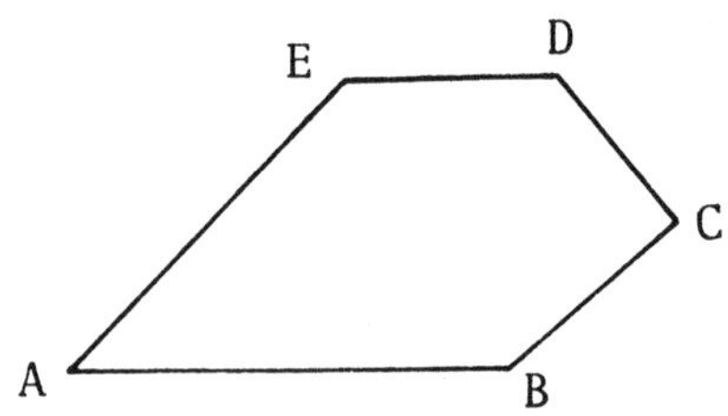

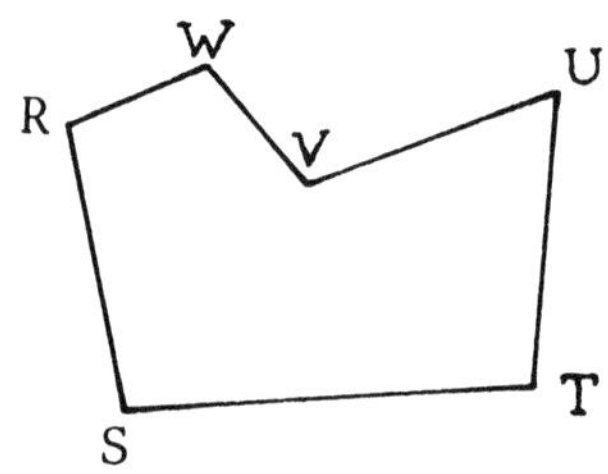

polyhedron: A three dimensional figure formed by polygons which intersect in such a way that vertices and sides are common. The polygonal figures are called the faces of the polyhedron, the common sides are called the edges of the polyhedron, and the common vertices are called the vertices of the polyhedron.

primary colors: In pigments, yellow, red and blue. These are the essential colors from which all others are derived.

prime number: A natural number, different from one, that is divisible only by one and itself. There are an infinite number of prime numbers. The sequence of primes is as follows:

$$2, 3, 5, 7, 11, 13, 17, 19, 23, 29, \cdots$$

proportion: A pair of ratios set equal to each other.

$$\frac{a}{b} = \frac{c}{d}$$

In this proportion a and d are called the extremes and b and c are called the means of the proportion. These words refer only to the position of the terms.

proportional spiral: See spiral (logarithmic spiral).

protractor:

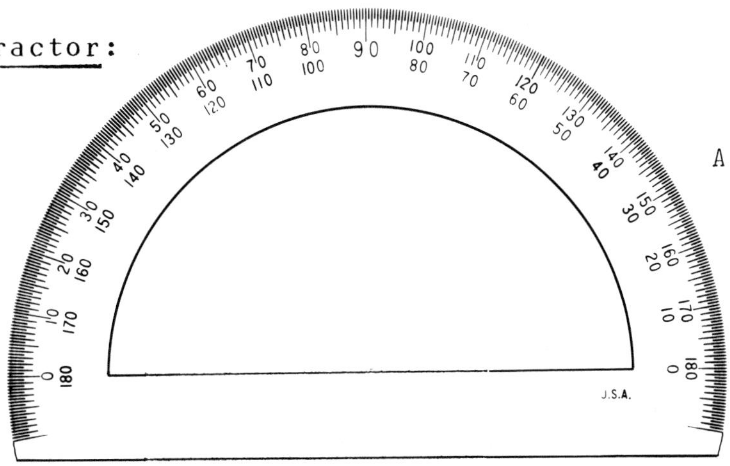

A tool used for measuring the number of degrees in an angle.

pyramid: A three dimensional figure set on a polygonal base whose lateral faces are triangles formed by segments joining the vertices of the polygon to a single point (called the vertex) not in the same plane.

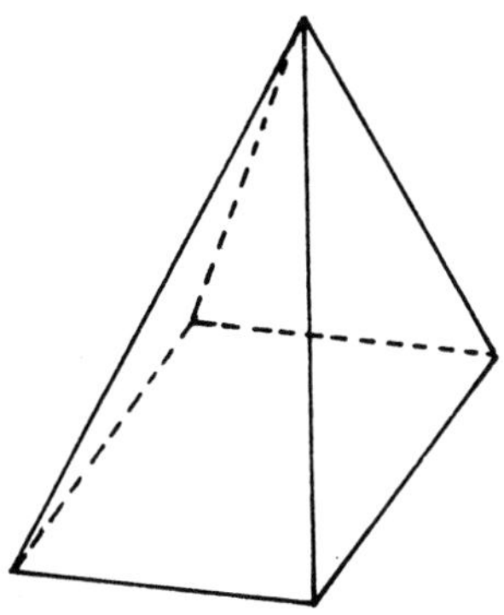

If the base is regular and the segment joining the vertex to the center of the base is perpendicular to the plane of the base, the pyramid is a right pyramid.

Pythagorean Theorem: $a^2 + b^2 = c^2$, where a and b name the lengths of the legs of a right triangle and c names the length of the hypotenuse.

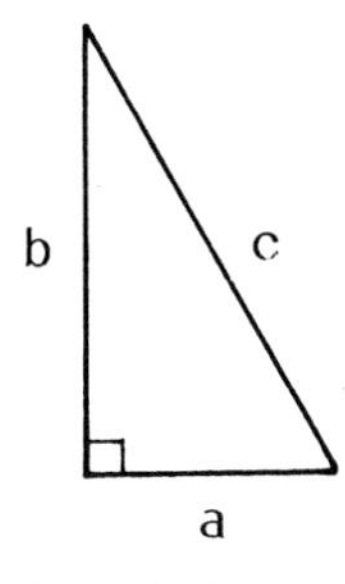

Pythagorean triple or triad: Three integers that have the property that the sum of the squares of two of them equals the square of the third.

eg. 3, 4, 5 $\quad 3^2 + 4^2 = 5^2 \quad 9 + 16 = 25$

5, 12, 13 $\quad 5^2 + 12^2 = 13^2 \quad 25 + 144 = 169$

Q

quadratic equation: An equation of the form $ax^2 + bx + c = 0$, where a, b and c are real numbers and $a \neq 0$.

quadratic formula: $x = \frac{-b \pm \sqrt{b^2 - 4ac}}{2a}$ where a, b and c are the coefficients of a quadratic equation $ax^2 + bx + c = 0$.

quadrilateral: A four sided polygon.

quadrilaterals

R

radicand: The quantity under the radical sign.

eg. in $\sqrt{15}$, 15 is the radicand.

In $\sqrt[3]{3xy}$, 3xy is the radicand.

radius: A line segment that joins the center of a circle to any point on the circle is a radius of the circle. The measure of any of these segments (all are congruent) is the radius of the circle.

$\overline{OP}$ is a radius of ⊙O.

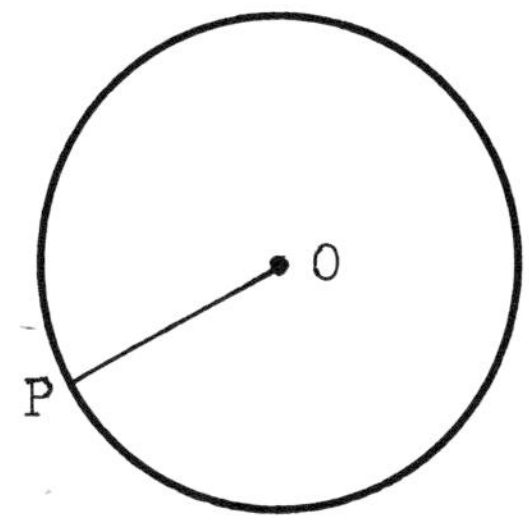

radius vector: Any ray emanating from the pole of a spiral and intersecting the spiral curve.

$\overline{PQ}$ is a radius vector.

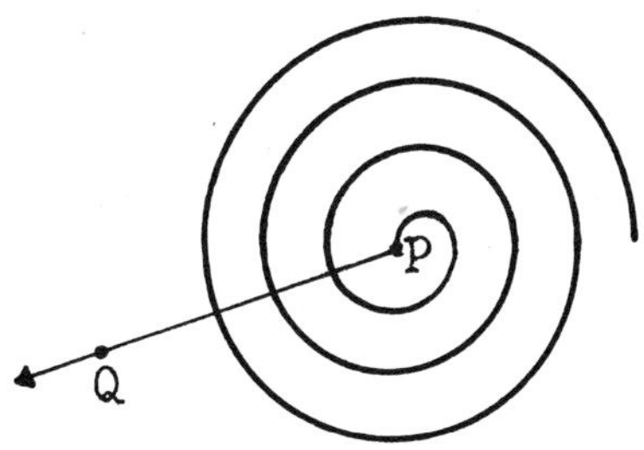

radius of a regular polygon: If a regular polygon is inscribed in a circle, the radius of the polygon is the radius of the circumscribed circle, or the line segment from the "center" of the polygon to a vertex.

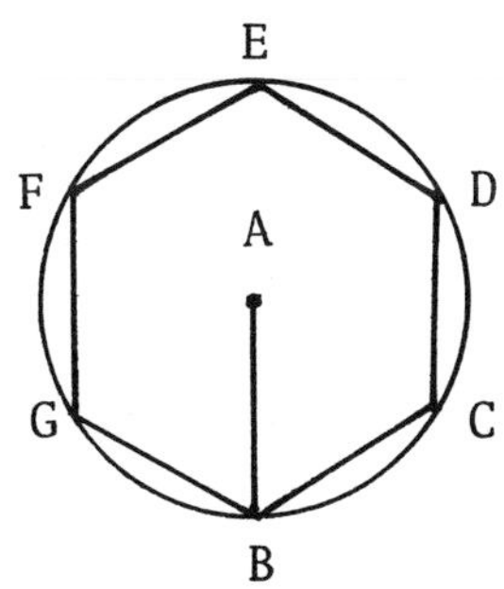

$\overline{AB}$ is a radius of BCDEFG.

rational number: Any number that can be put into the form $\frac{p}{q}$ where p and q are integers and $q \neq 0$.

ratio: A comparison of numbers to each other.

eg. the ratio of 2 to 3 may be expressed as $\frac{2}{3}$ or 2:3 or simply 2 to 3.

ray: The figure formed by a point on a line and all the points on the same side of that point. The point is called the endpoint of the ray, and the ray is named by its endpoint and any other point on the ray.

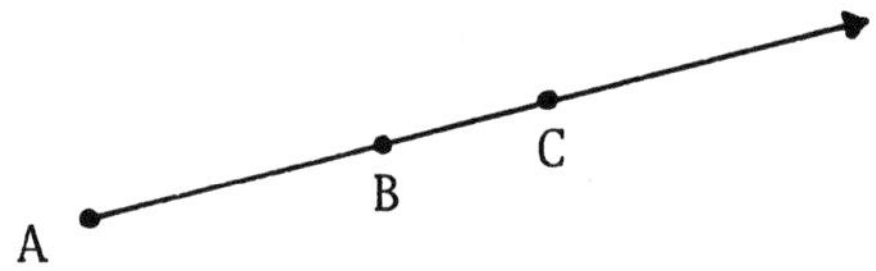

$\overrightarrow{AB} = \overrightarrow{AC}$

reciprocal: Of a rectangle: a similar rectangle cut from the parent rectangle whose length is the width of the parent rectangle.

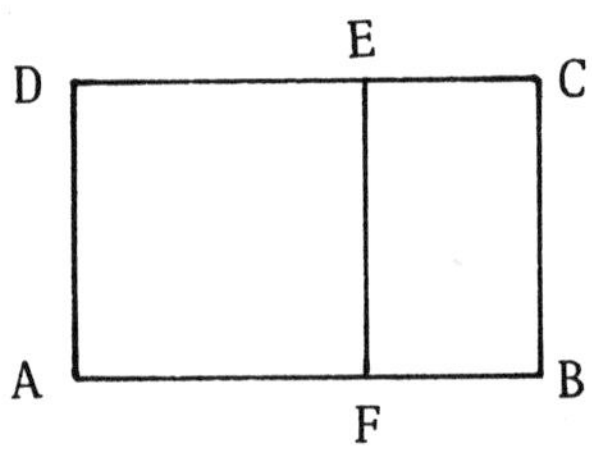

ABCD ∼ BCEF where BCEF is the reciprocal of ABCD.

Of a number: if n is any number then $\frac{1}{n}$ is its reciprocal.

Note: the product of any number and its reciprocal is always one. $n \cdot \frac{1}{n} = 1.$

rectangle: A parallelogram whose angles are right angles.

ABCD is a rectangle.

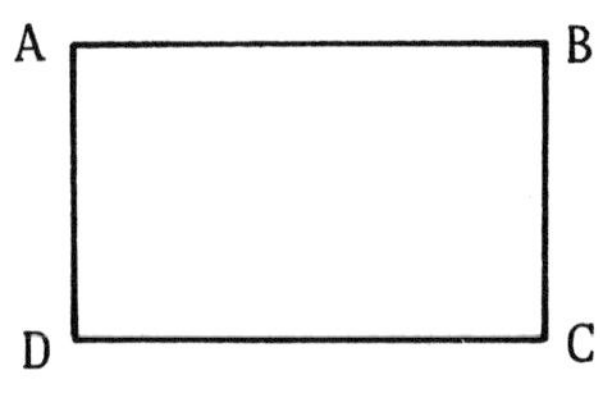

Rectangle of Price: See $\sqrt{\emptyset}$ Rectangle.

reflection: A symmetry operation in which a figure is flipped about an axis, inverting right and left.

regular: Regular polygon: a polygon having all sides and angles congruent.

Regular polyhedron: A polyhedron whose faces are all congruent regular polygons.

rhombus: A parallelogram in which all sides are congruent.

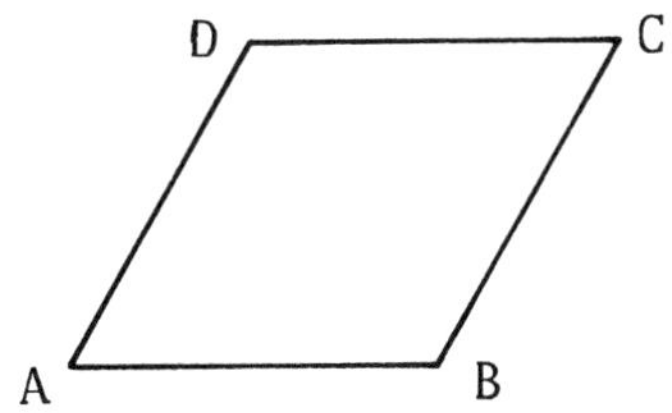

ABCD is a rhombus.

right angle: An angle that measures 90°.

right angles

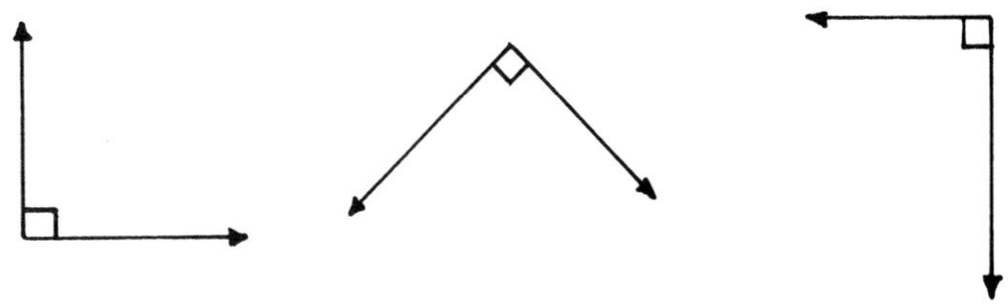

right triangle: A triangle having one right angle.

right triangles

root: See solution.

root rectangles: The rectangles derived from the square such that the width measures one, and the length measures the square root of a natural number. These rectangles are members of the family of Dynamic Rectangles.

Rope Knotter's Triangle: The right triangle whose legs measure 3 and 4 respectively, and whose hypotenuse measures 5 (or 3-4-5 Right Triangle). It was so named because it was discovered by the ancient Egyptian rope knotters who prepared ropes for purposes of measuring.

rotation: A symmetry operation which involves rotating a figure about a point.

S

<u>scalene triangle</u>: A triangle having no congruent sides.

scalene triangles

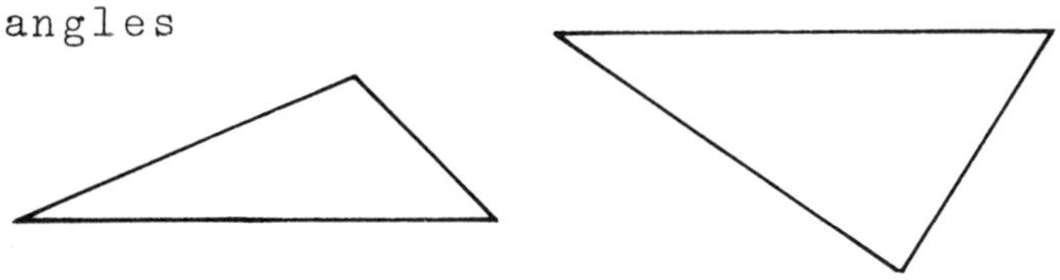

<u>secant</u>: A line that intersects a circle in exactly two points.

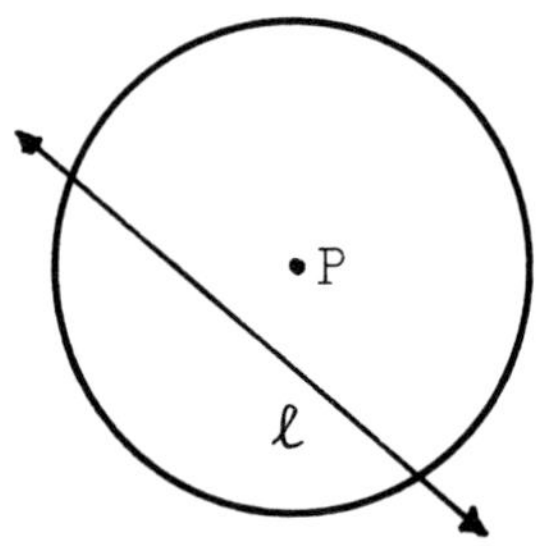

Line l is a secant with respect to ⊙P.

<u>second</u>: One-sixtieth of a minute in angle measure.

60" = 1' or 3600" = 1°

18°32'21" reads 18 degrees, 32 minutes, 21 seconds.

<u>septagon</u>: A polygon having seven sides, sometimes called a heptagon.

septagons

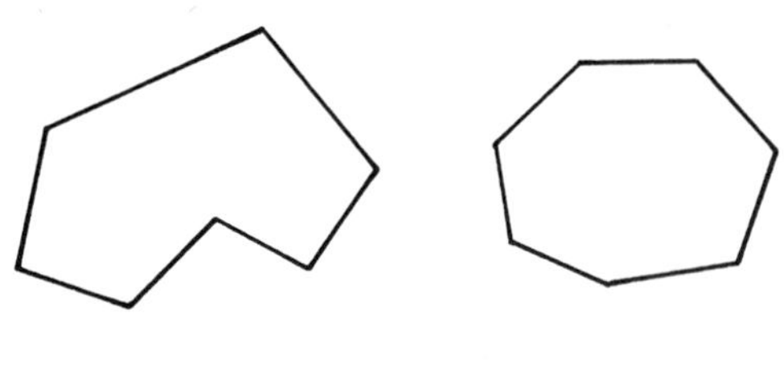

sequence: An ordered progression of terms. A sequence may have a finite number of terms, eg. 1,2,3,4,5,6, or an infinite number of terms, eg. 1,1,2,3,5,8,13,···

If the notation $S_1, S_2, S_3, \cdots, S_n, \cdots$ is used to denote a sequence, the subscript names the position of the term. eg. S_{10} would be the tenth term.

side: Of an angle: see angle.
Of a line: either of the half planes formed by a line in the plane.

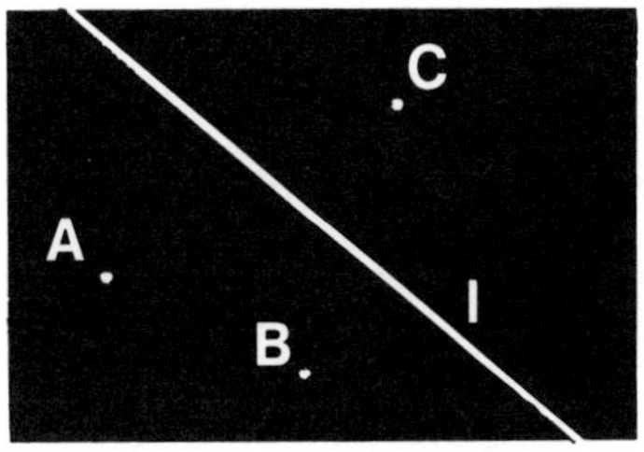

The shaded portions represent the two half planes.
A and B lie on the same side of line l. A and C lie on opposite sides of line l.

Of a polygon: see polygon.

similar: Having the same shape. Two polygons are similar if corresponding angles are congruent and corresponding sides are in proportion.

$$\triangle ABC \sim \triangle DEF$$

$$\frac{AB}{DE} = \frac{BC}{EF} = \frac{AC}{DF}$$

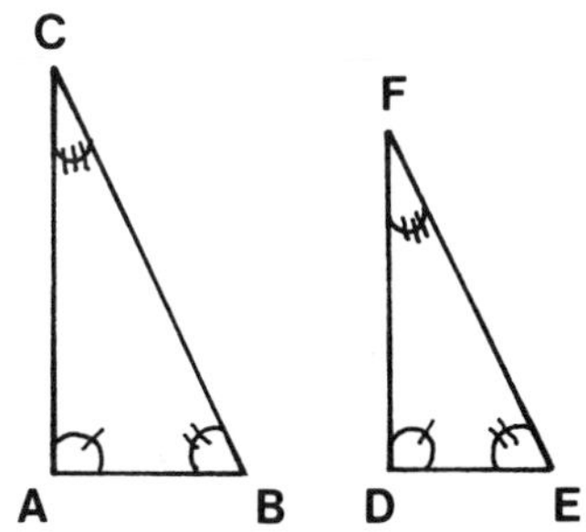

simple closed curve: A plane curve that can be traced without lifting pen from paper in such a way that the starting point and ending point are the same and no point is traced more than once.

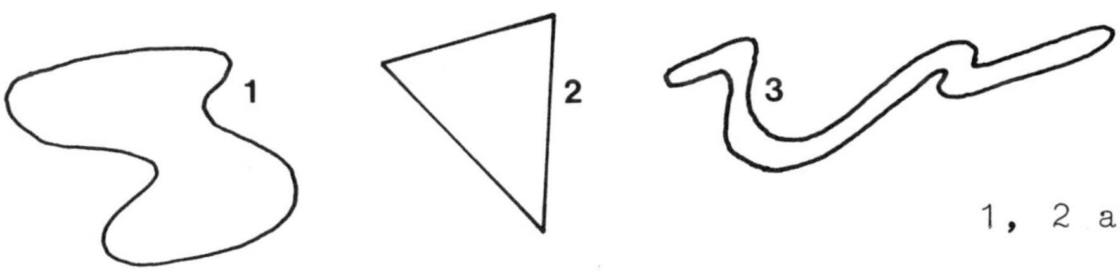

1, 2 and 3 are simple closed curves.

4 and 5 are not simple closed curves.

skew lines: Lines that lie in different planes and do not intersect.

slide reflection: See glide reflection.

solution: Any quantity which may be substituted for a variable in an equation to produce a true statement.

eg. 3 is the solution for the equation $x + 1 = 4$.

space lattice: A uniform three dimensional grid which is confined to compositions of triangles, squares, rhombuses and hexagons. A space lattice may be extended infinitely in all directions in a continuous pattern.

spiral: A curve which moves around a central point, called the pole, while continually increasing its distance from the pole.

Archimedean spiral: A spiral whose successive whorls are spaced at equal intervals along any radius vector.

Archimedean spiral

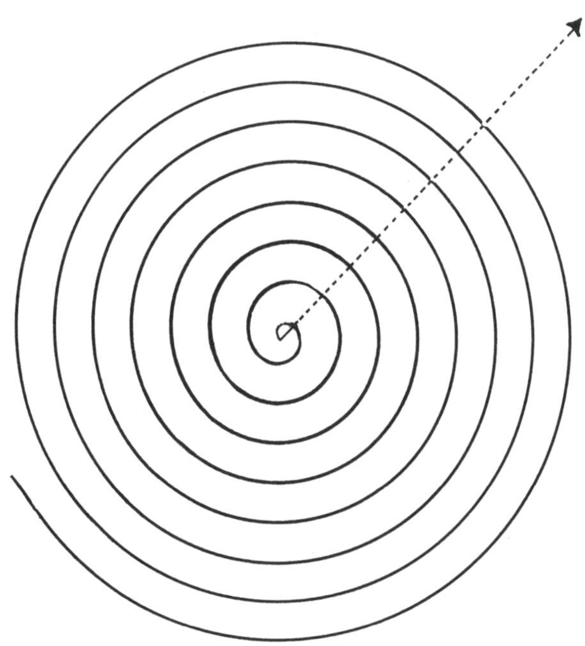

logarithmic spiral: A spiral whose successive whorls form congruent angles with any radius vector and are spaced in geometric progression along the radius vector. Also known as equiangular spiral, proportional spiral, Spira Mirabilis.

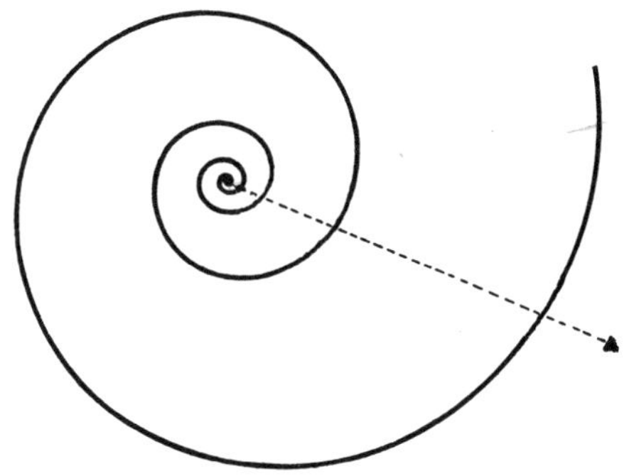

logarithmic spiral

Spira Mirabilis: See spiral (logarithmic spiral).

square: Noun: a rectangle all of whose sides are congruent.

ABCD is a square.

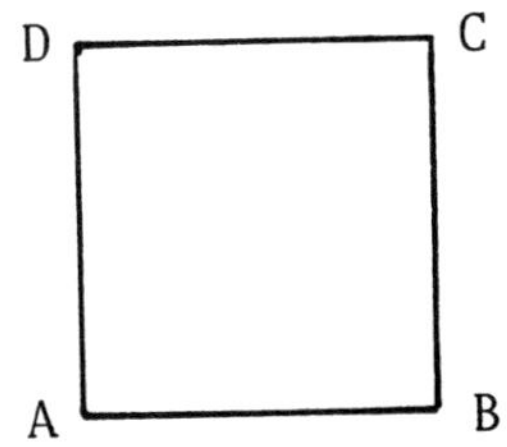

Verb: to multiply a quantity by itself. The number 2 used as an exponent indicates the operation of squaring.

eg. $a^2 = a \cdot a$

square root: A number that when multiplied by itself produces a given number. The operation of taking the square root of a number is indicated by a radical sign ($\sqrt{\ }$) and the given number, under the sign, is called the radicand.

eg. $\sqrt{16} = 4$ since $4 \times 4 = 16$.

$\sqrt{\phi}$ Rectangle (Square Root of Phi Rectangle): Any rectangle in which the ratio of the length to the width is $\sqrt{\phi}$ to 1 or approximately 1.271.

straight angle: An angle measuring 180°.

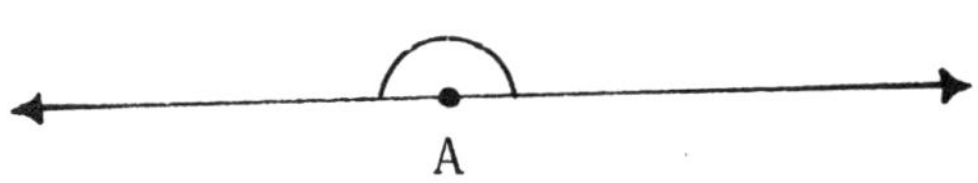

$\angle A$ is a straight angle, so called because its sides form a line.

Sublime Triangle: See Golden Triangle.

summation sequence: A sequence whose terms are arranged such that any term is the sum of the two preceding terms.

$u_1, u_2, u_3, \cdots, u_n, \cdots$ is a summation sequence if

$u_i + u_{i+1} = u_{i+2}$ for all integers i.

supplementary angles: Two angles the sum of whose measures is 180°. The angles are said to be supplements of each other.

$\angle ABC$ and $\angle CBD$ are supplementary.

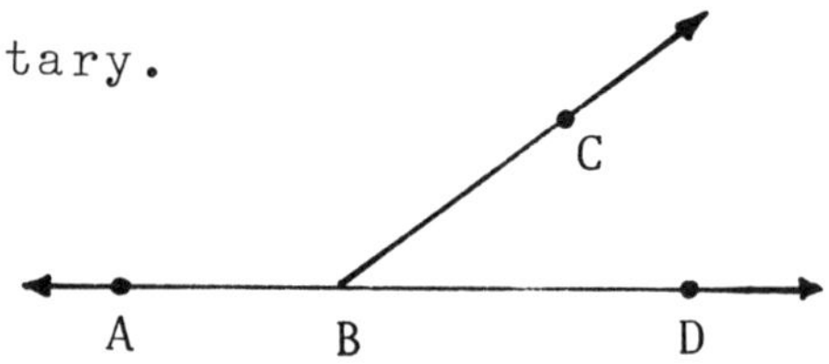

symmetry: With measure. See the symmetry operations of: translation, rotation, reflection and glide reflection.

T

tangent: A line lying in the same plane as a circle that intersects the circle in exactly one point.

Line m is tangent to $\odot$P at A.

A is called the point of tangency.

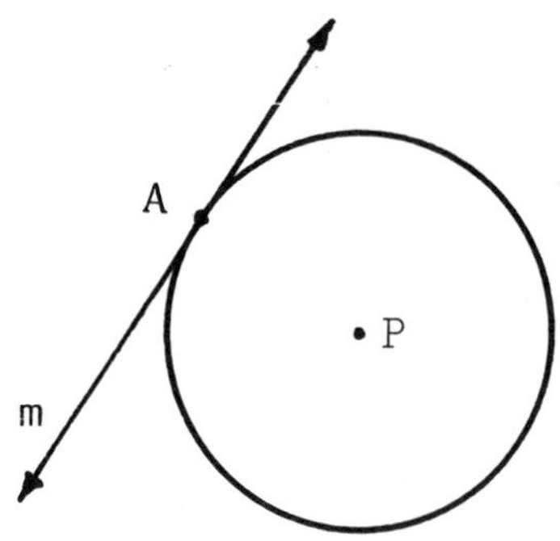

tetrahedron: A polyhedron having four triangular faces.

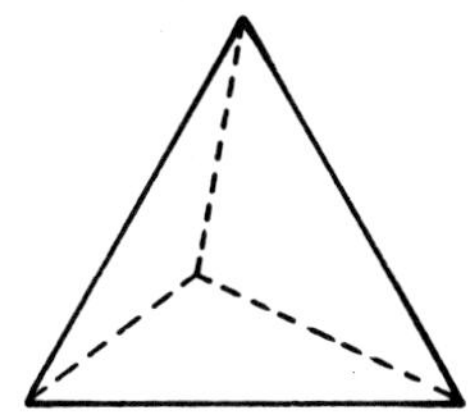

translation: A symmetry operation that involves a slide or a glide.

transversal: A line that intersects two or more other lines. The concept of transversal is most frequently used in connection with parallel lines.

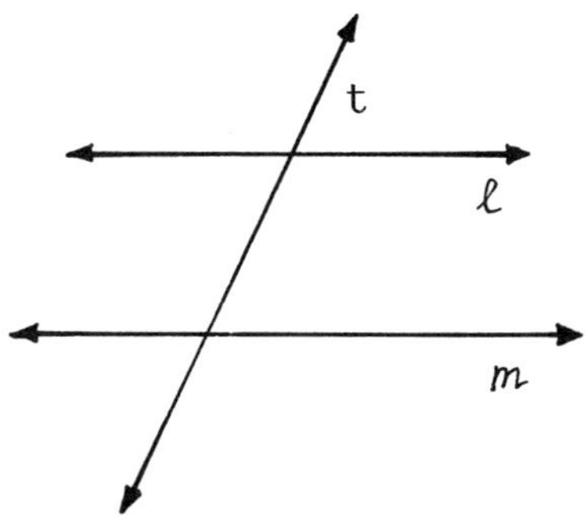

l ||m cut by transversal t.

trapezoid: A quadrilateral with exactly one pair of parallel sides. The parallel sides are called the bases of the trapezoid and the angles including either base are called the base angles. There are two pairs of base angles in a trapezoid.

trapezoids

$\overline{AB} \parallel \overline{DC}$

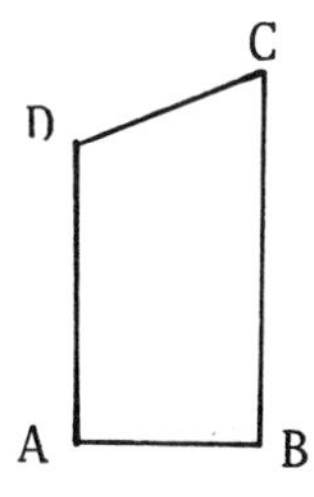

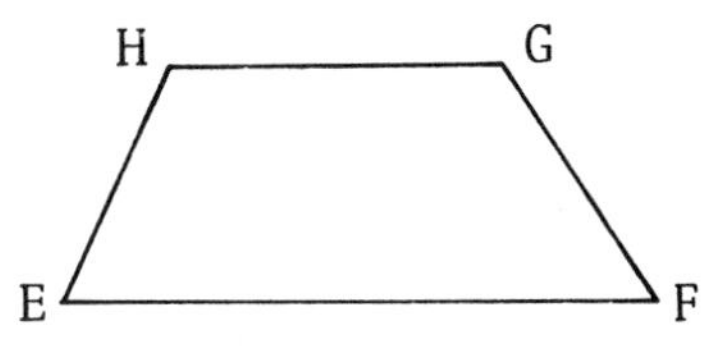

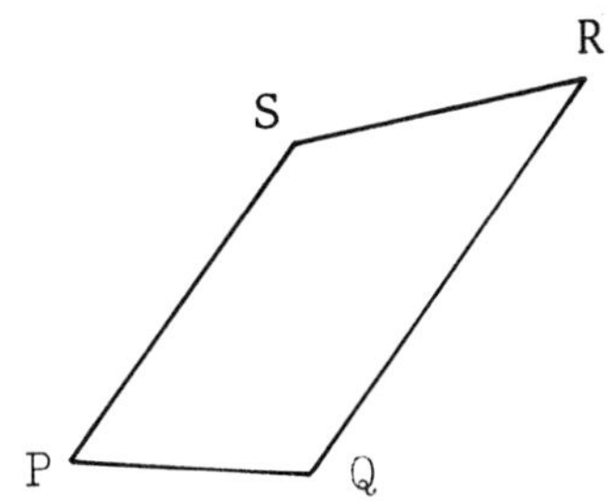

$\overline{EF} \parallel \overline{HG}$

$\overline{PS} \parallel \overline{QR}$

triangle: A polygon having three sides. Individual types of triangles are listed under their respective names.

Triangle of Price: Any right triangle whose sides are in the ratio of $1:\sqrt{\phi}:\phi$. Also called the Egyptian Triangle.

Triangle of the Pentalpha: See Golden Triangle.

triangulation: The process of dividing an irregular shape into triangular shapes for purposes of finding area.

truncate: To cut off.

U

unit measure: The length taken to represent the number one in a given set of circumstances.

V

vertex: Of an angle: see angle.

Of a polygon: see polygon.

volume: The measure in cubic units of any three dimensional object.

W

Whirling Square Rectangle: Another name for the Golden Rectangle.

whorl: A circular or spiral arrangement of like things such as leaves, petals, hair, etc.

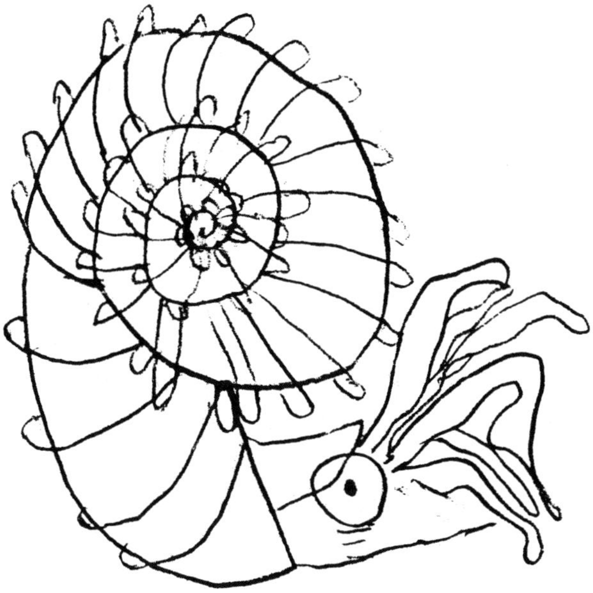

Chambered Nautilus

Graham Boles
Age 6